Organic Redox Chemistry

Organic Redox Chemistry

Chemical, Photochemical and Electrochemical Syntheses

Edited by

Jun-Ichi Yoshida
Frédéric W. Patureau

WILEY-VCH

Editors

Prof. Jun-Ichi Yoshida
Kyoto University
School of Engineering
Nishikyo-ku
Kyoto University
606-8510 Kyoto
Japan

Prof. Frédéric W. Patureau
Institut für Organische Chemie
RWTH Aachen University
Landoltweg 1
52074 Aachen
Germany

Cover
Cover Image: © Netfalls Remy Musser/Shutterstock

Library of Congress Card No.:
applied for

British Library Cataloguing-in-Publication Data
A catalogue record for this book is available from the British Library.

Bibliographic information published by the Deutsche Nationalbibliothek
The Deutsche Nationalbibliothek lists this publication in the Deutsche Nationalbibliografie; detailed bibliographic data are available on the Internet at <http://dnb.d-nb.de>.

Print ISBN: 978-3-527-34487-1
ePDF ISBN: 978-3-527-81570-8
ePub ISBN: 978-3-527-81568-5
oBook ISBN: 978-3-527-81567-8

Typesetting Straive, Chennai, India
Printing and Binding CPI Group (UK) Ltd, Croydon CR0 4YY

Printed on acid-free paper

C085623_070921

Contents

Biography *xi*
Preface *xiii*

1 **Chemical Oxidative C—C Bond Formation** *1*
Koji Hirano
1.1 Introduction *1*
1.2 Oxidative Aryl–Alkenyl Bond Formation *1*
1.2.1 Oxidative Mizoroki–Heck Reaction with Arylmetal Reagents *2*
1.2.2 Direct Oxidative Mizoroki–Heck Reaction with Arene C–Hs (Fujiwara–Moritani Reaction) *4*
1.3 Oxidative Aryl–Aryl Bond Formation *8*
1.3.1 Oxidative C–H/C–M Biaryl Cross-Coupling *10*
1.3.2 Oxidative C–H/C–H Biaryl Cross-Coupling *12*
1.4 Oxidative Aryl–Alkynyl Bond Formation *15*
1.5 Oxidative C—C Bond Formation at $C_{sp}{}^3$ Center *18*
1.6 Conclusion *22*
References *23*

2 **Electrochemical Oxidative C—C Bond Formation** *29*
Sebastian Lips and Siegfried R. Waldvogel
2.1 Electrochemical Oxidative Aryl–Aryl Cross-Coupling Reaction *29*
2.2 Electrochemical Oxidative Benzyl–Aryl Cross-Coupling Reaction *35*
2.3 Electrochemical Oxidative Arylation of Olefins *36*
2.4 Electrochemical Oxidative Arylation of Alkynes *39*
2.5 Electrochemical Oxidative Cross-Dehydrogenative Coupling of $C(sp^3)$—H and $C(sp^2)$—H Bonds *39*
References *39*

3 **Fundamentals of Photochemical Redox Reactions** *45*
Daniel A. Corbin, Nicholas A. Swisher, and Garret M. Miyake
3.1 Introduction: A Brief History of Photochemistry *45*
3.2 Photochemistry: Background and Theory *50*
3.2.1 The Electromagnetic Spectrum *50*

3.2.2 Allowed and Forbidden Transitions *51*
3.2.3 Photophysical Processes *52*
3.2.3.1 Jablonski Diagrams *52*
3.2.3.2 Absorption *53*
3.2.3.3 Vibrational Relaxation *55*
3.2.3.4 Internal Conversion *56*
3.2.3.5 Fluorescence *56*
3.2.3.6 Intersystem Crossing *57*
3.2.3.7 Phosphorescence *58*
3.2.4 Electron Transfers *58*
3.2.4.1 Photoinduced Electron Transfer *58*
3.2.4.2 Mechanisms of Electron Transfer *59*
3.2.4.3 Marcus Theory *60*
3.2.5 Laboratory Techniques for Studying Photoredox Processes *61*
3.2.5.1 sUV–Visible Spectroscopy *61*
3.2.5.2 Emission Spectroscopy *63*
3.2.5.3 Transient Absorption Spectroscopy *65*
3.2.5.4 Cyclic Voltammetry *67*
3.2.6 Practical Considerations for Performing Photochemical Reactions *68*
3.2.6.1 Factors Influencing Bimolecular Reactions *68*
3.2.6.2 Photoreactor Design *68*
3.2.6.3 Choice of Light Source *69*
3.3 Photoredox Catalysis *69*
3.3.1 General Mechanisms of Photocatalysis *69*
3.3.2 Design Principles for Effective Photoredox Catalysts *70*
3.3.2.1 Effective Absorption of Light *70*
3.3.2.2 High Quantum Yield of Desired Excited State *71*
3.3.2.3 Long-Lived Excited State *71*
3.3.2.4 Favorable Thermodynamics *71*
3.3.2.5 Redox Reversibility *72*
3.3.3 Inorganic Photocatalysts *72*
3.3.4 Organic Excited-State Oxidants *75*
3.3.5 Organic Excited-State Reductants *78*
3.3.6 Open-Shell Photoredox Catalysts *82*
3.4 Photochemistry of Electron Donor–Acceptor Complexes *85*
3.4.1 Background and Theory *85*
3.4.1.1 What Is an EDA Complex? *85*
3.4.1.2 How do EDA Complexes Interact with Light? *85*
3.4.1.3 Electron Transfer in EDA Complexes *86*
3.4.1.4 Environmental Factors Affecting EDA Complexes *86*
3.4.2 Early Examples of EDA Photochemistry *87*
3.4.3 Recent Examples of EDA Photochemistry *87*
3.4.3.1 Rediscovering EDA Complexes through Photoredox Catalysis *87*
3.4.3.2 Stoichiometric EDA Reactions *88*
3.4.3.3 Use of Sacrificial Donors and Acceptors *89*

3.4.3.4 Redox Auxiliaries to Expand Donor and Acceptor Scope *90*
3.4.3.5 Catalytic EDA Reactions *91*
3.4.3.6 Enantioselective Reactions of EDA Complexes *91*
3.5 Concluding Thoughts *92*
Suggested Additional Reading *92*
Photochemistry and Photophysical Processes *92*
Electrochemical Methods *93*
Photoredox Catalysis *93*
Earth Abundant Metal Photoredox Catalysis *93*
EDA Complexes *93*
References *93*

4 C–H Bond Functionalization with Chemical Oxidants *103*
Jia-Xiang Xiang, Pooja Vemuri, and Frédéric W. Patureau
4.1 Introduction *103*
4.1.1 A Shift in the Rate-Determining Step *103*
4.1.2 The Nature of the Oxidant *103*
4.2 Metal-Based Oxidants and Other Inorganic Oxidants *104*
4.2.1 Silver Salt Oxidants *105*
4.2.2 Copper Salt Oxidants *108*
4.2.3 Other Inorganic Oxidants *109*
4.3 Organic Oxidants *109*
4.3.1 Organic Peroxides *110*
4.3.2 Quinones *112*
4.4 Internal Oxidants (DG^{ox}) *115*
4.5 Use of O_2 as an Oxidant *119*
4.6 Dehydrogenative Couplings with No Oxidant at All *124*
4.7 Conclusion *125*
References *125*

5 Electrochemical Reductive Transformations *129*
Mahito Atobe and Toshio Fuchigami
5.1 General Characteristics of Electrochemical Reactions *129*
5.2 Mechanism of Organic Electrochemical Reductions *130*
5.3 Characteristics of Organic Electrochemical Reductions *131*
5.3.1 Umpolung *131*
5.3.2 Selectivity *132*
5.3.2.1 Chemoselectivity *133*
5.3.2.2 Reaction Pathway Selectivity *133*
5.3.2.3 Regioselectivity *133*
5.3.2.4 Stereoselectivity *134*
5.3.2.5 Selectivity Depending on Electrode Materials *134*
5.4 Electroauxiliaries *135*
5.4.1 Electroauxiliaries Based on Readily Electron-Transferable Functional Groups *135*

5.4.2 Electroauxiliaries Based on Coordination Effects *136*
5.5 Reaction Pattern of Organic Electrochemical Reductions *137*
5.5.1 Transformation Type of Functional Group *137*
5.5.2 Addition Type *138*
5.5.3 Insertion Type *138*
5.5.4 Substitution Type *139*
5.5.5 Substitutive Exchange Type *139*
5.5.6 Elimination Type *139*
5.5.7 Dimerization Type *139*
5.5.8 Crossed Dimerization *140*
5.5.9 Cyclization Type *140*
5.5.10 Polymorphism Formation Type *140*
5.5.11 Polymerization Type *141*
5.5.12 Cleavage Type *141*
5.5.13 Metalation Type *141*
5.5.14 Asymmetric Synthesis Type *141*
5.6 Electrochemically Generated Reactive Species *141*
5.6.1 Cathodically Generated Carbon Species *142*
5.6.1.1 Reduction of Alkyl Halides *142*
5.6.1.2 Reduction of Ketone and Imine *142*
5.6.1.3 Reduction of Activated Olefin and Conjugated Olefin *142*
5.6.1.4 Reduction of Active Hydrogen Compounds *143*
5.6.1.5 Reduction of *gem*- and *vic*-Dihalogeno Compounds *143*
5.6.2 Cathodically Generated Heteroatom Species *143*
5.6.2.1 Cathodically Generated Nitrogen Species *143*
5.6.2.2 Reduction of Alcohol and Carboxylic Acid *143*
5.6.2.3 14-Family and 15-Family Element Species *144*
5.7 Advanced Methodology for Electrochemical Reductive Transformations *144*
5.7.1 Electrocatalysis for Reductive Transformations *144*
5.7.1.1 Direct and Indirect Electrochemical Reductions *144*
5.7.1.2 Kinds of Mediators for Reductive Transformations *145*
5.7.1.3 Electrorechemical Reductive Transformations Using Mediators *146*
5.7.2 Electrogenerated Bases *148*
5.8 Conclusions *150*
References *150*

6 Electrochemical Redox-Mediated Polymer Synthesis *153*
Naoki Shida and Shinsuke Inagi
6.1 Introduction *153*
6.2 Synthesis of Conducting Polymers by Electrochemical Redox *154*
6.2.1 Electrochemical Redox Behavior of Conducting Polymers *154*
6.2.2 Oxidative Electropolymerization of Aromatic Monomers *154*
6.2.3 Electrochemical Copolymer Synthesis *155*
6.2.4 Reductive Electropolymerization of Aromatic Monomers *157*

6.2.5 Polysilane Synthesis by Cathodic Reduction *157*
6.2.6 Electropolymerization Under Nonconventional Conditions *158*
6.3 Post-Functionalization of Conducting Polymers by Electrochemical Redox *159*
6.3.1 Functionalization of Conducting Polymers by Anodic Substitution *159*
6.3.2 Cathodic Reduction and Paired Reactions *162*
6.3.3 Functionalization of Polyaniline by the CRS Method *162*
6.3.4 Oxidation-Induced Intramolecular Cyclization of Conducting Polymer *163*
6.3.5 Electrogenerated Transition-Metal Catalysts for Post-Functionalization *164*
6.4 Synthesis of Nonconjugated Polymers by Electrochemical Redox *164*
6.4.1 Electropolymerization of Electroactive Polymers *164*
6.4.2 Electrochemical Redox-Controlled Polymerization *165*
6.4.3 Electrochemically Induced Film Formation via Crosslinking *167*
6.5 Conclusion *167*
References *168*

7 Chemical Paired Transformations *171*
Eiji Shirakawa
7.1 Introduction *171*
7.2 Direct Arylation of Arenes with Aryl Halides *173*
7.3 Electron-Catalyzed Cross-Coupling Reactions of Aryl Halides *178*
7.4 Conclusions *182*
References *183*

8 Photochemical Paired Transformations *187*
Takashi Koike and Munetaka Akita
8.1 Introduction *187*
8.2 Basic Concepts for Photochemical Hydrogen Atom Transfer (HAT) Process *188*
8.2.1 Concept 1: Direct HAT by the Excited Photocatalyst *188*
8.2.2 Concept 2: Indirect HAT Triggered by Photocatalysis *188*
8.3 Asymmetric Radical Functionalization Associated with Direct HAT by Photocatalysts *189*
8.3.1 Photocatalytic Functionalization of $C(sp^3)$—H Bonds Based on Concept 1 *189*
8.3.2 Asymmetric Transformations Based on Concept 1 *194*
8.4 Asymmetric Radical Functionalization Associated with Indirect HAT Triggered by Photocatalysis *195*
8.4.1 Photocatalytic Functionalization of $C(sp^3)$—H Bonds Through 1,5-Hydrogen Atom Transfer Processes *197*
8.4.2 Asymmetric Transformations Based on Concept 2 *200*
8.5 Summary and Outlook *201*
References *202*

9 **Paired Electrolysis** *209*
Kouichi Matsumoto and Toshiki Nokami
9.1 Introduction *209*
9.2 Paired Electrolysis for Sequential Reactions at both Electrodes *210*
9.2.1 Using an Undivided Cell *210*
9.2.2 Using a Flow Cell *211*
9.3 Paired Electrolysis with Two Different Reactions at both Electrodes *213*
9.3.1 Using an Undivided Cell *213*
9.3.2 Using a Divided Cell *214*
9.3.3 Using a Flow Cell *215*
9.4 Paired Electrolysis for Generation of Two Intermediates to Afford a Final Product by the Sequential Reaction *216*
9.4.1 Using an Undivided Cell *216*
9.4.2 Using a Divided Cell *219*
9.4.3 Using a Flow Cell *220*
9.5 Conclusion *221*
References *221*

Index *225*

Biography

Jun-Ichi Yoshida (1952–2019)

Jun-Ichi Yoshida was born in Osaka, Japan, on 13 November 1952. He graduated from Kyoto University in 1975 and obtained his PhD under the supervision of Professor Makoto Kumada in 1981. He joined the faculty at Kyoto Institute of Technology as an Assistant Professor of the research group of Professor Nariyoshi Kawabata in 1979. In the meantime, he visited the University of Wisconsin in 1982–1983, where he joined the research group of Professor Barry M. Trost. In 1985, he moved to the research group of Professor Sachihiko Isoe of Osaka City University, wherein he was promoted to an Associate Professor position in 1992. He returned to Kyoto University as a full professor in 1994 and served there until 2018. After his retirement, he was appointed as President of the National Institute of Technology, Suzuka College, and the supervisor of the research program "Innovative reactions" of Japan Science and Technology Agency. At the same time, he was chairperson of the Group for Research on Automated Flow and Microreactor Synthesis (GRAMS), the Kinka Chemical Society, Japan, since its founding in 1996. He served also as president of the Society of Synthetic Organic Chemistry, Japan, from 2017 to 2019.

His research interests included flash chemistry, integrated organic synthesis, and organic electron transfer chemistry. He received numerous awards, including the Progress Award of Synthetic Organic Chemistry, Japan (1987), the Chemical Society of Japan Award for Creative Work (2001), the Nagoya Silver Medal of Organic Chemistry (2006), the Humboldt Research Award (2007), the Dougane Award (2010), the Green and Sustainable Chemistry Award (2010), the Ta-shue Chou Lectureship Award (2013), the Chemical Society Japan Award (2013), the Manuel. M. Baizer Award (2014), and the Medal of Honor with Purple Ribbon (2015). His tremendous passion toward chemical research, education, and society lasted until the day before he passed away on 14 September 2019.

13th October 2020 *Toshiki Nokami*

Preface

"*Because this method* [Metal- and chemical-oxidant-free C–H/C–H cross-coupling of aromatic compounds: the use of radical-cation pools] *consists of two sequential steps, namely the generation and accumulation of a radical cation of an aromatic compound under oxidative conditions and then the coupling of the radical cation with another aromatic substrate under nonoxidative conditions, nonselective oxidation of the starting materials and oxidation of the products are avoided.*" (The Yoshida scenario for a controlled dehydrogenative hetero-coupling reaction, T. Morofuji, A. Shimizu, J. Yoshida, *Angew. Chem. Int. Ed.* 2012, *51*, 7259). The problem presented in this milestone paper and its solution are characteristic of the quest for (radical) hetero-cross-dehydrogenative couplings.

This present book is dedicated to its original editor, the late Prof. Yoshida, and his achievements in the field of the title topic.

The late Prof. Yoshida, originally the principal and only editor of this book, sadly passed away in 2019, leaving this important *Organic Redox Chemistry* book project unfinished. As one of its authors, I accepted in November 2019 to finish assembling it, as a tribute to his inspiring scientific achievements. In his memory, I kept as much as possible the original structure laid out by Prof. Yoshida. I had to make nevertheless some necessary adjustments, such as in the order of chapters and their contents. Moreover, in view of these exceptional circumstances, much topical freedom was given to all authors. Some chapters had already been received by Prof. Yoshida before his passing. Some were received in the winter and spring of 2020. The very last chapter to be submitted, which I absolutely wanted for this book, was received in September 2020. I thank all authors for their expert contributions and for their patience. It is now time to move forward.

Beyond the tribute to Prof. Yoshida, one of the reasons I volunteered to finish assembling this book project is that I very much agree with its topical importance and scope. This also coincided well with my recent appointment as Professor at the RWTH Aachen University as well as with the general objectives of my ongoing ERC project ("2O2ACTIVATION": Development of Direct Dehydrogenative Couplings mediated by Dioxygen, grant agreement 716136). I would like to thank at this point Dr. Rajaa Benchouaia and M. Sc. Pooja Vemuri from our research group here in Aachen for their editorial assistance.

The broad and inclusive title *Organic Redox Chemistry, Chemical, Photochemical, and Electrochemical Syntheses* is the original one proposed by Prof. Yoshida. The chapters herein highlight the increasing importance of redox aspects in numerous classes of organic reactions. Indeed, the latter are often governed – at least to some degree – by redox events. Where did the electron go? Its displacement can change to a stark degree the reactivity and life times of key reaction intermediates, as well as their relative philicity (electrophilicity or nucleophilicity). In other words, redox events often control reactivity and selectivity in chemical reactions. Understanding and controlling those processes will thus continue to be at the heart of twenty first century synthetic methodology. I hope the present collection of chapters will provide the reader, both novice and expert, with a vision into this imminent future, as well as with some inspiration to shape it.

13 October 2020

Frédéric W. Patureau
RWTH Aachen University

1

Chemical Oxidative C–C Bond Formation

Koji Hirano

Department of Applied Chemistry, Graduate School of Engineering, Osaka University, Yamadaoka 2-1, Suita, Osaka 565-0871, Japan

1.1 Introduction

Efficient and selective C—C bond formation has been one of the longstanding central topics in synthetic organic chemistry because it is the indispensable methodology for the construction of organic skeletons. In general, an overall redox-neutral process using a carbon electrophile and a carbon nucleophile is employed owing to preferable polarity of two coupling fragments (Scheme 1.1a). On the other hand, the C—C bond-forming reaction with two different nucleophiles in the presence of suitable chemical oxidants (chemical oxidative C—C bond formation) can provide a good alternative to the above overall redox-neutral process particularly when the corresponding carbon electrophile is difficult to prepare (Scheme 1.1b). Moreover, the ultimate direct C—C bond-forming reaction of two simple C—H fragments without any stoichiometric preactivations (e.g. halogenation and metalation) is also theoretically possible. Additionally, the oxidative strategy often enables otherwise challenging C—C bond formations with uniquely high chemo-, regio-, and stereoselectivity. Such complementary features have prompted synthetic chemists to develop numerous strategies for the oxidative C—C bond formation. In this chapter, the recently developed oxidative C—C bond formations are categorized according to the carbon hybridization state of the coupling fragments, and their scope, limitations, and mechanisms are briefly summarized.

1.2 Oxidative Aryl–Alkenyl Bond Formation

Since the aryl–alkenyl π-conjugation frequently occurs in many pharmaceuticals, biologically active compounds, and functional materials, the aromatic $C_{sp}{}^2$–alkenyl $C_{sp}{}^2$ bond formation has been widely explored by many synthetic chemists. The most famous and standard approach is the Mizoroki–Heck reaction with aryl halides and alkenes, in conjunction with a suitable palladium catalyst and base [1, 2]. This

Organic Redox Chemistry: Chemical, Photochemical and Electrochemical Syntheses, First Edition.
Edited by Jun-Ichi Yoshida and Frédéric W. Patureau.

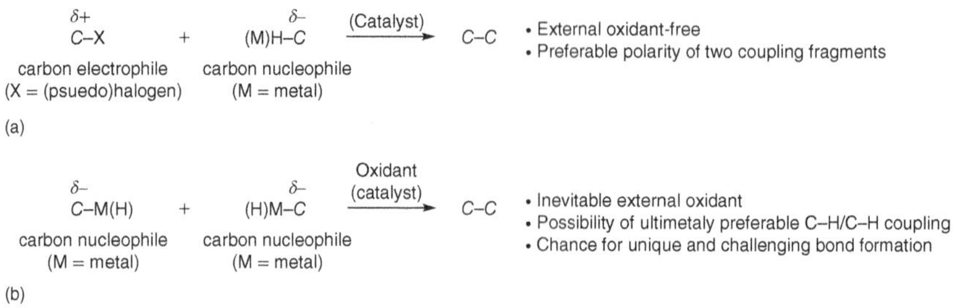

Scheme 1.1 Overall redox-neutral C–C bond formation (a) vs. oxidative C–C bond formation (b).

reaction is the overall redox-neutral process containing oxidative addition and reductive elimination in the catalytic cycle (Scheme 1.2). Although numerous efforts for development of palladium catalysts and their supporting ligands have allowed various aryl halides, including unactivated aryl chlorides, to be adopted, the alkene fragments are still largely limited to electronically activated α,β-unsaturated carbonyls and styrenes. Moreover, the preparation of the corresponding aryl halides from the parent arenes (stoichiometric halogenation) is an additional drawback to be addressed. The chemical oxidative coupling approach can be a good solution to the above problems inherent in the classical Mizoroki–Heck reaction. In this section, the oxidative Mizoroki–Heck reaction with arylmetal reagents as aromatic $C_{sp}{}^2$ fragments and direct aromatic $C_{sp}{}^2$–alkenyl $C_{sp}{}^2$ coupling (Fujiwara–Moritani reaction) are mainly presented.

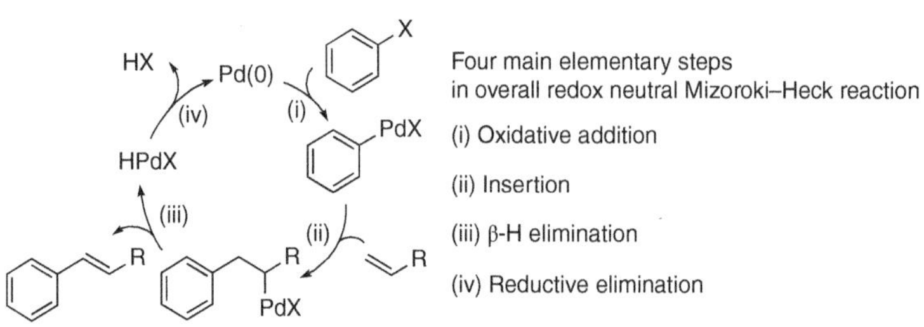

Scheme 1.2 General mechanism of palladium-catalyzed overall redox-neutral Mizoroki–Heck reaction of aryl halides with alkenes.

1.2.1 Oxidative Mizoroki–Heck Reaction with Arylmetal Reagents

As mentioned in the above introduction part, the redox-neutral Mizoroki–Heck reaction still suffers from the relatively narrow scope of alkenes. The oxidative Mizoroki–Heck reaction can address the problem probably because of the formation of more reactive, coordinately unsaturated arylpalladium species through transmetalation between PdX_2 and arylmetal reagents rather than the oxidative addition of aryl halides (Scheme 1.3).

In 2008, White and coworkers reported the $Pd(OAc)_2$/bis(sulfoxide) catalyst for the oxidative Mizoroki–Heck reaction with arylboronic acids [3]. In the presence

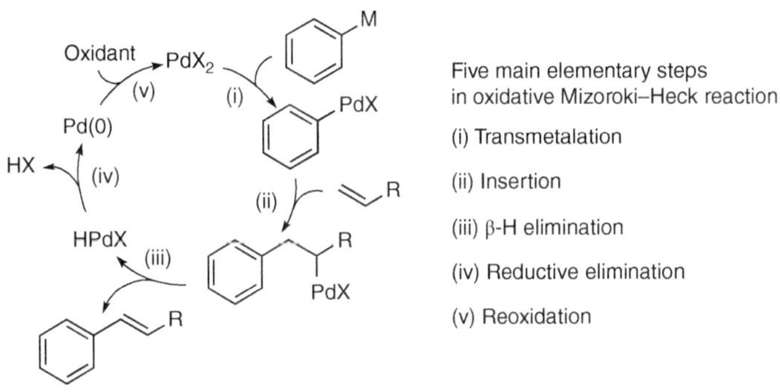

Scheme 1.3 General mechanism of palladium-catalyzed oxidative Mizoroki–Heck reaction of arylmetal reagents with alkenes.

of a benzoquinone (BQ) terminal oxidant, unactivated aliphatic terminal alkenes undergo the Mizoroki–Heck-type arylation (Scheme 1.4). Milder reaction conditions are compatible with somewhat labile point chirality derived from α-amino acids as well as functional groups such as a free carboxylic acid. The regioselectivity (internal/terminal) is also well controlled in most cases, but the olefinic position of product (styrenyl/allylic) is highly dependent on the substrate structure and its control still remains a challenging task. A related Pd(I*i*Pr)$(OTs)_2$ catalysis was reported by Sigman and Werner in 2010 (Scheme 1.5): the beneficial point is the use of molecular oxygen as an terminal oxidant, where $Cu(OTf)_2$ is added as a co-oxidant [4]. Also in this case, the reaction proceeds without erosion-of-point chirality. Particularly notable is the high styrenyl/allylic selectivity as well as internal/terminal selectivity in almost all cases.

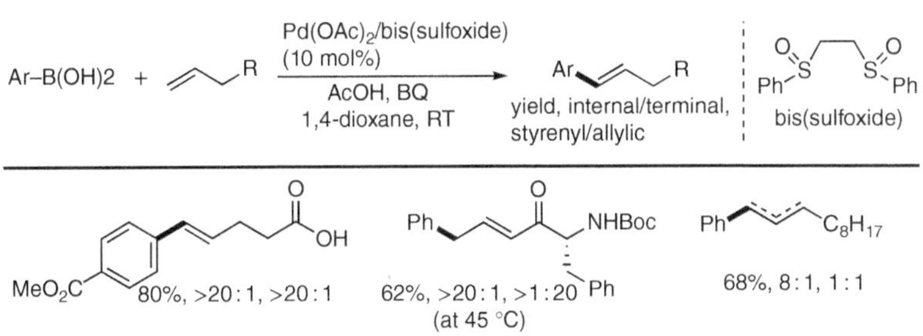

Scheme 1.4 $Pd(OAc)_2$/bis(sulfoxide)-catalyzed oxidative Mizoroki–Heck reaction of unactivated terminal alkenes. BQ = benzoquinone.

While not aryl–alkenyl bond formation, the group of Sigman subsequently developed the enantioselective oxidative Mizoroki–Heck reaction of internal alkenyl alcohols by using the chiral pyridine-oxazoline hybrid ligand, PyrOx (Scheme 1.6) [5]. The key to success is the redox-relay process: the alkene is migrated toward the alcohol via an iterative β-H elimination and insertion, and finally converted to the carbonyl functionality by the formal oxidation event. As a result, the regioselective and enantioselective remote arylation of carbonyl compound is possible. This

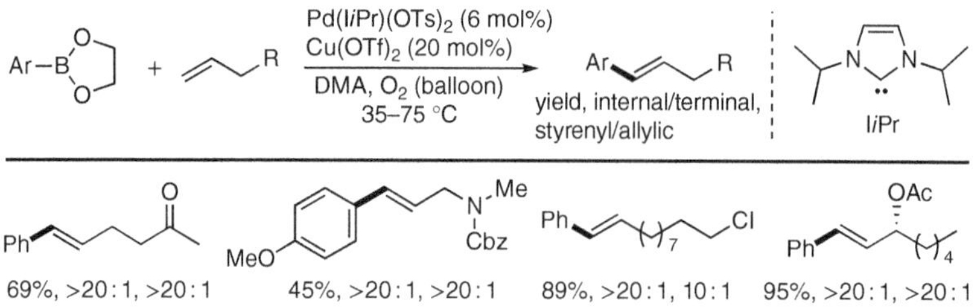

Scheme 1.5 Pd(I*i*Pr)(OTs)-catalyzed oxidative Mizoroki–Heck reaction of unactivated terminal alkenes. Ts = *p*-toluenesulfonyl, Tf = trifluoromethanesulfonyl.

strategy is also applicable to alkenyl aldehydes and enelactams to deliver the remotely arylated enantioenriched α,β-unsaturated aldehydes and α,β-unsaturated δ-lactams, respectively (Scheme 1.7) [6].

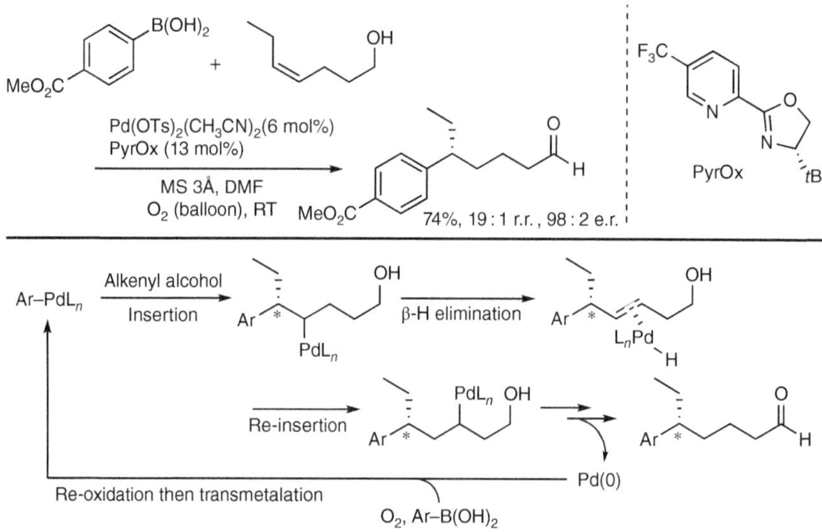

Scheme 1.6 Palladium-catalyzed enantioselective redox-relay oxidative Mizoroki–Heck reaction of internal alkenyl alcohols and its redox-relay mechanism. Source: Modified from Mei et al. [5].

1.2.2 Direct Oxidative Mizoroki–Heck Reaction with Arene C–Hs (Fujiwara–Moritani Reaction)

One of the biggest drawbacks in the above Mizoroki–Heck reaction with aryl halides or aryl boronic acids is their tedious preparation from the parent simple arenes. In 1969, Fujiwara et al. reported seminal work on the palladium-catalyzed coupling reaction of simple arenes and alkenes, in the presence of $Cu(OAc)_2$ or AgOAc terminal oxidant, to form the corresponding alkenylarenes directly (Scheme 1.8) [7]. This protocol received significant attention from the viewpoint of organic synthesis because the arene C–H and alkene C–H are directly cross-coupled without any preactivation steps of both starting substrates. Since then, such a

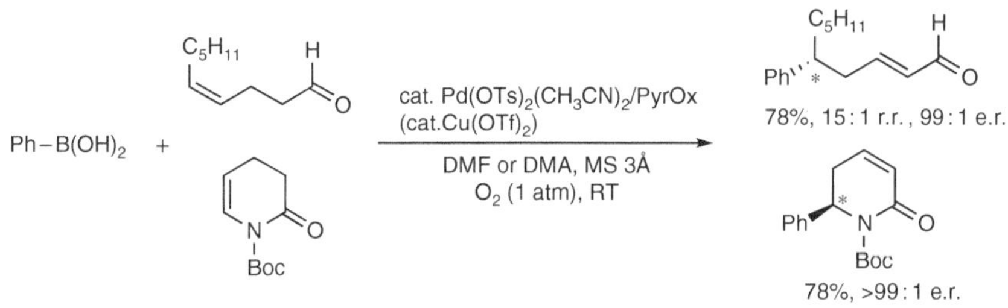

Scheme 1.7 Palladium-catalyzed enantioselective redox-relay oxidative Mizoroki–Heck reaction of alkenyl aldehydes and enelactams. Source: Modified from Zhang et al. [6a], Yuan and Sigman [6b].

transition-metal-promoted "C–H activation" strategy has greatly and rapidly progressed by efforts of many research groups, and the Fujiwara–Moritani reaction is now a powerful synthetic tool for the construction of aryl–alkenyl π-conjugation. However, the disadvantage of early studies on the Fujiwara–Moritani reaction is the inevitable use of excess arene substrates (in many cases solvent amount).

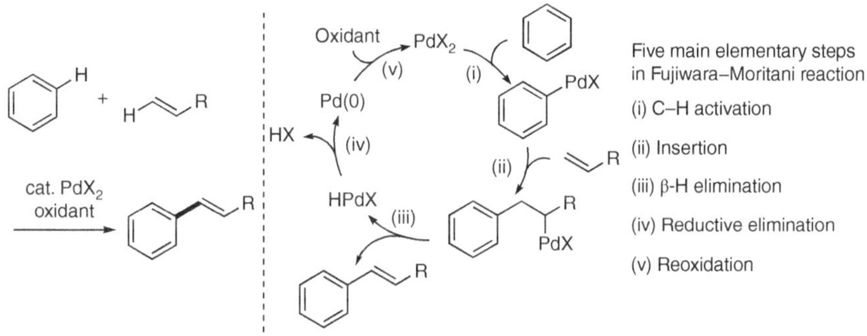

Scheme 1.8 General reaction scheme and mechanism of palladium-catalyzed Fujiwara–Moritani reaction. Source: Modified from Fujiwara et al. [7a], Jia et al. [7b].

In this context, Yu and coworkers reported the Fujiwara–Moritani reaction with the simple arene as the limiting reagent (1.0 equiv) under Rh(II)/PCy$_3$/Cu(TFA)$_2$/V$_2$O$_5$ oxidative catalysis (Scheme 1.9) [8]. Although the exact role of PCy$_3$ ligand as well as copper and vanadium combined oxidation system still remains to be elucidated, the reaction proceeds smoothly even in the presence of 1 equiv of simple arenes. More recently, the same research group developed the well-defined and robust Pd(OAc)$_2$/hydroxypyridine catalyst for the reaction with much broader simple arenes, including benzene derivatives, heteroaromatics, and even more challenging complex drug-like molecules (Scheme 1.10) [9]. Also in this case, the arene substrate works well even at 1 equiv loading. The well-designed hydroxypyridine ligand is key to success, and its pivotal role in the C–H activation step of otherwise unreactive simple arene is also uncovered by computational studies with density functional theory (DFT) calculation. Although the alkene coupling partners are still limited to electronically activated acrylates and styrenes, the above work successfully overcomes the big issue in the conventional Fujiwara–Moritani reaction.

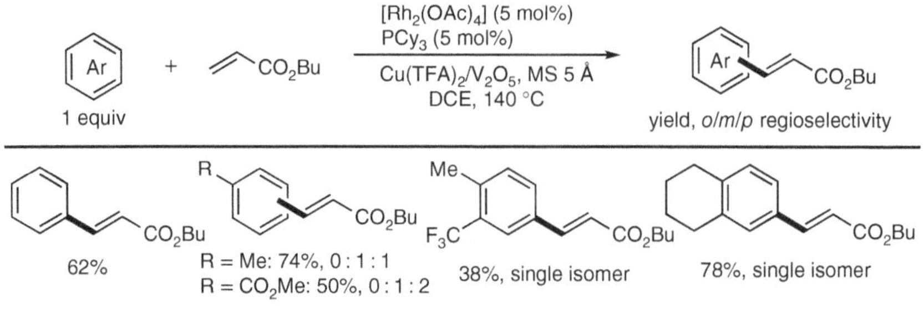

Scheme 1.9 Rhodium(II)-catalyzed Fujiwara–Moritani reaction of simple arenes as limiting reagents. DCE = 1,2-dichloroethane. Source: Modified from Vora et al. [8].

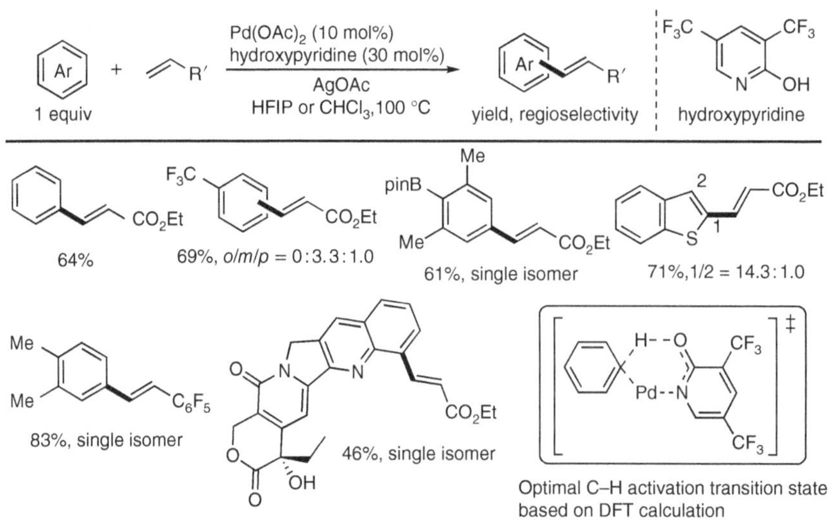

Scheme 1.10 Palladium/hydroxypyridine-catalyzed Fujiwara–Moritani reaction of simple arenes as limiting reagents and proposed transition state based on DFT calculation. HFIP = 1,1,1,3,3,3-hexafluoroisopropyl alcohol. Source: Modified from Wang et al. [9].

Additional inherent problem of the Fujiwara–Moritani reaction is the regioselectivity: for example, when the mono-substituted arene is employed, there are three possible reaction sites, namely ortho, meta, and para C–Hs, and its control is also of great importance in chemical synthesis. A good solution of such a regioselective issue is the introduction of coordinating functional group (directing group): a suitable functional group coordinates to the metal center to promote the C–H activation at the proximal ortho position. The formed five- or six-membered metalacycle intermediate undergoes the insertion reaction with the alkene substrate, eventually leading to the ortho-alkenylated product with high regioselectivity. Since Murai's et al. milestone work on the ruthenium-catalyzed ortho-alkylation of ketones [10], numerous research groups joined this field to develop various directing groups containing nitrogen, oxygen, sulfur, phosphorus, and even less polar C—C π-bond [11]. A representative example of indole carboxylic acids is shown in Scheme 1.11. In this case, the carboxyl group works as a unique "traceless" ortho-directing group: 2- and 3-indole carboxylic acids react with the acrylate with concomitant decarboxylation under palladium catalysis to furnish 3- and 2-alkenylated indoles, regioselectively [12].

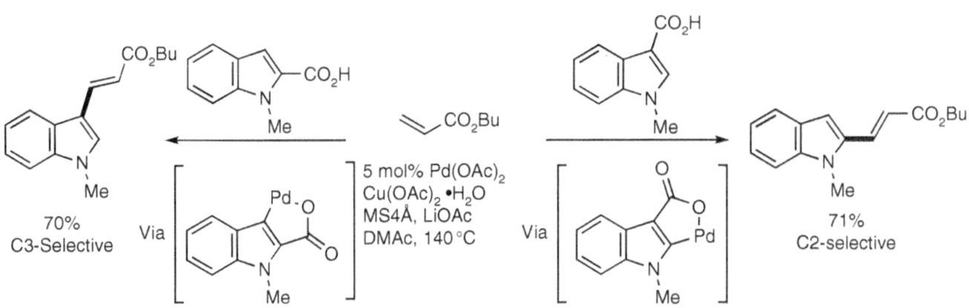

Scheme 1.11 Carboxylic-acid-directed ortho-selective Fujiwara–Moritani reaction of indole carboxylic acids with concomitant decarboxylation.

In addition to the palladium salts, Cp*Rh(III) catalysts also show high performance in the directed Fujiwara–Moritani reaction. In 2007, Satoh and coworkers reported the pioneering work on the Cp*Rh(III)-catalyzed ortho-selective alkenylation of benzoic acids with acrylates (Scheme 1.12) [13]. The initially formed alkenylated product spontaneously undergoes intramolecular Michael-type addition of carboxylic acid directing group to form the observed lactone derivative. Additional beneficial point of the Cp*Rh(III) catalysis is the reoxidation system: the most environmentally benign atmospheric molecular oxygen works as a terminal oxidant in the presence of a catalytic amount of internal oxidant, $Cu(OAc)_2{\cdot}H_2O$. Thus, the formed byproduct is only nontoxic water. Since then, more and more

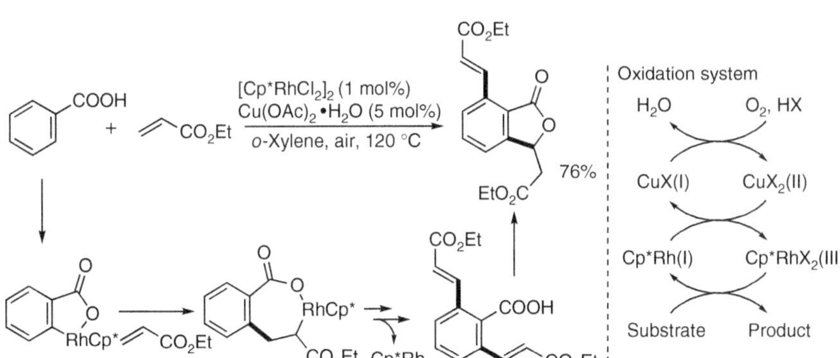

Scheme 1.12 Cp*Rh(III)-catalyzed ortho-selective Fujiwara–Moritani reaction of benzoic acids with acrylates. Cp* = 1,2,3,4,5-pentamethylcyclopentadienyl. Source: Modified from Ueura et al. [13].

oxidative coupling reactions between aromatic compounds and alkenes or alkynes have been explosively developed under the Rh(III)/Cu(II) or related Rh(III)/Ag(I) oxidative catalysis [14].

The directed meta- or para-alkenylation is much more challenging than the ortho-alkenylation because of formation of kinetically less favored medium- and large-sized metalacycle intermediate. However, some seminal works recently appeared. In 2012, Yu and coworkers elegantly designed the nitrile-based, U-shaped template to direct the meta-selective Fujiwara–Moritani reaction under Pd/Ag catalysis (Scheme 1.13) [15]. Owing to the end-on coordinating nature of nitrile, the relatively large and unique Pd/Ag bimetallic metalacycle is formed as the key intermediate, which is supported by computational studies in the follow-up article [16]. This work prompted several researchers to develop the related directing groups for the meta-C–H alkenylation of various aromatic compounds, but all of them still rely on the nitrile functionality [17]. The same strategy is also effective for the rhodium-catalyzed meta-C–H alkenylation (Scheme 1.14) [18].

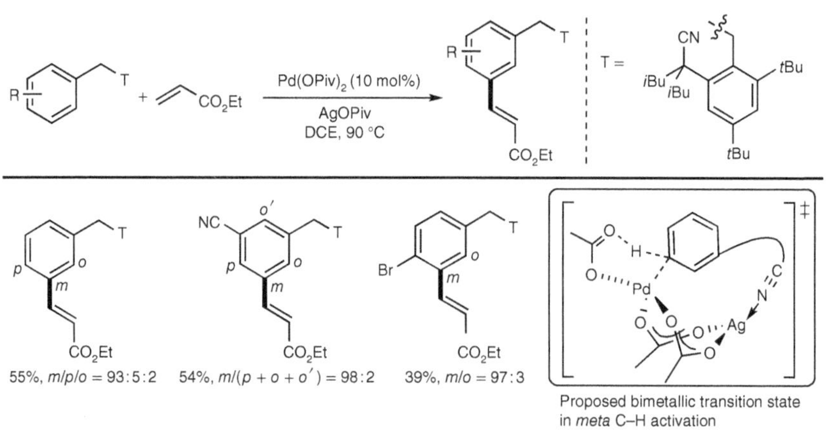

Scheme 1.13 Palladium/silver-catalyzed meta-selective Fujiwara–Moritani reaction assisted by nitrile-based U-shaped template and proposed bimetallic transition state in C–H activation. Piv = *tert*-butylcarbonyl. Source: Modified from Leow et al. [15].

In 2015, Maiti and coworkers developed a similar nitrile-based D-shaped template for the para-selective Fujiwara–Moritani reaction (Scheme 1.15) [19]. The biggest feature is the long biphenyl template ligated with flexible Si tether of sp^3 hybridization. Additionally, the positive Thorpe–Ingold effect is successfully promoted by two bulky isopropyl groups on Si. In the presence of a $Pd(OAc)_2$ catalyst and amino acid ligand, Ac-Phe-OH, a variety of benzene derivatives undergo the alkenylation selectively at the para position beyond their innate electronic biases.

1.3 Oxidative Aryl–Aryl Bond Formation

Due to the ubiquity of biaryl structure in biologically active compounds, natural products, pharmaceutical targets, and organic functional materials, the aromatic

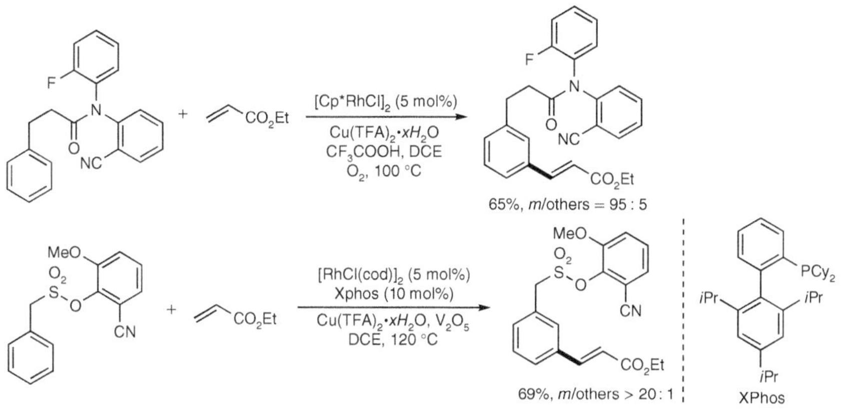

Scheme 1.14 Rhodium-catalyzed meta-selective Fujiwara–Moritani reaction assisted by nitrile-based U-shaped template. Source: Modified from Xu et al. [18a], Bera et al. [18b].

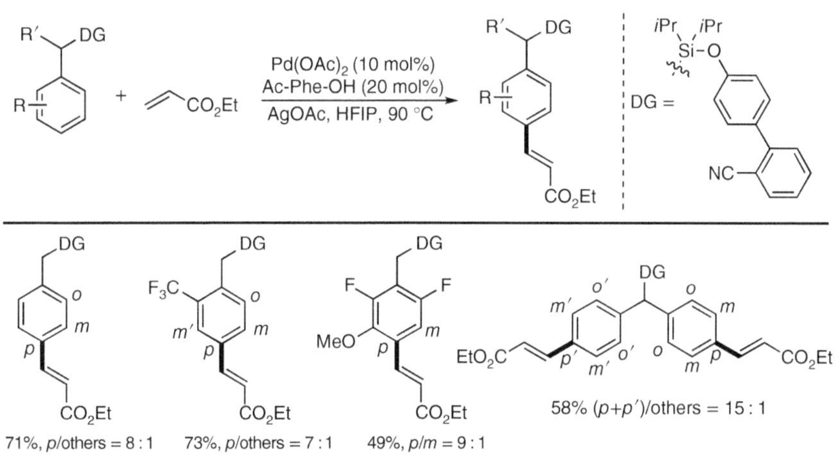

Scheme 1.15 Palladium-catalyzed para-selective Fujiwara–Moritani reaction assisted by nitrile-based D-shaped template. Ac-Phe-OH = *N*-acetylphenylalanine. Source: Modified from Bag et al. [19a], Patra et al. [19b].

$C_{sp}{}^2$–aromatic $C_{sp}{}^2$ bond-forming reaction is always one of the hot research topics in synthetic organic chemistry. The Nobel Prize–winning palladium-catalyzed overall redox-neutral cross-coupling reaction of aryl halides with arylmetal reagents is now the most powerful and reliable approach to the above biaryl linkage (Scheme 1.16, left). On the other hand, the oxidative aryl–aryl bond-forming reaction can replace the aryl halide electrophiles with the simple and readily accessible arenes (C–H/C–M cross-coupling; Scheme 1.16, right). Such an oxidative coupling protocol often enables the challenging biaryl coupling under the redox-neutral conditions. Moreover, an ideal oxidative cross-coupling of two simple arenes (C–H/C–H cross-coupling) is potentially possible. In this section, the recent advances in the above two types of oxidative biaryl coupling are demonstrated.

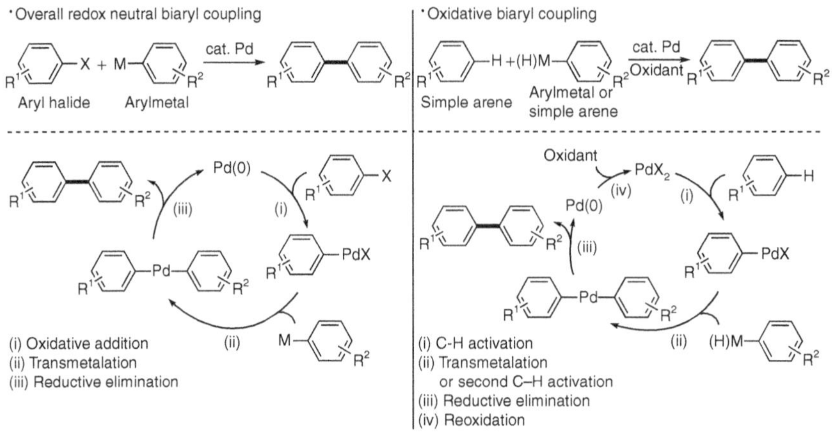

Scheme 1.16 General mechanisms of overall redox-neutral biaryl cross-coupling of aryl halide and arylmetal (left) and oxidative biaryl cross-coupling of simple arene and arylmetal or another simple arene (right).

1.3.1 Oxidative C–H/C–M Biaryl Cross-Coupling

Electron-rich heterocycles such as indole, thiazole, and pyrrole relatively easily participate in the oxidative biaryl cross-coupling reaction with arylmetals (Scheme 1.17). In the case of indole, the simple $Pd(OAc)_2$-catalyzed C–H arylation with phenylboronic acid often proceeds even at room temperature by using the molecular oxygen as the sole oxidant [20]. Owing to its experimental simplicity, the reaction is also easily scaled up to a gram quantity. The thiazole is also arylated in the presence of a $Pd(OAc)_2$/1,10-phenanthroline (phen) catalyst and 2,2,6,6-tetramethylpiperidine-1-oxyl (TEMPO) oxidant [21]. The additional unique feature is the otherwise challenging C4 selectivity. Although generally less reactive, the abundant first-row transition metals also promote similar reactions. For example, $Cu(TFA)_2$ mediates the multiple oxidative arylation of pyrrole with excess phenylboronic acid to form the tetraphenylpyrrole in one shot [22].

The 1,3-azoles have relatively acidic C–Hs at the C2 position, and thus they are more reactive under somewhat basic conditions. The oxazole and imidazole are directly arylated selectively at the C2 position under cost-effective nickel catalysis (Scheme 1.18) [23]. Notably, in the latter case, the less reactive arylsilane also works well as the arylation reagent.

The benzene derivatives are less reactive than heteroarenes mentioned above, and the suitable coordinating functional groups, namely directing groups, are generally essential for obtaining satisfactory reactivity as well as control of regioselectivity (Scheme 1.19). Similar to the Fujiwara–Moritani reaction (Schemes 1.11 and 1.12), the carboxyl group is the effective ortho-directing group under oxidative Pd(II)/BQ catalysis to deliver the corresponding arylated product in a good yield [24]. With the assistance of well-designed N,N-bidentately coordinating group, the abundant copper salt also catalyzes the C–H arylation with the arylboronate in the presence of Ag_2O oxidant [25].

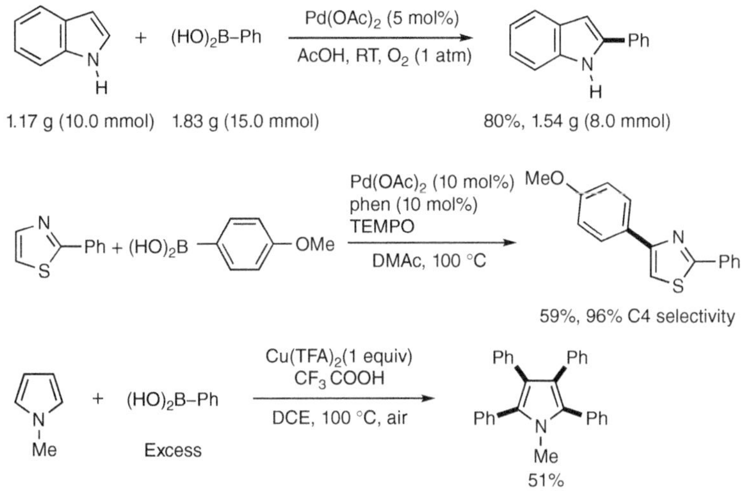

Scheme 1.17 Oxidative C–H/C–M biaryl coupling of electron-rich indole, thiazole, and pyrrole.

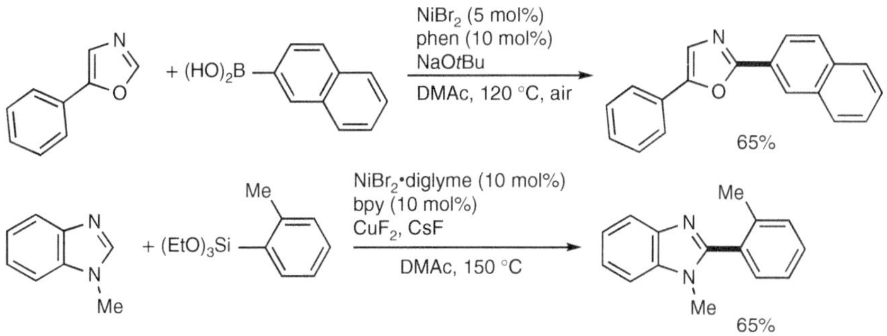

Scheme 1.18 Nickel-catalyzed oxidative C–H/C–M biaryl coupling of relatively acidic C2 C–Hs of 1,3-azoles. bpy = 2,2′-bipyridyl. Source: Modified from Hachiya et al. [23].

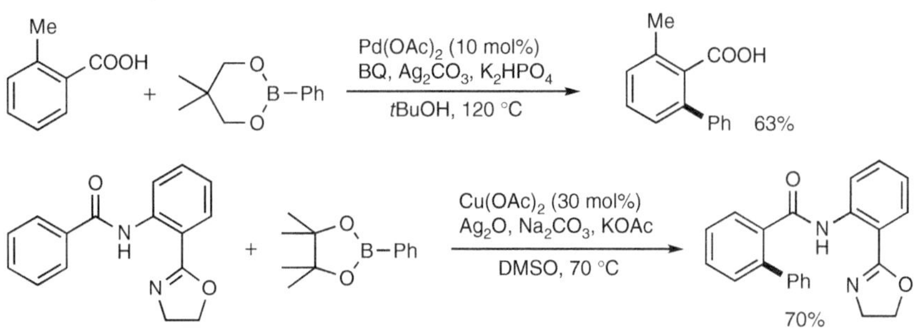

Scheme 1.19 Oxidative C–H/C–M biaryl coupling of benzene derivatives with assistance of directing groups.

Exceptionally, the highly fused benzene derivatives show a remarkably high reactivity under the oxidative C–H arylation conditions because of their lower aromaticity, i.e. alkene-like reactivity (Scheme 1.20). In 2011, Itami and coworkers reported the Pd(II)/*o*-chloranil oxidative catalyst for the C–H arylation of pyrene derivatives with arylboroxines [26]. The reaction is apparently unique to the higher-fused aromatics but of great interest in the bottom-up synthesis of π-extended polyaromatic hydrocarbons (PAHs), which have received tremendous attention in the field of material science. Actually, application to the rapid and concise synthesis of extended PAHs was demonstrated by the C–H arylation/Scholl reaction sequence. Subsequently, the same catalyst system was successfully applied to the K-region-selective one-shot π-extension with the dibenzosiloles [27]. The *o*-chloranil is the key of catalysis, and its multitask nature as the ligand, oxidant, and base was recently uncovered by computational studies with DFT calculation.

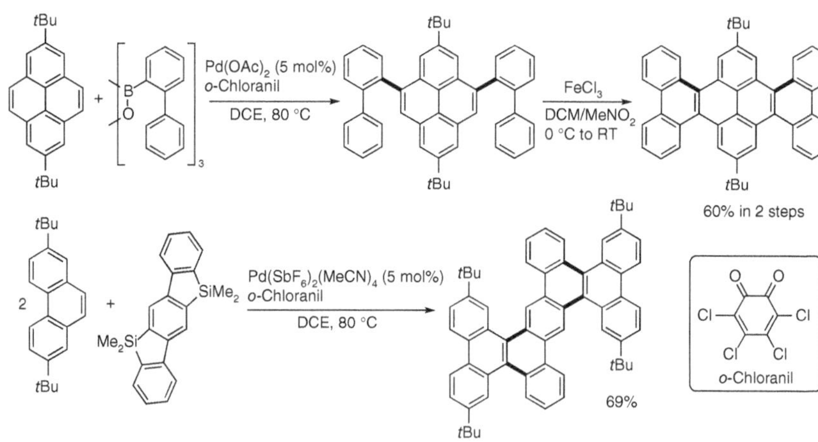

Scheme 1.20 Pd/*o*-chloranil-catalyzed oxidative C–H/C–M biaryl coupling of higher fused benzene derivatives. DCM = dichloromethane.

As mentioned in Schemes 1.17–1.20, the air-stable and easy-to-handle arylboronic acids and arylsilanes are frequently employed as the arylmetals in the oxidative C–H/C–M biaryl coupling reaction. Although still limited in scope, more readily available benzoic acids can also couple with some arene C–Hs, via decarboxylation, to afford the corresponding biaryls under appropriate oxidative conditions (Scheme 1.21). The decarboxylative C–H arylation of indole and oxazole derivatives efficiently occurs in the presence of Pd(II)/Ag(I) or Cu(II) oxidative catalysts [28]. The copper salt alone also mediates a similar reaction of benzamide substrate that bears the suitable 8-aminoquinoline-based N,N-bidentate coordination [29].

1.3.2 Oxidative C–H/C–H Biaryl Cross-Coupling

The direct oxidative coupling of two different arenes via dual C—H bond cleavage is the most attractive and ideal approach to the biaryl structure from the viewpoint of

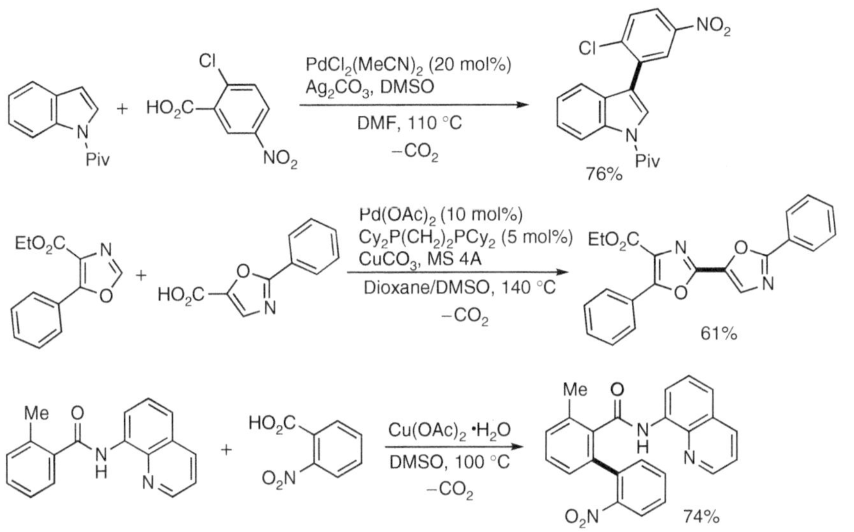

Scheme 1.21 Oxidative biaryl coupling via decarboxylative C–H arylation.

atom and step economy because the preactivation step of both starting substrates can be skipped. The biggest challenge is chemoselectivity for the desired cross-coupling product over the homo-coupling byproducts. The first breakthrough was reported by Fagnou and coworker in 2007: the oxidative $Pd(TFA)_2$/3-nitropyridine/$Cu(OAc)_2$ system promotes the highly selective C–H/C–H biaryl coupling of indole and simple benzene to form the C3-arylated indole selectively (Scheme 1.22). The undesired homo-coupling products from both indole and benzene are not detected at all, and the selectivity switching is thus perfectly operated in the first and second C–H activation steps. Subsequently, related oxidative couplings of benzofurans and arylpyridines were reported. Recently, the direct coupling of benzoxazole with electron-deficient pyridines was also achieved under Pd(II)/organic bromide oxidative catalysis, where the C2/C3 regioselectivity was controlled by the judicious choice of organic halide oxidant. However, all procedures rely on excess of simple benzene derivatives (c. 30 equiv to solvent amount), which still remains to be improved [30].

By replacing the simple benzenes with more reactive heteroarenes or suitably functionalized arenes such as indazole, furan, and polyfluoroarene, the loading can be dramatically decreased to the practical amount (~3.0 equiv) (Scheme 1.23). The former reaction is applied to the discovery of photostable near-infrared (NIR) probe for mitochondria. In the latter case, the highly selective cross-dehydrogenative polycondensation of bithiophene and octafluoronaphthalene is possible to form the corresponding polymer with a good yield, molecular weight, and M_w/M_n [31]. Thus, the C–H/C–H biaryl coupling now provides an opportunity to develop new biaryl-based functional materials, which are otherwise difficult to access by the conventional methods.

In addition to the palladium, other transition metals also promote the direct biaryl coupling (Scheme 1.24). The Cp*Rh(III) complex catalyzes the C–H/C–H

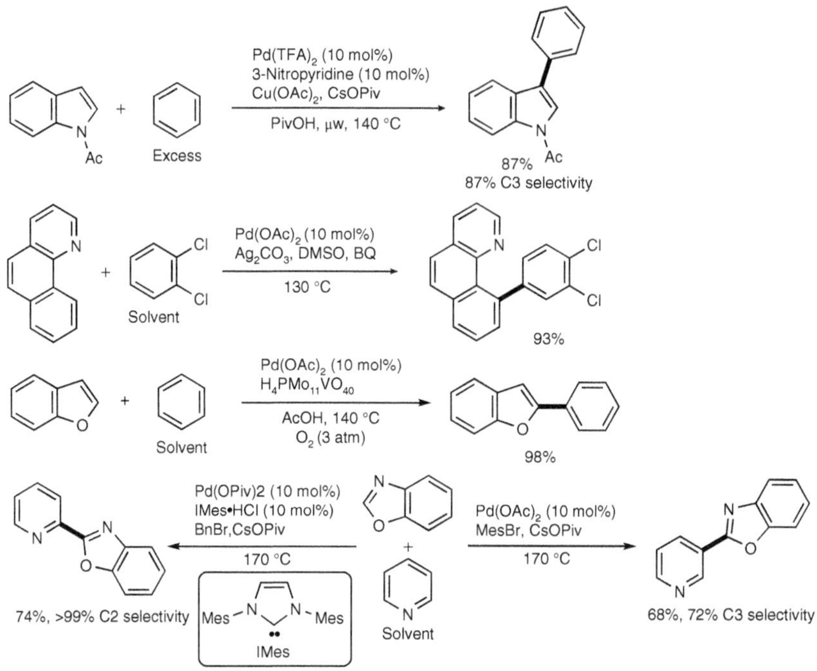

Scheme 1.22 Oxidative C–H/C–H biaryl coupling with simple benzene and pyridine substrates. Mes = 2,4,6-trimethylphenyl.

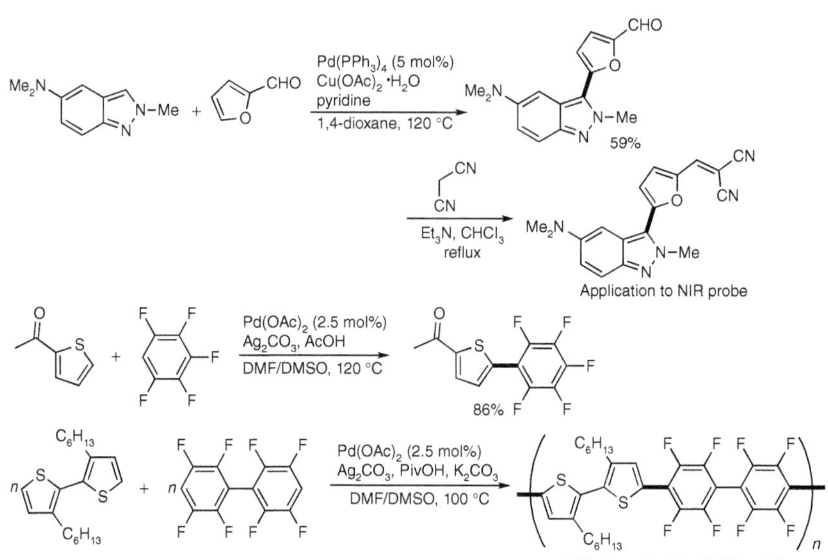

Scheme 1.23 Oxidative C–H/C–H biaryl coupling with relatively activated heteroarenes or functionalized arenes. M_n = the number average of molecular weight; M_w = the weight average of molecular weight.

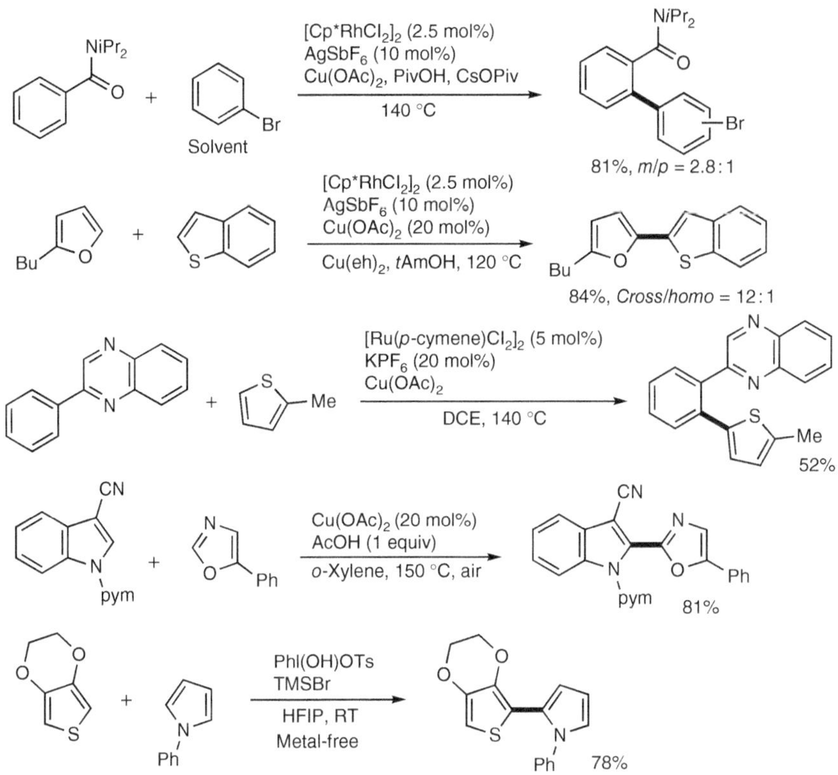

Scheme 1.24 Oxidative C–H/C–H biaryl couplings promoted by transition metals other than palladium or I(III) reagents. pym = 2-pyrimidyl.

coupling of benzamides with a solvent amount of bromobenzene in preference to the conceivably more reactive aryl C—Br bond of bromobenzene. Additionally, the same catalyst couples simple furans with thiophenes chemoselectively even in the absence of any directing groups [32]. The less expensive Ru(*p*-cymene)Cl_2 is also efficient for the direct coupling of arylpyridines and thiophenes [33]. The direct coupling between *N*-pyrimidylindoles and 1,3-azoles proceeds in the presence of $Cu(OAc)_2$ catalyst alone. In this case, the molecular oxygen works as the terminal oxidant, thus providing an environmentally benign process with H_2O as the sole byproduct [34]. Furthermore, metal-free coupling between thiophenes and pyrroles can also be achieved by using a well-defined hypervalent-iodine(III) reagent [35]. The stepwise addition of thiophene followed by pyrrole is essential, but the reaction occurs even at room temperature with high cross-coupling selectivity.

1.4 Oxidative Aryl–Alkynyl Bond Formation

Arylacetylenes are among the most fundamental and important π-conjugated systems in various areas of organic chemistry. The most powerful and reliable approach to these molecules is the palladium/copper-cocatalyzed cross-coupling of aryl halides with terminal alkynes, also known as Sonogashira coupling (Scheme 1.25, top). Given the preparation of starting aryl halides from the parent simple arenes, the oxidative direct coupling between arenes and terminal alkynes via twofold C—H bond cleavage of both substrates is an ideal protocol (Scheme 1.25, bottom). However, the C_{sp}–H at the alkyne termini readily reacts with transition metals, and thus the biggest challenge is the selectivity control toward cross-coupling over alkyne homo-coupling under oxidative conditions.

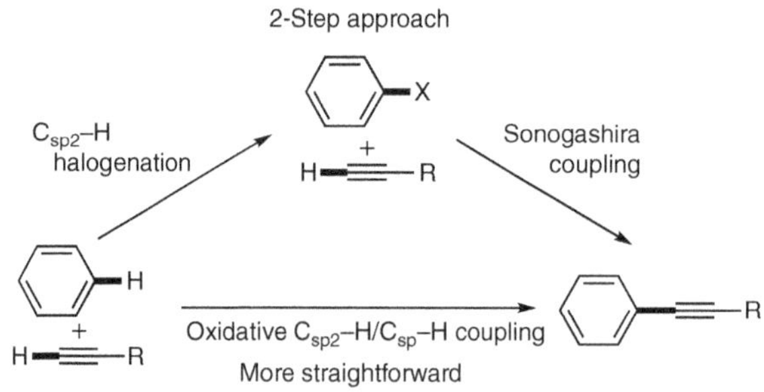

Scheme 1.25 Approaches to arylacetylenes via aryl–alkynyl coupling.

The first successful attempt of 1,3-azoles using $CuCl_2$ as a catalyst was reported by Miura and coworkers in 2010 (Scheme 1.26) [36]. Although the relatively high catalyst loading is essential, the pioneering work has stimulated further discoveries of other transition metal catalysts. These include the use of nickel [37] and palladium [38] complexes for the direct alkynylation of several acidic 1,3-azoles with the assistance of relatively strong base, LiO*t*Bu. It is noteworthy that molecular oxygen is the sole oxidant in these transformations.

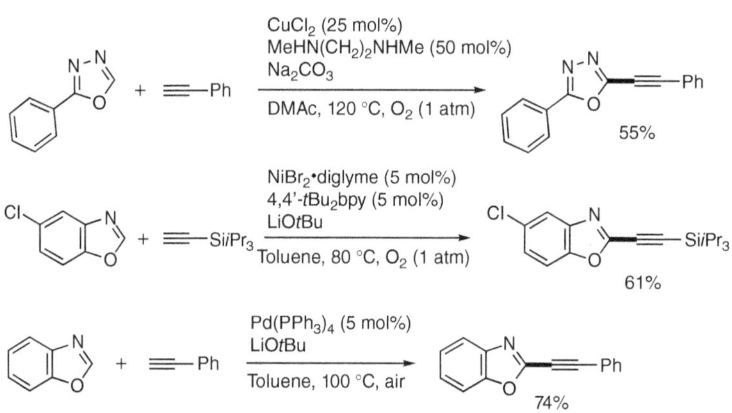

Scheme 1.26 Copper-, nickel-, and palladium-catalyzed oxidative aryl–alkynyl couplings of relatively acidic 1,3-azoles. Source: Modified from Kitahara et al. [36a], Wei et al. [36b].

The alkynylation of electron-rich indoles and pyrroles is achieved by a Ph_3PAuCl catalyst (Scheme 1.27) [39]. However, the use of $PhI(OAc)_2$ as the oxidant is critical for success. Subsequently, the more robust palladium-based catalyst system was reported, in which electron-rich thiophene, furan, oxazole, thiazole, and pyrazole as well as indole and pyrrole were accommodated [40].

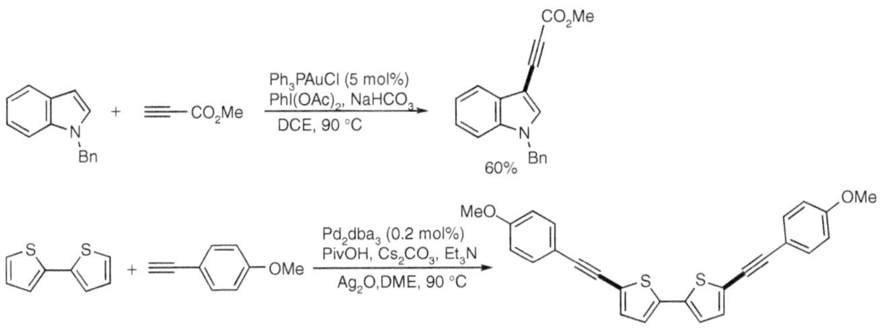

Scheme 1.27 Gold- and palladium-catalyzed oxidative aryl–alkynyl couplings of electron-rich heteroaromatics. dba = dibenzylideneacetone. Source: Modified from de Haro et al. [39a], Jie et al. [40].

The benzene derivatives can also be employed by using suitable directing groups (Scheme 1.28). The aniline derivatives are oxidatively coupled with tri(isopropyl)silylacetylene in the presence of a $Pd(acac)_2$ catalyst and BQ terminal oxidant [41]. However, to suppress the undesired homo-coupling of terminal acetylene, the somewhat tedious slow addition technique is inevitable. The well-designed N,N-bidentately coordinating groups allow the benzoic acid derivatives to be adopted in this reaction [42]. The phenyloxazine-containing benzamides undergo

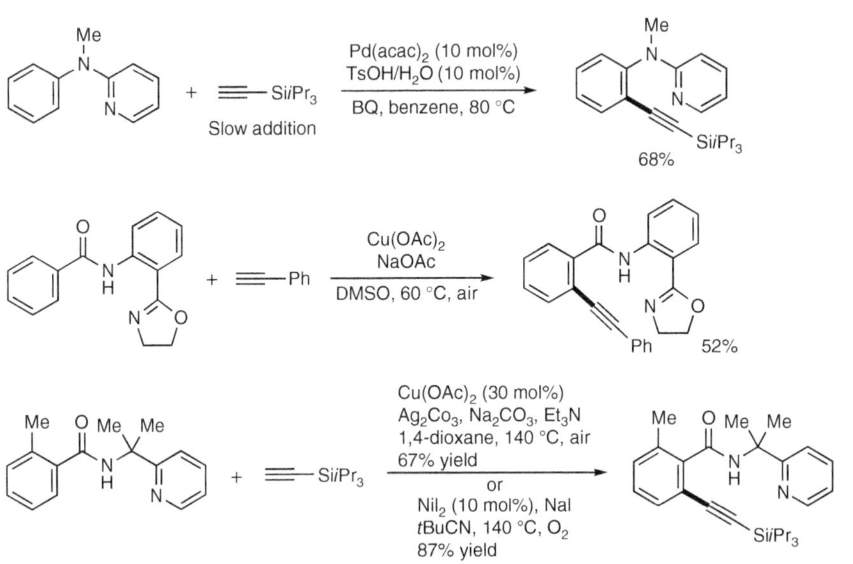

Scheme 1.28 N,N-bidentately coordinating group-promoted oxidative aryl–alkynyl couplings of benzene derivatives. acac = acetylacetonate.

the Cu(II)-mediated C–H/C–H coupling with various terminal alkynes, including alkylacetylenes and even more challenging arylacetylenes. While limited to the tri(isopropyl)silylacetylene, the $Cu(OAc)_2/Ag_2CO_3$ system promotes the oxidative coupling of benzamides derived from the 2-(pyridine-2-yl)isopropylamine. The same reaction occurs with better reaction efficiency under more environmentally benign $Ni(II)/O_2$ oxidative conditions.

1.5 Oxidative C–C Bond Formation at $C_{sp}{}^3$ Center

The oxidative cross-coupling reaction at the saturated $C_{sp}{}^3$ center still remains a great challenge because of the lack of π bond coordination to the reactive metal catalyst. However, the $C_{sp}{}^3$—H bond adjacent to heteroatoms (e.g. O, N, S) or π-bonds (e.g. alkene and benzene) shows exceptionally high reactivity under oxidative conditions: it can be easily oxidized to the cationic intermediate by the action of electron-donating nature of proximal heteroatoms or π-bonds to readily participate in the addition reaction with external nucleophiles (Scheme 1.29). Since the stimulating work on the oxidative coupling between *N*-methylanilines and terminal alkynes was reported by Li and coworkers in 2004, such an oxidative coupling at the $C_{sp}{}^3$ center was called cross-dehydrogenative coupling (CDC) and extensively explored by many research groups. The comprehensive review article was already published [43], and selected important examples are thus presented in this section.

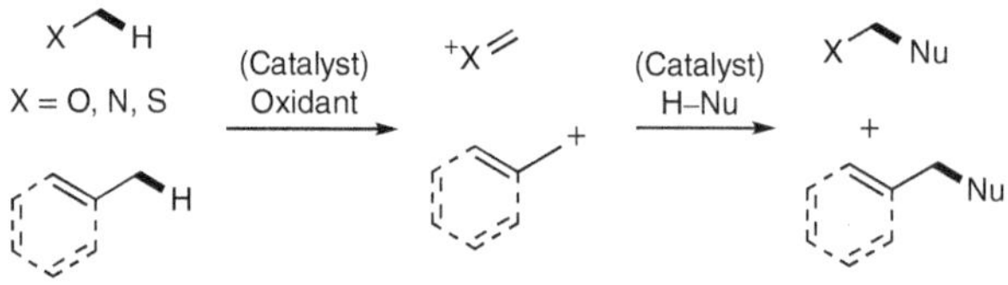

Scheme 1.29 General mechanism of cross-dehydrogenative coupling (CDC) of $C_{sp}{}^3$–H bonds adjacent to heteroatoms or π-bonds.

The tetrahydroisoquinoline is one of the most frequently employed substrates in the CDC reaction because the cationic intermediate can be doubly stabilized by proximal nitrogen atom and benzene ring. Additionally, the suitable chiral catalyst renders the reaction enantioselective (Scheme 1.30) [44]. Li developed the Cu(I)/Ph-pybox-catalyzed enantioselective CDC reaction of *N*-phenyltetrahydroisoquinoline with terminal alkynes. The enantioselectivity is still moderate (up to 87 : 13 e.r.), but the point chirality can be readily constructed from the unfunctionalized simple starting materials. In this case, *t*BuOOH works well as the oxidant. Sodeoka also reported the Pd(II)/DM-Segphos-catalyzed enantioselective coupling of *in situ*-generated *N*-Boc-tetrahydroisoquinolines and malonate nucleophiles in the presence of 2,3-dichloro-5,6-dicyano-1,4-benzoquinone (DDQ) terminal oxidant. The reaction proceeds smoothly even at room temperature, and the maximum 93 : 7 enantiomeric ratio was obtained.

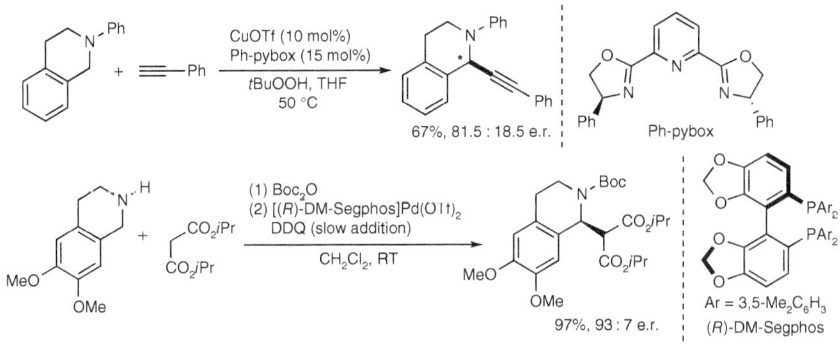

Scheme 1.30 Enantioselective CDC reactions of tetrahydroisoquinoline derivatives. DDQ = 2,3-dichloro-4,5-dicyanobenzoquinone. Source: Modified from Li and Li [44a], Dubs [44b].

The simple benzylic position is also reactive under the typical CDC conditions (Scheme 1.31). Shi and coworkers reported the Fe(II)/DDQ catalyst system for the oxidative coupling of diarylmethane and electron-rich arenes to form the corresponding triarylmethanes directly [45]. Additionally notable is the compatibility with the otherwise reactive aryl C–I moiety. The isoelectronic allylic system is also a viable substrate and can be directly coupled with active methylene compounds in the presence of a Pd(II) catalyst and modified BQ, 2,6-dimethylbenzoquinone (DMBQ) [46]. In this case, the π-allyl palladium species is formed as the putative intermediate, and the reaction can be thus regarded as the direct oxidative Tsuji–Trost reaction.

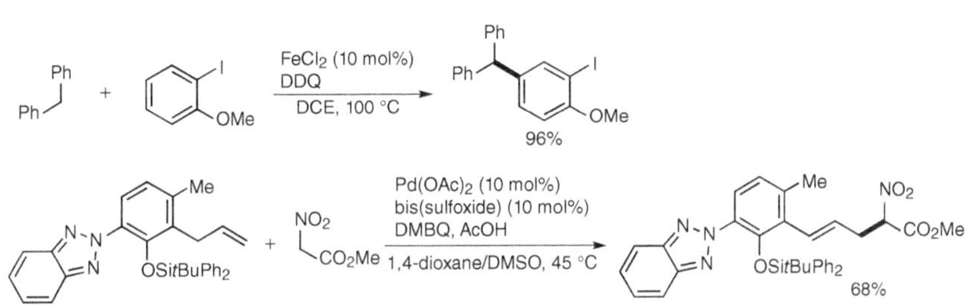

Scheme 1.31 CDC reactions of benzyl- and allyl-type substrates.

The completely unfunctionalized simple alkane also participated in the CDC reaction, although a solvent amount of alkane is inevitable (Scheme 1.32). Under the oxidative Ru/$(t\text{BuO})_2$ system, cyclohexane directly reacts with acetophenone to deliver the cross-coupling product with high para selectivity [47]. Due to the regioselective issue, almost all reported examples use the cyclic alkane as the $C_{sp}{}^3$ source.

The $C_{sp}{}^3$—H bond adjacent to electron-withdrawing groups such as carbonyl is usually reluctant to be oxidized. Thus, a protocol different from the above CDC reaction is necessary for the oxidative coupling at the relatively electron-deficient $C_{sp}{}^3$

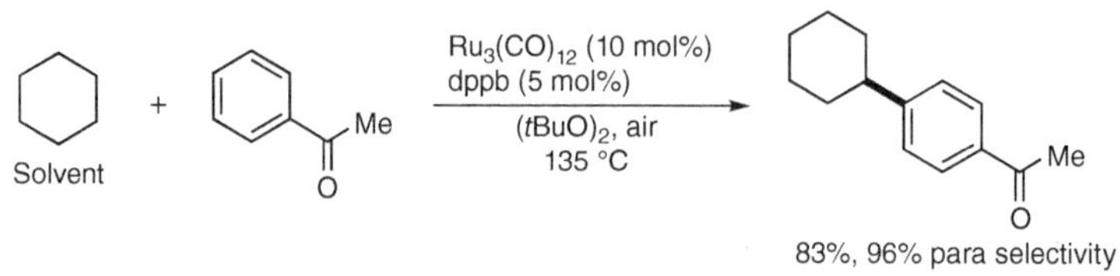

Scheme 1.32 CDC reaction of simple alkane.

center. In the case of carbonyl compounds, the generation of radical species via enolization and one-electron oxidation is the promising strategy (Scheme 1.33).

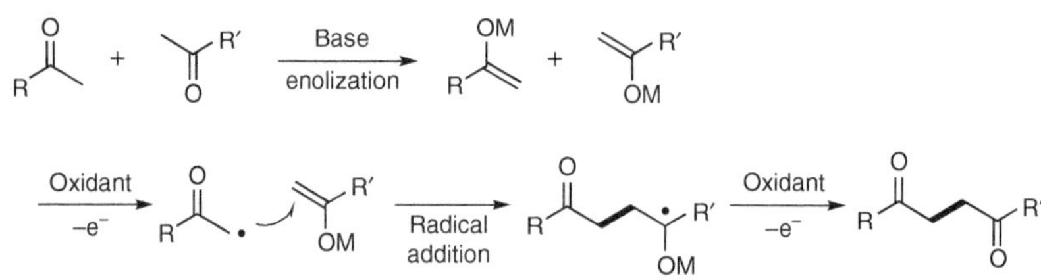

Scheme 1.33 General mechanism of oxidative coupling of two carbonyl compounds via enolization and one-electron oxidation.

Baran and coworkers reported the $Cu(eh)_2$- or $Fe(acac)_3$-mediated oxidative cross-coupling of two different lithium enolates derived from the Evans amides and simple ketones or esters (Scheme 1.34) [48]. A similar strategy is applicable to the oxidative coupling between ketones and indoles. The enolate–enolate oxidative cross-coupling reaction can be extended to the reaction of silyl enolates and *in-situ*-generated enamines, in which the suitable chiral organocatalyst successfully induces the enantioselectivity [49]. Recently, Amaya et al. succeeded in the related V(V)-mediated highly selective oxidative coupling of boron enolates and silyl enolates despite that both enolates originate from ketones at the same oxidation level [50].

Another approach to the oxidative C—C bond formation at the position α to carbonyl is the copper-catalyzed decarboxylative oxidative coupling of malonic acid half esters with arylboronic acids (Scheme 1.35) [51]. The mild reaction conditions (at room temperature under air) are compatible with a wide range of functional groups such as halides, ethers, carbonyls, and heterocycles. The 8-aminoquinoline-derived benzamide is also a viable aryl-coupling partner under similar copper-based oxidative conditions.

The suitable directing groups allow more general and otherwise unactivated $C_{sp}{}^3$–H to be adopted in the oxidative C—C bond formation. The highly fluorinated amide, pyridine, and sulfonamide groups direct the Fujiwara–Moritani-type coupling at the $C_{sp}{}^3$–H with electron-deficient alkenes under oxidative Pd(II) catalysis (Scheme 1.36) [52]. In all cases, the initially formed alkenylated intermediates spontaneously undergo the intramolecular Michael addition to form the cyclized products, same as shown in Scheme 1.12. Notably, the second reaction uniquely

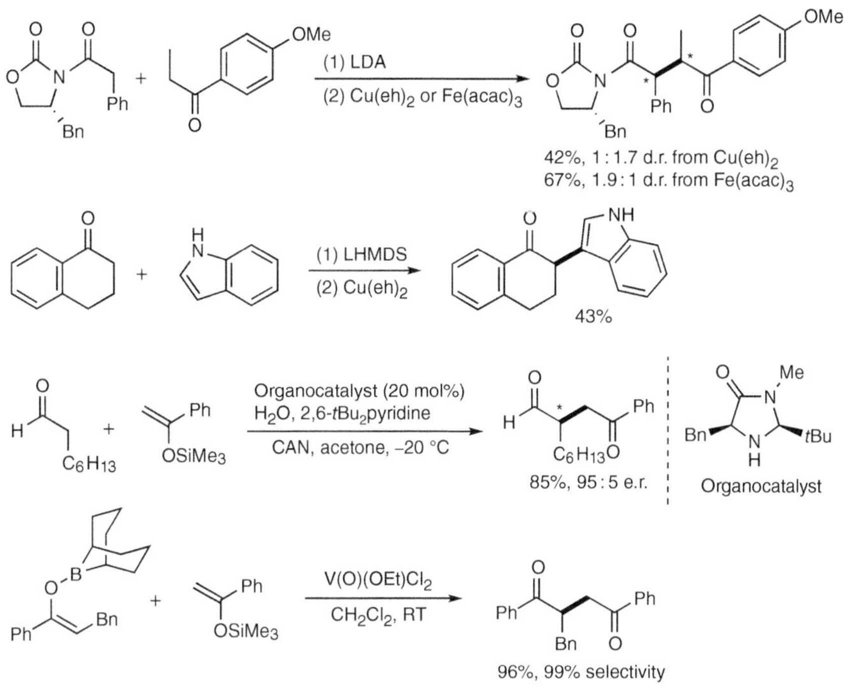

Scheme 1.34 Cu(II)-, Fe(III)-, Ce(IV)-, and V(V)-mediated oxidative cross-coupling of carbonyl compounds via enolization. eh = 2-ethylhexanoate, LHMDS = lithium hexamethyldisilazide, CAN = $(NH_4)_2Ce(NO_3)_6$. Source: Modified from DeMartino et al. [48a], Richter et al. [48b], Jang et al. [49], Amaya et al. [50].

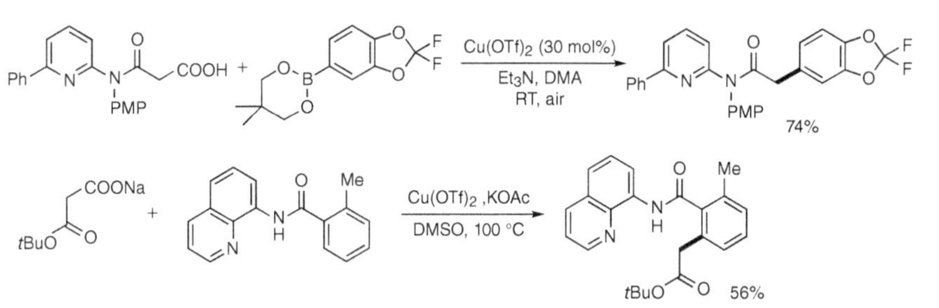

Scheme 1.35 Oxidative decarboxylative coupling of malonic acid half esters. PMP = *p*-methoxyphenyl. Source: Modified from Moon et al. [51a], Takamatsu et al. [51b].

necessitates the polyoxometalate co-oxidant, $H_4[PMo_{11}VO_{40}]$, for the acceptable catalyst turnover.

The 8-aminoquinoline-based N,N-bidentately coordinating group is also frequently employed in the oxidative couplings of unactivated $C_{sp}{}^3$–H (Scheme 1.37). Although limited to the highly electron-deficient polyfluoroarenes, the Cu(II)/$(t\text{BuO})_2$ system mediates direct $C_{sp}{}^3$–H/aromatic $C_{sp}{}^2$–H coupling [53]. The oxidative alkynylation with tri(isopropyl)silylacetylene is also possible under Ni(II)/Cu(I) cooperative catalysis [54]. Furthermore, probably the most challenging

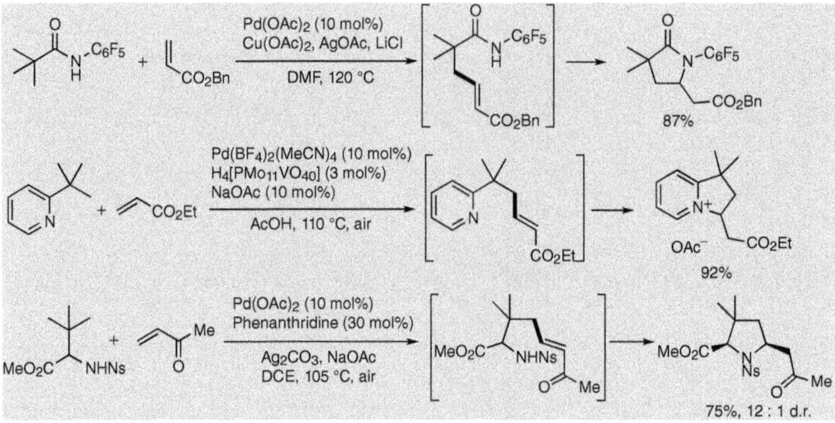

Scheme 1.36 Directed Fujiwara–Moritani-type reaction at unactivated C_{sp}^3–H. Ns = *o*-nitrobenzenesulfonyl. Source: Modified from Wasa et al. [52a], Stowers et al. [52b], and Jiang et al. [52c].

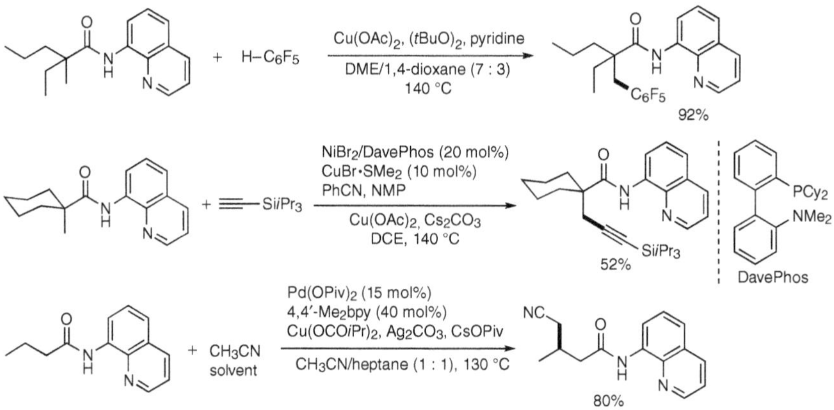

Scheme 1.37 Bidentately coordinating group directed oxidative C–C bond formations at unactivated C_{sp}^3–H with aromatic C_{sp}^2–H, alkynyl C_{sp}–H, and aliphatic C_{sp}^3–H.

dehydrogenative C_{sp}^3–H/C_{sp}^3–H cross-coupling was recently achieved albeit with solvent amount MeCN as the coupling partner [55]. Additionally, in the last case, more sterically crowding methylene C_{sp}^3–H can be functionalized, whereas the former two reactions are restricted at the sterically most accessible methyl C_{sp}^3–H.

1.6 Conclusion

An oxidative strategy can make the C—C bond potentially more effective than the conventional overall redox process using organic halide electrophiles because the parent simple organic materials can be directly used via C–H activation. Thus, it deserves significant attention from the viewpoint of atom- and step-economies, and more and more oxidative catalysts as well as stoichiometric conditions have been developed by synthetic chemists. Additionally, even in the case that organometallic coupling partners are still necessary, the oxidative protocol often shows the uniquely high functional group compatibility, chemoselectivity, and stereoselectivity, which are difficult to achieve under the overall redox-neutral process. The oxidative C—C bond formation is already one of the indispensable synthetic tools in chemical synthesis, but further development of oxidative catalysis and discovery/design of new chemical oxidants will improve the reaction efficiency and selectivity much more and realize the creation of complex functional molecules inaccessible by means of the redox-neutral reaction.

References

1 For seminal works: (a)Mizoroki, T., Mori, K., and Ozaki, A. (1971). Arylation of olefin with aryl iodide catalyzed by palladium. *Bull. Chem. Soc. Jpn.* 44: 581. (b) Heck, R.F. and Nolley, J.P. (1972). Palladium-catalyzed vinylic hydrogen substitution reactions with aryl, benzyl, and styryl halides. *J. Organomet. Chem.* 37: 2320–2322.

2 Selected reviews: (a)Beletskaya, I.P. and Cheprakov, A.V. (2000). The Heck reaction as a sharpening stone of palladium catalysis. *Chem. Rev.* 100: 3009–3066. (b) Dounay, A.B. and Overman, L.E. (2003). The asymmetric intramolecular Heck reaction in natural product total synthesis. *Chem. Rev.* 103: 2945–2964. (c) Mc Cartney, D. and Guiry, P.J. (2011). The asymmetric Heck and related reactions. *Chem. Soc. Rev.* 40: 5122–5150.

3 Delcamp, J.H., Brucks, A.P., and White, M.C. (2008). A general and highly selective chelate-controlled intermolecular oxidative Heck reaction. *J. Am. Chem. Soc.* 130: 11270–11271.

4 Werner, E.W. and Sigman, M.S. (2010). A highly selective and general palladium catalyst for the oxidative Heck reaction of electronically nonbiased olefins. *J. Am. Chem. Soc.* 132: 13981–13983.

5 Mei, T.-S., Werner, E.W., Burckle, A.J., and Sigman, M.S. (2013). Enantioselective redox-relay oxidative Heck arylations of acyclic alkenyl alcohols using boronic acids. *J. Am. Chem. Soc.* 135: 6830–6833.

6 (a) Zhang, C., Santiago, C.B., Kou, L., and Sigman, M.S. (2015). Alkenyl carbonyl derivatives in enantioselective redox relay Heck reactions: accessing α,β-unsaturated systems. *J. Am. Chem. Soc.* 137: 7290–7293. (b) Yuan, Q. and Sigman, M.S. (2018). Palladium-catalyzed enantioselective relay Heck arylation of enelactams: accessing α,β-unsaturated δ-lactams. *J. Am. Chem. Soc.* 140: 6527–6530.

7 (a) Fujiwara, Y., Moritani, I., Danno, S. et al. (1969). Aromatic substitution of olefins. VI. Arylation of olefins with palladium(II) acetate. *J. Am. Chem. Soc.* 91: 7166–7169. (b) Jia, C., Kitamura, T., and Fujiwara, Y. (2001). Catalytic functionalization of arenes and alkanes via C—H bond activation. *Acc. Chem. Res.* 34: 633–639.

8 Vora, H.U., Silvestri, A.P., Engelin, C.J., and Yu, J.-Q. (2014). Rhodium(II)-catalyzed nondirected oxidative alkenylation of arenes: arene loading at one equivalent. *Angew. Chem. Int. Ed.* 53: 2683–2686.

9 Wang, P., Verma, P., Xia, G. et al. (2017). Ligand-accelerated non-directed C–H functionalization of arenes. *Nature* 551: 489–494.

10 Murai, S., Kakiuchi, F., Sekine, S. et al. (1993). Efficient catalytic addition of aromatic carbon–hydrogen bonds to olefins. *Nature* 366: 529–531.

11 For recent comprehensive reviews: Chen, Z., Wang, B., Zhang, J. et al. (2015). Transition metal-catalyzed C—H bond functionalizations by the use of diverse directing groups. *Org. Chem. Front.* 2: 1107–1295.

12 Maehara, A., Tsurugi, H., Satoh, T., and Miura, M. (2008). Regioselective C–H functionalization directed by a removable carboxyl group: palladium-catalyzed vinylation at the unusual position of indole and related heteroaromatic rings. *Org. Lett.* 10: 1159–1162.

13 Ueura, K., Satoh, T., and Miura, M. (2007). An efficient waste-free oxidative coupling via regioselective C—H bond cleavage: Rh/Cu-catalyzed reaction of benzoic acids with alkynes and acrylates under air. *Org. Lett.* 9: 1407–1409.

14 A leading review: Satoh, T. and Miura, M. (2010). Oxidative coupling of aromatic substrates with alkynes and alkenes under rhodium catalysis. *Chem. Eur. J.* 16: 11212–11222.

15 Leow, D., Li, G., Mei, T.-S., and Yu, J.-Q. (2012). Activation of remote meta-C—H bonds assisted by an end-on template. *Nature* 486: 518–522.

16 Yang, Y.-F., Cheng, G.-J., Liu, P. et al. (2014). Palladium-catalyzed *meta*-selective C—H bond activation with a nitrile-containing template: computational study on mechanism and origins of selectivity. *J. Am. Chem. Soc.* 136: 344–355.

17 (a) Li, S., Cai, L., Ji, H. et al. (2016). Pd(II)-catalysed *meta*-C–H functionalizations of benzoic acid derivatives. *Nat. Commun.* 7: 10443. (b) Modak, A., Mondal, A., Watile, R. et al. (2016). Remote *meta* C—H bond functionalization of 2-phenethylsulphonic acid and 3-phenylpropanoic acid derivatives. *Chem. Commun.* 52: 13916–13919. (c) Yang, L., Fu, L., and Li, G. (2017). Incorporation of carbon dioxide into carbamate directing groups: palladium-catalyzed *meta*-C–H olefination and acetoxylation of aniline derivatives. *Adv. Synth. Catal.* 359: 2235–2240. (d) Deng, Y. and Yu, J.-Q. (2015). Remote *meta*-C–H olefination of phenylacetic acids directed by a versatile U-shaped template. *Angew. Chem. Int. Ed.* 54: 888–891.

18 (a) Xu, H.-J., Lu, Y., Farmer, M.E. et al. (2017). Rh(III)-catalyzed *meta*-C–H olefination directed by a nitrile template. *J. Am. Chem. Soc.* 139: 2200–2203. (b) Bera, M., Agasti, S., Chowdhury, R. et al. (2017). Rhodium-catalyzed *meta*-C–H functionalization of arenes. *Angew. Chem. Int. Ed.* 56: 5272–5276.

19 (a) Bag, S., Patra, T., Modak, A. et al. (2015). Remote para-C–H functionalization of arenes by a D-shaped biphenyl template-based assembly. *J. Am. Chem. Soc.* 137: 11888–11891. (b) Patra, T., Bag, S., Kancherla, R. et al. (2016). Palladium-catalyzed directed *para* C–H functionalization of phenols. *Angew. Chem. Int. Ed.* 55: 7751–7755.

20 Yang, S.-D., Sun, C.-L., Fang, Z. et al. (2008). Palladium-catalyzed direct arylation of (hetero)arenes with aryl boronic acids. *Angew. Chem. Int. Ed.* 47: 1473–1476.

21 Kirchberg, S., Tani, S., Ueda, K. et al. (2011). Oxidative biaryl coupling of thiophenes and thiazoles with arylboronic acids through palladium catalysis: otherwise difficult C4-selective C–H arylation enabled by boronic acids. *Angew. Chem. Int. Ed.* 50: 2387–2391.

22 Ban, I., Sudo, T., Taniguchi, T., and Itami, K. (2008). Copper-mediated C—H bond arylation of arenes with arylboronic acids. *Org. Lett.* 10: 3607–3609.

23 (a) Hachiya, H., Hirano, K., Satoh, T., and Miura, M. (2010). Oxidative nickel–air catalysis in C–H arylation: direct cross-coupling of azoles with arylboronic acids using air as sole oxidant. *ChemCatChem* 2: 1403–1406. (b) Hachiya, H., Hirano, K., Satoh, T., and Miura, M. (2010). Nickel-catalyzed direct C–H arylation and alkenylation of heteroarenes with organosilicon reagents. *Angew. Chem. Int. Ed.* 49: 2202–2205.

24 Giri, R., Maugel, N., Li, J.-J. et al. (2007). Palladium-catalyzed methylation and arylation of sp^2 and sp^3 C—H bonds in simple carboxylic acids. *J. Am. Chem. Soc.* 129: 3510–3511.

25 Shang, M., Sun, S.-Z., Dai, H.-X., and Yu, J.-Q. (2014). $Cu(OAc)_2$-catalyzed coupling of aromatic C—H bonds with arylboron reagents. *Org. Lett.* 16: 5666–5669.

26 Mochida, K., Kawasumi, K., Segawa, Y., and Itami, K. (2011). Direct arylation of polycyclic aromatic hydrocarbons through palladium catalysis. *J. Am. Chem. Soc.* 133: 10716–10719.

27 (a) Ozaki, K., Kawasumi, K., Shibata, M. et al. (2015). One-shot K-region-selective annulative p-extension for nanographene synthesis and functionalization. *Nat. Commun.* 6: 6251. (b) Shibata, M., Ito, H., and Itami, K. (2018). C–H arylation of phenanthrene with trimethylphenylsilane by Pd/*o*-chloranil catalysis: computational studies on the mechanism, regioselectivity, and role of *o*-chloranil. *J. Am. Chem. Soc.* 140: 2196–2205.

28 (a) Cornella, J., Lu, P., and Larrosa, I. (2009). Intermolecular decarboxylative direct C-3 arylation of indoles with benzoic acids. *Org. Lett.* 11: 5506–5509. (b) Xie, K., Yang, Z., Zhou, X. et al. (2010). Pd-catalyzed decarboxylative arylation of thiazole, benzoxazole, and polyfluorobenzene with substituted benzoic acids. *Org. Lett.* 12: 1564–1567. (c) Zhang, F. and Greaney, M.F. (2010). Decarboxylative C–H cross-coupling of azoles. *Angew. Chem. Int. Ed.* 49: 2768–2771.

29 Takamatsu, K., Hirano, K., and Miura, M. (2017). Copper-mediated decarboxylative coupling of benzamides with *ortho*-nitrobenzoic acids by directed C–H cleavage. *Angew. Chem. Int. Ed.* 56: 5353–5357.

30 (a) Stuart, D.R. and Fagnou, K. (2007). The catalytic cross-coupling of unactivated arenes. *Science* 316: 1172–1175. (b) Hull, K.L. and Sanford, M.S. (2007). Catalytic and highly regioselective cross-coupling of aromatic C–H substrates. *J. Am. Chem. Soc.* 129: 11904–11905. (c) Dwight, T.A., Rue, N.R., Charyk, D. et al. (2007). C—C bond formation via double C–H functionalization: aerobic oxidative coupling as a method for synthesizing heterocoupled biaryls. *Org. Lett.* 9: 3137–3139. (d) Yamada, S., Murakami, K., and Itami, K. (2016).

Regiodivergent cross-dehydrogenative coupling of pyridines and benzoxazoles: discovery of organic halides as regio-switching oxidants. *Org. Lett.* 18: 2415–2418.

31 (a) Cheng, Y., Li, G., Liu, Y. et al. (2016). Unparalleled ease of access to a library of biheteroaryl fluorophores via oxidative cross-coupling reactions: discovery of photostable NIR probe for mitochondria. *J. Am. Chem. Soc.* 138: 4730–4738. (b) He, C.-Y., Fan, S., and Zhang, X. (2010). Pd-catalyzed oxidative cross-coupling of perfluoroarenes with aromatic heterocycles. *J. Am. Chem. Soc.* 132: 12850–12852. (c) Aoki, H., Saito, H., Shimoyama, Y. et al. (2018). Synthesis of conjugated polymers containing octafluorobiphenylene unit via Pd-catalyzed cross-dehydrogenative-coupling reaction. *ACS Macro Lett.* 7: 90–94.

32 (a) Wencel-Delord, J., Nimphius, C., Patureau, F.W., and Glorius, F. (2012). [Rh^{III}Cp*]-catalyzed dehydrogenative aryl–aryl bond formation. *Angew. Chem. Int. Ed.* 51: 2247–2251. (b) Kuhl, N., Hopkinson, M.N., and Glorius, F. (2012). Selective rhodium(III)-catalyzed cross-dehydrogenative coupling of furan and thiophene derivatives. *Angew. Chem. Int. Ed.* 51: 8230–8234.

33 Dong, J., Long, Z., Song, F. et al. (2013). Rhodium or ruthenium-catalyzed oxidative C–H/C–H cross-coupling: direct access to extended π-conjugated systems. *Angew. Chem. Int. Ed.* 52: 580–584.

34 Nishino, M., Hirano, K., Satoh, T., and Miura, M. (2012). Copper-mediated and copper-catalyzed cross-coupling of indoles and 1,3-azoles: double C–H activation. *Angew. Chem. Int. Ed.* 51: 6993–6997.

35 Kita, Y., Morimoto, K., Ito, M. et al. (2009). Metal-free oxidative cross-coupling of unfunctionalized aromatic compounds. *J. Am. Chem. Soc.* 131: 1668–1669.

36 (a) Kitahara, M., Hirano, K., Tsurugi, H. et al. (2010). Copper-mediated direct cross-coupling of 1,3,4-oxadiazoles and oxazoles with terminal alkynes. *Chem. Eur. J.* 16: 1772–1775. (b) Wei, Y., Zhao, H., Kan, J. et al. (2010). Copper-catalyzed direct alkynylation of electron-deficient polyfluoroarenes with terminal alkynes using O_2 as an oxidant. *J. Am. Chem. Soc.* 132: 2522–2523.

37 Matsuyama, N., Kitahara, M., Hirano, K. et al. (2010). Nickel- and copper-catalyzed direct alkynylation of azoles and polyfluoroarenes with terminal alkynes under O_2 or atmospheric conditions. *Org. Lett.* 12: 2358–2361.

38 Kim, S.H., Yoon, J., and Chang, S. (2011). Palladium-catalyzed oxidative alkynylation of heterocycles with terminal alkynes under air conditions. *Org. Lett.* 13: 1474–1477.

39 (a) de Haro, T. and Nevado, C. (2010). Gold-catalyzed ethynylation of arenes. *J. Am. Chem. Soc.* 132: 1512–1513. stoichiometric version: (b)Fuchita, Y., Utsonomiya, Y., and Yasutake, M. (2001). Synthesis and reactivity of arylgold(III) complexes from aromatic hydrocarbons *via* C—H bond activation. *J. Chem. Soc., Dalton Trans.*: 2330–2334.

40 Jie, X., Shang, Y., Hu, P., and Su, W. (2013). Palladium-catalyzed oxidative cross-coupling between heterocycles and terminal alkynes with low catalyst loading. *Angew. Chem. Int. Ed.* 52: 3630–3633.

41 Kim, S.H., Park, S.H., and Chang, S. (2012). Palladium-catalyzed oxidative alkynylation of arene C—H bond using the chelation-assisted strategy. *Tetrahedron Lett.* 68: 5162–5166.

42 (a) Shang, M., Wang, H.-L., Sun, S.-Z. et al. (2014). Cu(II)-mediated ortho C–H alkynylation of (hetero)arenes with terminal alkynes. *J. Am. Chem. Soc.* 136: 11590–11593. (b) Liu, Y.-J., Liu, Y.-H., Yin, X.-S. et al. (2015). Copper/silver-mediated direct *ortho*-ethynylation of unactivated (hetero)aryl C—H bonds with terminal alkyne. *Chem. Eur. J.* 21: 205–209. (c) Liu, Y.-H., Liu, Y.-J., Yan, S.-Y., and Shi, B.-F. (2015). Ni(II)-catalyzed dehydrative alkynylation of unactivated (hetero)aryl C—H bonds using oxygen: a user-friendly approach. *Chem. Commun.* 51: 11650–11653.

43 (a) Li, Z. and Li, C.-J. (2004). CuBr-catalyzed efficient alkynylation of sp^3 C–H bonds adjacent to a nitrogen atom. *J. Am. Chem. Soc.* 126: 11810–11811. (b) Girard, S.A., Knauber, T., and Li, C.-J. (2014). The cross-dehydrogenative coupling of C_{sp3}—H bonds: a versatile strategy for C—C bond formations. *Angew. Chem. Int. Ed.* 53: 74–100.

44 (a) Li, Z. and Li, C.-J. (2004). Catalytic enantioselective alkynylation of prochiral sp^3 C—H bonds adjacent to a nitrogen atom. *Org. Lett.* 6: 4997–4999. (b) Dubs, C., Hamashima, Y., Sasamoto, N. et al. (2008). Mechanistic studies on the catalytic asymmetric Mannich-type reaction with dihydroisoquinolines and development of oxidative Mannich-type reactions starting from tetrahydroisoquinolines. *J. Organomet. Chem.* 73: 5859–5871.

45 Li, Y.-Z., Li, B.-J., Lu, X.-Y. et al. (2009). Cross dehydrogenative arylation (CDA) of a benzylic C—H bond with arenes by iron catalysis. *Angew. Chem. Int. Ed.* 48: 3817–3820.

46 Young, A.J. and White, M.C. (2008). Catalytic intermolecular allylic C–H alkylation. *J. Am. Chem. Soc.* 130: 14090–14091.

47 Guo, X. and Li, C.-J. (2011). Ruthenium-catalyzed para-selective oxidative cross-coupling of arenes and cycloalkanes. *Org. Lett.* 13: 4977–4979.

48 (a) DeMartino, M.P., Chen, K., and Baran, P.S. (2008). Intermolecular enolate heterocoupling: scope, mechanism, and application. *J. Am. Chem. Soc.* 130: 11546–11560. (b) Richter, J.M., Whitefield, B.W., Maimone, T.J. et al. (2007). Scope and mechanism of direct indole and pyrrole couplings adjacent to carbonyl compounds: total synthesis of acremoauxin A and oxazinin 3. *J. Am. Chem. Soc.* 129: 12857–12869.

49 Jang, H.-Y., Hong, J.-B., and MacMillan, D.W.C. (2007). Enantioselective organocatalytic singly occupied molecular orbital activation: the enantioselective r-enolation of aldehydes. *J. Am. Chem. Soc.* 129: 7004–7005.

50 Amaya, T., Maegawa, Y., Masuda, T. et al. (2015). Selective intermolecular oxidative cross-coupling of enolates. *J. Am. Chem. Soc.* 137: 10072–10075.

51 (a) Moon, P.J., Yin, S., and Lundgren, R.J. (2016). Ambient decarboxylative arylation of malonate half-esters via oxidative catalysis. *J. Am. Chem. Soc.* 138: 13826–13829. (b) Takamatsu, K., Hirano, K., and Miura, M. (2018). Copper-mediated decarboxylative coupling of benzamides with potassium malonate monoesters via directed C–H cleavage. *Chem. Lett.* 47: 450–453.

52 (a) Wasa, M., Engle, K.M., and Yu, J.-Q. (2010). Pd(II)-catalyzed olefination of *sp*^{3} C—H bonds. *J. Am. Chem. Soc.* 132: 3680–3681. (b) Stowers, K.J., Fortner, K.C., and Sanford, M.S. (2011). Aerobic Pd-catalyzed sp^3 C–H olefination: a route to both N-heterocyclic scaffolds and alkenes. *J. Am. Chem. Soc.* 133: 6541–6544. (c) Jiang, H., He, J., Liu, T., and Yu, J.-Q. (2016). Ligand-enabled γ-C(sp^3)–H olefination of amines: en route to pyrrolidines. *J. Am. Chem. Soc.* 138: 2055–2059.

53 Wu, X., Zhao, Y., and Ge, H. (2015). Pyridine-enabled copper-promoted cross dehydrogenative coupling of C(sp^2)—H and unactivated C(sp^3)—H bonds. *Chem. Sci.* 6: 5978–5983.

54 Luo, F.-X., Cao, Z.-C., Zhao, H.-W. et al. (2017). Nickel-catalyzed oxidative coupling of unactivated C(sp^3)—H bonds in aliphatic amides with terminal alkynes. *Organometallics* 36: 18–21.

55 Liu, Y., Yang, K., and Ge, H. (2016). Palladium-catalyzed ligand-promoted site-selective cyanomethylation of unactivated C(sp^3)—H bonds with acetonitrile. *Chem. Sci.* 7: 2804–2808.

2

Electrochemical Oxidative C—C Bond Formation

Sebastian Lips and Siegfried R. Waldvogel

Johannes Gutenberg University Mainz, Department of Chemistry, Duesbergweg 10-14, 55128 Mainz, Germany

2.1 Electrochemical Oxidative Aryl–Aryl Cross-Coupling Reaction

The selective bond formation between two carbon atoms is an essential transformation in modern organic chemistry. The development of transition-metal-catalyzed coupling reactions enabled the synthesis of many complex structures, natural products, and large organic scaffolds for the first time [1–4]. In particular, aryl–aryl coupling reactions are of huge significance for the design and construction of modern organic architectures.

Biaryls can be synthesized by transition metal catalysis in combination with prefunctionalized aryls or by application of chemical redox agents to activate C—H or C—X bonds. Nevertheless, besides all advantages, these methods produce a significant amount of reagent waste or lack selective bond formation. Considering electrochemistry as an alternative pathway, electrons can be directly used as a stoichiometric reagent-free substitute for classical chemical oxidizers and reducing agents. Therefore, costs can be tremendously reduced and toxic reagent waste may be completely avoided in many cases. Besides such factors, these electrochemical transformations often provide an unusual reactivity, which cannot be achieved by conventional methods and have the potential to shorten many conventional synthetic routes.

Scheme 2.1 provides an overview of different protocols allowing a selective homo-coupling reaction of aryls. The Waldvogel group is especially focused on employment of phenols within an anodic conversion. In particular, 2,2′-biphenols represent a structural feature of several natural products, active pharmaceutical ingredients, or as ligand system in catalysis [11–14].

The investigations in the Waldvogel group started with the anodic homo-coupling of simple, methyl-substituted phenols, which, despite their relatively simple substitution pattern, are susceptible to side reactions during oxidative treatment. The first selective homo-coupling of 2,4-dimethylphenol was achieved by using

Organic Redox Chemistry: Chemical, Photochemical and Electrochemical Syntheses, First Edition.
Edited by Jun-Ichi Yoshida and Frédéric W. Patureau.

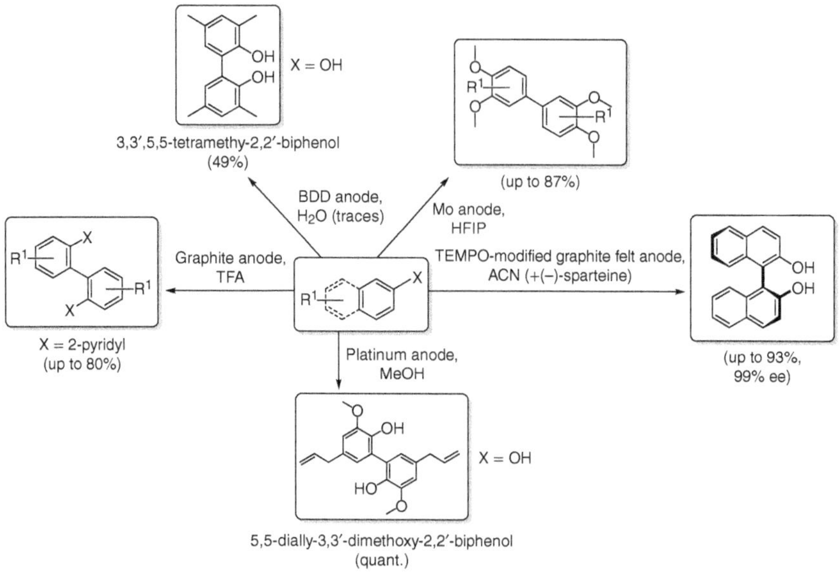

Scheme 2.1 Overview of some anodic C–C homo-coupling reactions. Source: Beil et al. [5], Malkowsky et al. [6, 7], Saito et al. [8], Osa et al. [9], Iguchi et al. [10].

boron-doped diamond electrodes [7]. A subsequent boron-based template strategy made it possible to carry out the synthesis on a larger scale [6]. In addition, the Waldvogel group established a protocol using a molybdenum anode for a C—C bond formation for the first time [5]. By using an active molybdenum electrode in 1,1,1,3,3,3-hexafluoroisopropanol (HFIP), a compact, conductive, and electroactive layer of higher-valent molybdenum species is formed allowing coupling reactions of anisoles and veratroles. Coupling reactions of these compounds were investigated much earlier by Erdtman and Parker. They realized the formation of hexamethoxytriphenylenes as well [15, 16].

Osa et al. reported a protocol for the enantioselective anodic coupling of 2-naphthol, 2-methoxynaphthalene, and 9-hydroxyphenanthrene at a 2,2,6,6-tetramethylpiperidinyloxyl (TEMPO)-modified graphite felt anode [9]. By using this anode in combination with (−)-sparteine, access to biaryls in high yields with high enantiomeric excess is possible. Furthermore, a selective homo-coupling reaction of 4-allyl-2-methoxyphenols in quantitative yields by using basic conditions was reported by Iguchi et al. [10]. When isoeugenol is applied as cross-coupling partner, diisoeugenol can be isolated in yields up to 57%. Therefore, the diastereoisomer α-diisoeugenol is formed exclusively due to a radical chain mechanism, where the initially formed radical intermediate faces a molecular orientation within the HFIP solvation cage [17].

Kakiuchi and coworkers reported a regioselective homo-coupling of aryl pyridines in an undivided cell [8]. The anodic compartment contains 10 mol% $Pd(OAc)_2$ and iodine as redox mediator. The regeneration of the catalyst is realized at the platinum anode. A selective homo-coupling reaction can also be achieved by using leaving groups on both coupling partners, similar to conventional coupling reactions. Therefore, the use of palladium catalysts is necessary again. In addition, TEMPO or para-benzoquinone as redox mediators are required. Such protocols were described by Tanaka, Amatore, and others [18–20].

Key for the selectivity of the following and some of the already-described conversions is the usage of HFIP as a specific solvent that is employed as a part of the electrolyte system. HFIP exerts a strong stabilizing effect on reactive intermediates such as radicals and radical cations. This unique stabilization is due to HFIP's ability to form a strong hydrogen-bonding network being responsible for stabilizing the individual compounds and reactive intermediates by their solvation [21]. In addition, it is chemically inert and offers an unusually wide potential window. Recent studies reveal an even more complex picture of HFIP as a solvent [22]. During epoxidation of olefins in HFIP, small-angle X-ray scattering experiments showed that the fluorous alcohol is capable of forming nano-domains. Therefore, a separation of hydroxyl groups, protic additives, and fluorous sections takes place. This unusual microheterogeneous feature can be attributed as the outstanding source of selectivity in the anodic coupling reactions.

When it comes to cross-coupling reactions, a direct oxidative C—C bond formation of two different arenes is challenging, since selective oxidation of one coupling partner has to occur. Otherwise, statistical formation of homo- and cross-coupling products will take place in nonselective oxidation processes. This would lead to moderate yields of the desired cross-coupling products in the best case. Additionally, overoxidation of the obtained biaryls will lower the overall yield and form oligomeric side products (Scheme 2.2).

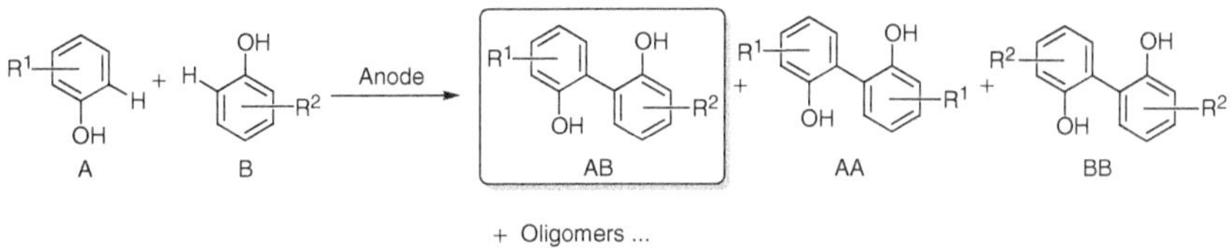

Scheme 2.2 Potential products for an anodic, electrochemical C–C coupling reaction of phenols.

A remarkable and modern example of such a conversion that overcomes these challenges is the "radical-cation pool" method by Yoshida and Suga [23, 24]. This method enables the generation and temporary accumulation of these particular reactive intermediates. The key step of this technique is electrolysis at very low temperatures in the absence of nucleophiles leading to a separation of the oxidation and the coupling event in time and space (Scheme 2.3). This requires sometimes specific supporting electrolytes such as BArF salts.

The "radical-cation pool" method enables, for example, the generation of ions of *N*-acyliminium [24, 26–28], alkoxycarbenium [29], diarylcarbenium [30–32], glycosyl [33–35], silyl [36], iodonium [37], alkoxysulfonium [38, 39], benzylaminosulfonium [40], arene [25], thioarenium [41, 42], and thionium [43]. After

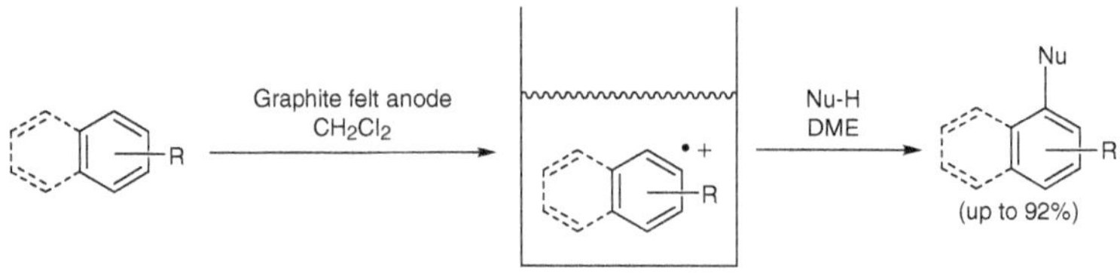

Scheme 2.3 Concept of "radical-cation pool" method. Source: Modified from Morofuji et al. [25].

generating a cation reservoir at low temperatures, these ions can be converted into corresponding compounds by adding a selected nucleophile. The range of possible nucleophiles includes organic and organometallic substances such as allylsilanes, silylenol ethers, Grignard compounds, 1,3-dicarbonyls, arenes as well as alkenes and alkynes, which can undergo [4+2] cycloadditions with the cations. For a successful formation of the final product, at least two equivalents of cations were required. For the "radical-cation pool" method, the use of protic substrates is not possible due to a potential proton abstraction.

After that, with a focus on cross-coupling reactions, the direct electrochemical coupling of phenols with electron-rich arenes under mild reaction conditions was reported (Scheme 2.4) [44].

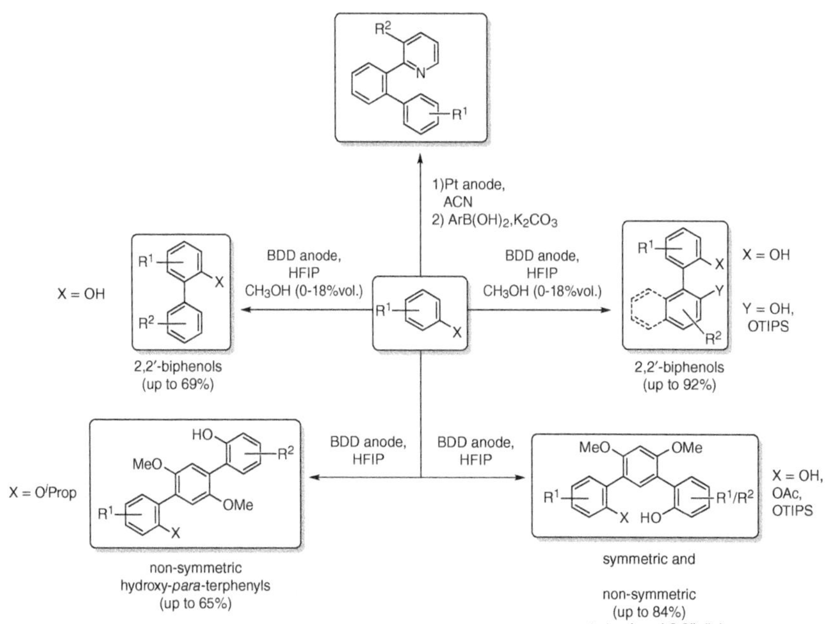

Scheme 2.4 Overview of anodic C–C cross-coupling reactions between phenols and arenes. Source: Kirste et al. [21, 44], Lips et al. [45], Aiso et al. [46], Elsler et al. [47].

In addition, further investigations provided first indications of a mechanistic rationale. Typically, the mechanism starts with an oxidation step to generate a phenoxyl radical, followed by a nucleophilic attack of the second component, the one with the higher oxidation potential [48]. In contrast to the homo-coupling reactions, a simple radical recombination of anodically generated phenoxyl radicals was not sufficient to describe the formation of these cross-coupling products. Instead, a selective oxidation of the component with the lower oxidation potential, followed by a nucleophilic attack and C—C bond formation of the component with the higher oxidation potential explains formation of these cross-coupling products [44]. By adding water or methanol to HFIP, an increase of yields up to 69% was possible.

The positive effect of protic additives was attributed to a decoupling of the nucleophilicity from the individual oxidation potential [49]. This effect was evaluated by different trends for the oxidation potentials of phenols and arenes when adding methanol. This can be rationalized by the formation of HFIP solvates for the individual coupling partners. These spheres can be manipulated by the incorporation of methanol, acting as base for hydrogen bonding.

Besides phenol-arene cross-coupling products, the first anodic cross-coupling reaction of different phenols to nonsymmetrical 2,2′-biphenols was realized as well. An improved protocol allowed an increase of yields from 63% to 92% by using a triisopropylsilyl-protecting group on the phenol with the higher oxidation potential (Scheme 2.4) [50]. Within a new synthetic approach, the electrochemical formation of OCO-pincer ligands has been developed (Scheme 2.4) [51]. This involves the direct coupling of phenols to one single arene moiety in a one-pot electrolysis, which provides symmetric OCO-pincer ligands, as well as a two-step protocol. The two-step procedure offers the possibility to design nonsymmetric OCO-pincer ligands. Here, onefold coupling of one phenol to a 1,3-disubstituted arene was carried out. Key step for a successful second coupling was the application of the previously developed protecting group strategy. Application of the cost-efficient acetyl group to protect the starting material resulted in yields up to 84% and an outstanding selectivity. This protocol was successfully transferred for the synthesis of the corresponding nonsymmetric hydroxyl-*para*-terphenyls, which were accessible by a similar protecting group strategy (Scheme 2.4) [45]. Kakiuchi and coworkers enabled C—C bond formation between two aryls by an electrochemical C,H-iodination with a subsequent arylation of arylpyridines within a one-pot synthesis [46].

Atobe and coworkers investigated C—C cross-coupling reactions of aromatic molecules realized by a parallel laminar flow in a two-inlet flow microreactor [52]. By this setup, the oxidation and subsequent coupling steps can be separated allowing a selective cross-coupling reaction of both compounds.

Besides anodic coupling reactions of phenols with different phenols or arenes, C—C bond formation between phenols and heterocycles was investigated by different working groups. These coupling products, so-called heterobiaryls, are important structural motifs in molecular electronics [53], natural products [1], pharmaceuticals [54], and catalysts [55].

Later, Lei and coworkers described an external oxidant-free electrooxidative [3+2] annulation between phenol and *N*-actetyl indole derivatives, which can be used for the synthesis of benzofuro[3,2-*b*]- and benzofuro[2,3-*b*]indolines in good-to-excellent yields (Scheme 2.5) [56]. The Waldvogel group enabled the selective cross-coupling reaction of phenols with thiophenes (Scheme 2.5) [57]. The reaction was optimized to give the phenol-thiophene biaryls in high selectivity and yields up to 60%. Additionally, the twofold coupling of phenols to one thiophene moiety had been observed. Reaction to these 2,5-bis(2-hydroxyphenyl)thiophenes can be simply favored by optimizing the ratio of phenol to thiophene, which gave the desired product in up to 53% yield in a one-pot electrochemical coupling. In addition, coupling of phenols with benzothiophenes is possible in yields up to 88%. Therefore, benzothiophenes bearing a substituent in either position 2 or 3 were used to obtain either 3-(hydroxyphenyl)- or 2-(hydroxyphenyl)benzothiophenes (Scheme 2.5) [58].

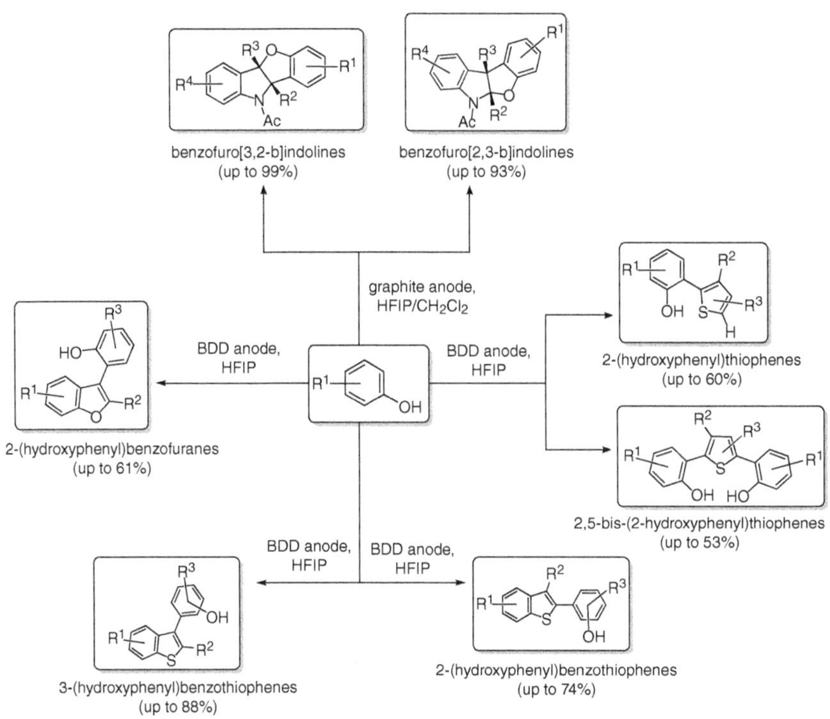

Scheme 2.5 Overview of anodic C—C bond formations between phenols and heterocycles. Source: Liu et al. [56], Wiebe et al. [57], Lips et al. [58, 59].

When using benzofurans within the anodic C—C coupling reaction, not only was a simple C—C bond formation observed, but also a furan metathesis leading to an exchange of the substituents of phenol and benzofuran takes place, thereby enabling the formation of products (Scheme 2.5) [59].

Besides phenols, also anilines were investigated as easily oxidizable motif. Due to their low oxidation potential, anodic treatment of substrates like these usually results in the formation of oligomeric byproducts and uncontrolled polymerization reactions. Nevertheless, by adjusting and increasing the oxidation potentials of such substrates by using protecting groups, such as acet-, trifluoroacet-, benz-, and pivalamides or BOC, a selective coupling of anilines to 2,2′-diaminobiaryls was possible in yields up to 74% (Scheme 2.6) [60]. In addition, the direct anodic cross- and homo-coupling of formanilides was enabled, respectively (Scheme 2.6) [61]. Furthermore, naphthylamines were used to access electrochemically synthesized *N*,*O*-biaryl motifs in yields up to 48% for the first time (Scheme 2.6) [62].

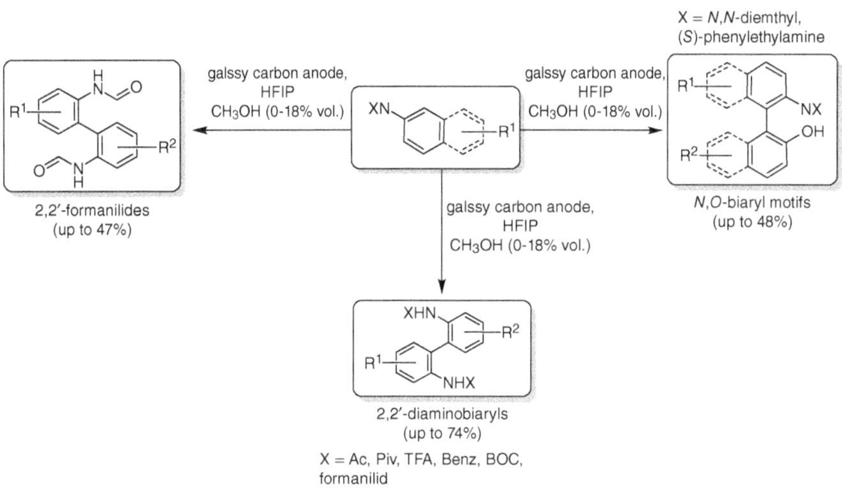

Scheme 2.6 Overview of anodic C–C bond formations using anilines or naphthylamines. Source: Schulz et al. [60, 61], Dahms et al. [62].

2.2 Electrochemical Oxidative Benzyl–Aryl Cross-Coupling Reaction

Besides a selective C—C bond formation between two aromatic compounds, mainly two different electrochemical approaches for a benzyl–aryl cross-coupling reaction yielding diarylmethanes are described. Such diarylmethanes are an important motif in biologically active compounds [63], medicinal chemistry [64], and material science [65, 66].

One protocol was reported by Waldvogel and coworkers, who showed a selective electrochemical functionalization of benzylic positions by HFIP [67]. The resulting ethers act as a molecular mask for the benzylic cation. A subsequent treatment with 2,2,2-trifluoroacetic acid led to the formation of an active benzylic cation. If this activation is carried out in the presence of an aromatic nucleophile, selective benzyl–aryl cross-coupling can be achieved (Scheme 2.7).

A broad substrate scope with various heterocycles such as thiophenes, pyrroles, indoles, benzofuranes, benzothiophenes, and also halosubstituents were tolerated. In addition, it was demonstrated that this method can be a powerful tool for late-stage functionalization.

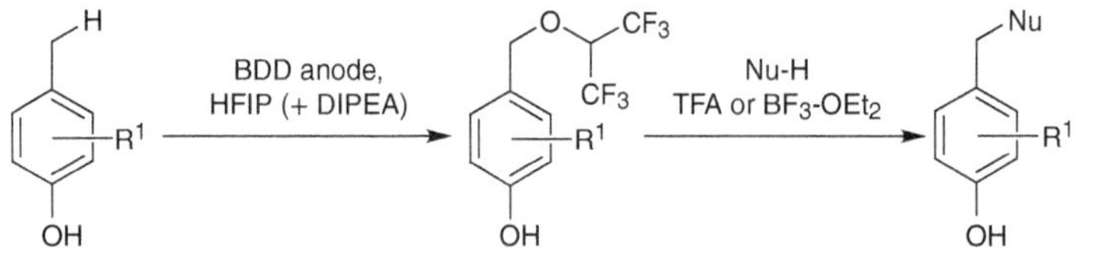

Scheme 2.7 Benzyl–aryl cross-coupling by anodic activation in HFIP. Source: Imada et al. [67].

Besides the benzyl–aryl coupling, these HFIP ethers can act as a leaving group within a nucleophilic substitution with cyanides as well leading to a broad scope of accessible 2-phenylacetonitriles in yields up to 90% [68].

Furthermore, Yoshida and coworkers used the already-mentioned "cation-pool" method to realize the electrochemical synthesis of such scaffolds [40].

To stabilize the electrochemically oxidized species in this procedure, the intermediate-generated benzylic cations had to be trapped with an additional reagent (Scheme 2.8). Afterwards, a subsequent elimination of the stabilizing reagent and coupling with aromatic nucleophiles is carried out yielding the final product.

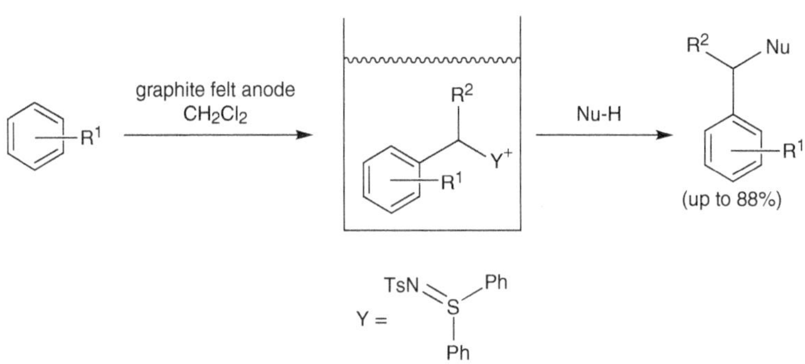

Scheme 2.8 Benzyl–aryl cross-coupling by stabilized "cation pool" method. Source: Modified from Hayashi et al. [40].

2.3 Electrochemical Oxidative Arylation of Olefins

To enable electrochemical oxidative arylation reactions of olefins, several different techniques were developed. Hence, one efficient approach is to regenerate the required reactants of conventional Heck reactions electrochemically.

For example, Jutand and coworkers reported a Pd-mediated C–H olefination of acetanilides by electrochemical oxidation. So a benzoquinone/hydroquinone redox pair is used as electron-transfer mediator. It is used to regenerate $Pd(OAc)_2$ as catalyst. *Para*-hydroquinone itself is afterwards oxidized at the anode (Scheme 2.9) [70]. The Moeller group used electrochemistry to reduce Pd(II) to the active Pd(0) catalyst (Scheme 2.9) [69].

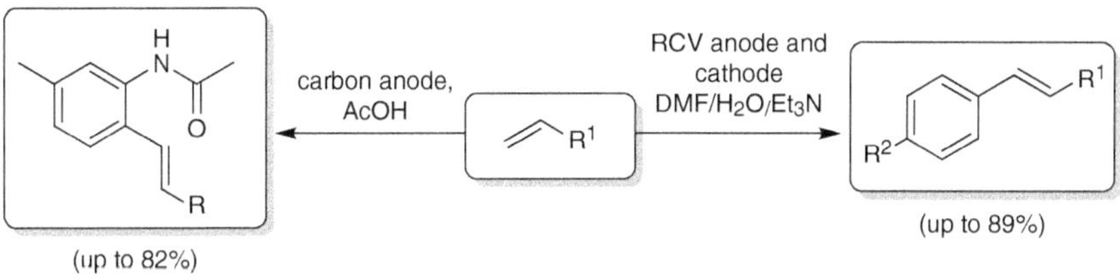

Scheme 2.9 Electrochemically assisted Heck reactions. Source: Tian et al. [69], Amatore et al. [70].

Shundo et al. instead developed a leaving group-free Mn^{3+}-mediated coupling cyclization of 5-arylpent-1-enes with active methylene compound by electrooxidation [71]. Hence, Mn^{3+} is generated in situ mediating the free-radical cyclization (Scheme 2.10).

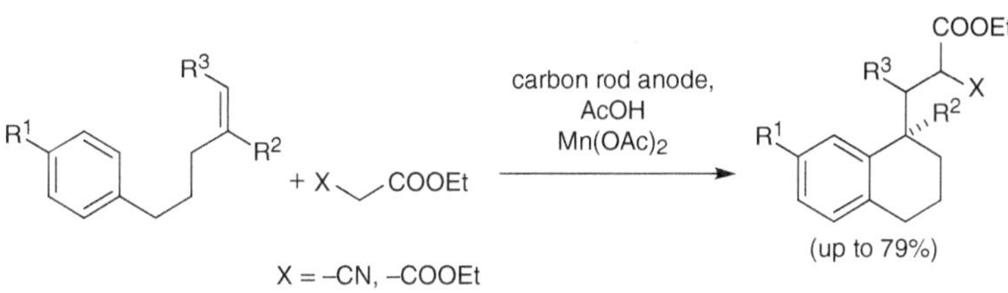

Scheme 2.10 Mn^{3+}-mediated intramolecular coupling reaction. Source: Modified from Shundo et al. [71].

However, conversions like these can also be conducted in a metal-free manner. Quite similar to the metal-free aryl–aryl cross-coupling reactions within a first oxidation step, phenoxyl species are generated and are afterwards attacked by the nucleophilic olefins.

When *para*-methoxyphenols are oxidized at a glassy carbon anode to the corresponding phenoxyl radicals, a subsequent conversion with different vinyl ethers leads to the formation of various dihydrobenzofurans (Scheme 2.11) [72–74]. Instead, when electron-rich styrene or propenylbenzene derivatives are used, the [3+2] cycloaddition delivers substituted dihydrobenzofuran derivatives again (Scheme 2.11) [75–77]. However, besides [3+2], also [5+2] cycloadditions are possible generating bridged bicyclic systems (Scheme 2.11) [78]. An optimized protocol for the synthesis of dihydrobenzofurans by [3+2] cycloaddition allowing the use of nonactivated alkyl-substituted alkenes was developed by Chiba et al. [79]. By using a PTFE-coated glassy carbon anode, dehydrodimerization reactions of the phenolic compound can be avoided.

The group of Yamamura reported an intramolecular protocol that can be used to obtain bioactive natural products like terpenoids or neolignans within a regio- and stereo-selective manner (Scheme 2.12) [74, 80–82].

Another intramolecular procedure for the synthesis of polycyclic ring systems was presented by the Moeller group (Scheme 2.13) [83, 84]. Remarkably, in this case, the first oxidation step involves the very electron-rich olefinic moiety, thereby

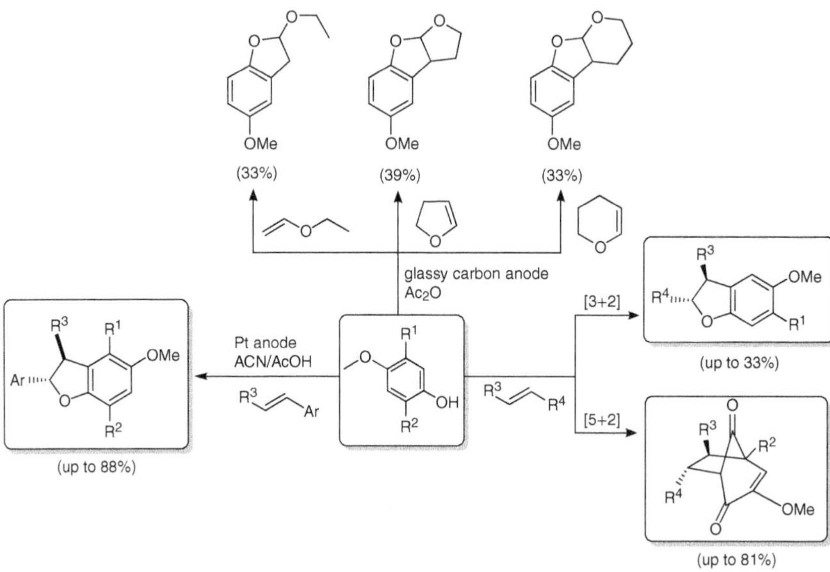

Scheme 2.11 Anodic conversions of phenols with olefins. Source: Shizuri et al. [72, 73], Yamamura et al. [74], Gates et al. [75], Kerns et al. [76], Swenton et al. [77] and Quideau et al. [78].

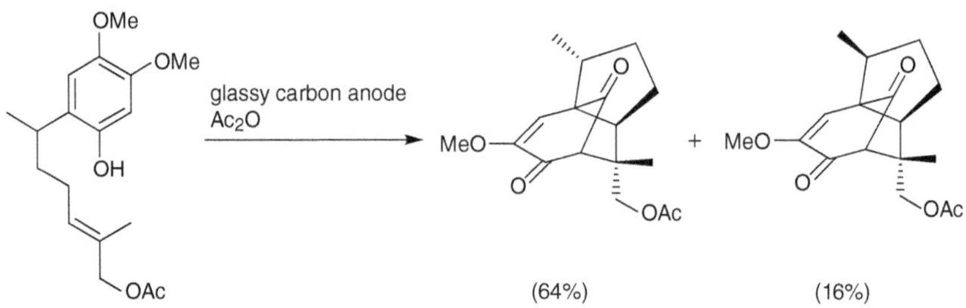

Scheme 2.12 Regio- and stereo-selective synthesis of aryl and olefin. Source: Yamamura and Niwa [80], Shizuri and Yamamura [81], Maki et al. [82].

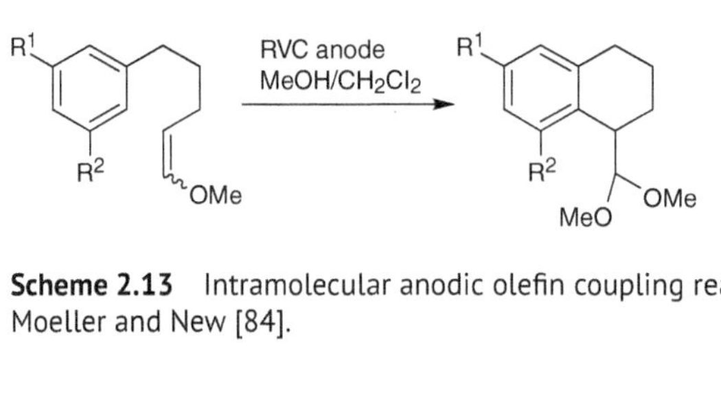

Scheme 2.13 Intramolecular anodic olefin coupling reaction. Source: New et al. [83] and Moeller and New [84].

preventing overoxidation of the final product. To ensure a primary oxidation of the side chain compared to the aromatic ring, phenol ethers are used as substrate.

2.4 Electrochemical Oxidative Arylation of Alkynes

Besides olefins, Xu and coworkers also realized an electrochemical arylation of alkynes [85, 86]. They conducted an intramolecular C–H/N–H [3+2] cyclization reaction of an aniline and alkyne moiety for the synthesis of indoles and azaindoles. In this case, Cp_2Fe is used as redox catalyst that is primarily oxidized at the anode. Cp_2Fe^+ oxidizes the corresponding substrate yielding amidyl radicals (Scheme 2.14).

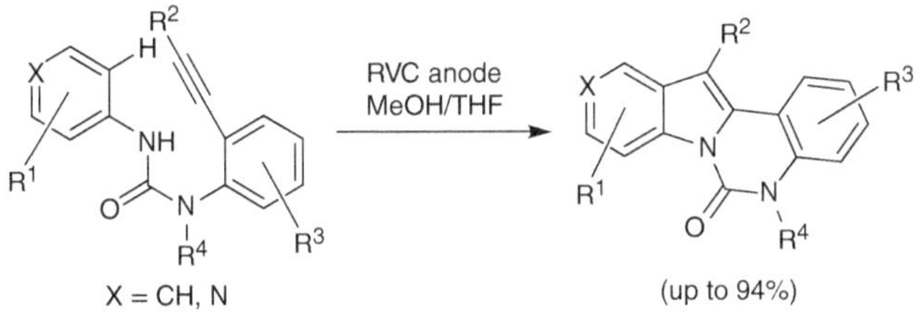

Scheme 2.14 Electrochemical intramolecular C–H/N–H [3+2] annulation mediated by Cp_2Fe. Source: Jiang et al. [85], Hou et al. [86].

2.5 Electrochemical Oxidative Cross-Dehydrogenative Coupling of C(sp^3)–H and C(sp^2)–H Bonds

The group of Xu was also able to present the first electrochemical oxidative cross-dehydrogenative coupling of C(sp^3)—H and C(sp^2)—H bonds. In the protocol, the synthesis of 3-fluorooxindoles with a broad substrate scope was reported (Scheme 2.15) [87]. Similar to the protocol of arylation of alkynes, Cp_2Fe is employed as redox catalyst.

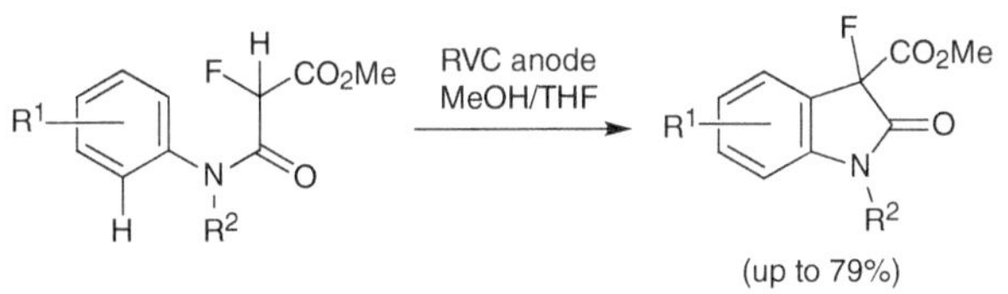

Scheme 2.15 Electrochemical synthesis of C3-fluorinated oxindoles. Source: Wu and Hu [87].

References

1 Nicolaou, K.C., Bulger, P.G., and Sarlah, D. (2005). *Angew. Chem. Int. Ed.* 44: 4442–4489; *Angew. Chem.* 117: 4516–4563.

2 Franke, R., Selent, D., and Börner, A. (2012). *Chem. Rev.* 112: 5675–5732.
3 Okamoto, K., Zhang, J., Housekeeper, J.B. et al. (2013). *Macromolecules* 46: 8059–8078.
4 von Nussbaum, F., Brands, M., Hinzen, B. et al. (2006). *Angew. Chem. Int. Ed.* 45: 5072–5129; *Angew. Chem.* 118: 5194–5254.
5 Beil, S.B., Müller, T., Sillart, S.B. et al. (2018). *Angew. Chem. Int. Ed.* 57: 2450–2454; *Angew. Chem.* 130: 2475–2479.
6 Malkowsky, I.M., Fröhlich, R., Griesbach, U. et al. (2006). *Eur. J. Inorg. Chem.* 2006: 1690–1697.
7 Malkowsky, I.M., Griesbach, U., Pütter, H., and Waldvogel, S.R. (2006). *Eur. J. Org. Chem.* 2006: 4569–4572.
8 Saito, F., Aiso, H., Kochi, T., and Kakiuchi, F. (2014). *Organometallics* 33: 6704–6707.
9 Osa, T., Kashiwagi, Y., Yanagisawa, Y., and Bobbitt, J.M. (1994). *J. Chem. Soc., Chem. Commun.* 54: 2535–2537.
10 Iguchi, M., Nishiyama, A., Terada, Y., and Yamamura, S. (1978). *Chem. Lett.* 7: 451–454.
11 Boudier, A., Breuil, P.-A.R., Magna, L. et al. (2015). *Dalton Trans.* 44: 12995–12998.
12 Vidal-Ferran, A., Mon, I., Bauzá, A. et al. (2015). *Chem. Eur. J.* 21: 11417–11426.
13 Liu, Y.-J., Zhang, Z.-Z., Yan, S.-Y. et al. (2015). *Chem. Commun.* 51: 7899–7902.
14 Chen, J.-S., Chen, Y.-L., Greenberg, A.S. et al. (2005). *J. Cell. Biochem.* 94: 1028–1037.
15 Ronlán, A., Bechgaard, K., Parker, V.D. et al. (1973). *Acta Chem. Scand.* 27: 2375–2382.
16 Erdtman, H.G.H. (1933). *Proc. R. Soc. London, Ser. A* 143: 191–222.
17 Yamamoto, T., Riehl, B., Naba, K. et al. (2018). *Chem. Commun.* 54: 2771–2773.
18 Amatore, C., Cammoun, C., and Jutand, A. (2008). *Eur. J. Org. Chem.* 2008: 4567–4570.
19 Mitsudo, K., Shiraga, T., and Tanaka, H. (2008). *Tetrahedron Lett.* 49: 6593–6595.
20 Mitsudo, K., Shiraga, T., Kagen, D. et al. (2009). *Tetrahedron* 65: 8384–8388.
21 Kirste, A., Elsler, B., Schnakenburg, G., and Waldvogel, S.R. (2012). *J. Am. Chem. Soc.* 134: 3571–3576.
22 Hollóczki, O., Berkessel, A., Mars, J. et al. (2017). *ACS Catal.* 7: 1846–1852.
23 Yoshida, J.-i. and Suga, S. (2002). *Chem. Eur. J.* 8: 2650.
24 Yoshida, J.-i., Suga, S., Suzuki, S. et al. (1999). *J. Am. Chem. Soc.* 121: 9546–9549.
25 Morofuji, T., Shimizu, A., and Yoshida, J.-i. (2012). *Angew. Chem.* 124: 7371–7374.
26 Suga, S., Tsutsui, Y., Nagaki, A., and Yoshida, J.-i. (2005). *Bull. Chem. Soc. Jpn.* 78: 1206–1217.
27 Suga, S., Nishida, T., Yamada, D. et al. (2004). *J. Am. Chem. Soc.* 126: 14338–14339.
28 Suga, S., Nagaki, A., Tsutsui, Y., and Yoshida, J.-i. (2003). *Org. Lett.* 5: 945–947.

29 Suga, S., Suzuki, S., Yamamoto, A., and Yoshida, J.-i. (2000). *J. Am. Chem. Soc.* 122: 10244–10245.

30 Nokami, T., Watanabe, T., Musya, N. et al. (2011). *Chem. Commun.* 47: 5575–5577.

31 Okajima, M., Soga, K., Watanabe, T. et al. (2009). *Bull. Chem. Soc. Jpn.* 82: 594–599.

32 Okajima, M., Soga, K., Nokami, T. et al. (2006). *Org. Lett.* 8: 5005–5007.

33 Saito, K., Saigusa, Y., Nokami, T., and Yoshida, J.-i. (2011). *Chem. Lett.* 40: 678–679.

34 Nokami, T., Shibuya, A., Tsuyama, H. et al. (2007). *J. Am. Chem. Soc.* 129: 10922–10928.

35 Suzuki, S., Matsumoto, K., Kawamura, K. et al. (2004). *Org. Lett.* 6: 3755–3758.

36 Nokami, T., Soma, R., Yamamoto, Y. et al. (2007). *Beilstein J. Org. Chem.* 3: 7.

37 Midorikawa, K., Suga, S., and Yoshida, J.-i. (2006). *Chem. Commun.* 2006: 3794–3796.

38 Ashikari, Y., Nokami, T., and Yoshida, J.-i. (2011). *J. Am. Chem. Soc.* 133: 11840–11843.

39 Ashikari, Y., Shimizu, A., Nokami, T., and Yoshida, J.-i. (2013). *J. Am. Chem. Soc.* 135: 16070–16073.

40 Hayashi, R., Shimizu, A., and Yoshida, J.-i. (2016). *J. Am. Chem. Soc.* 138: 8400–8403.

41 Matsumoto, K., Kozuki, Y., Ashikari, Y. et al. (2012). *Tetrahedron Lett.* 53: 1916–1919.

42 Fujie, S., Matsumoto, K., Suga, S. et al. (2010). *Tetrahedron* 66: 2823–2829.

43 Shimizu, A., Takeda, K., Mishima, S. et al. (2016). *Bull. Chem. Soc. Jpn.* 89: 61–66.

44 Kirste, A., Schnakenburg, G., Stecker, F. et al. (2010). *Angew. Chem. Int. Ed.* 49: 971–975; *Angew. Chem.* 122: 983–987.

45 Lips, S., Franke, R., and Waldvogel, S.R. (2019). *Synlett* 30: 1174–1177. https://doi.org/10.1055/s-0037-1611942.

46 Aiso, H., Kochi, T., Mutsutani, H. et al. (2012). *J. Org. Chem.* 77: 7718–7724.

47 Elsler, B., Schollmeyer, D., Dyballa, K.M. et al. (2014). *Angew. Chem. Int. Ed.* 53: 5210–5213; *Angew. Chem.* 126: 5311–5314.

48 Kirste, A., Schnakenburg, G., and Waldvogel, S.R. (2011). *Org. Lett.* 13: 3126–3129.

49 Elsler, B., Wiebe, A., Schollmeyer, D. et al. (2015). *Chem. Eur. J.* 21: 12321–12325.

50 Wiebe, A., Schollmeyer, D., Dyballa, K.M. et al. (2016). *Angew. Chem. Int. Ed.* 55: 11801–11805; *Angew. Chem.* 128: 11979–11983.

51 Lips, S., Wiebe, A., Elsler, B. et al. (2016). *Angew. Chem. Int. Ed.* 55: 10872–10876; *Angew. Chem.* 128: 11031–11035.

52 Arai, T., Tateno, H., Nakabayashi, K. et al. (2015). *Chem. Commun.* 51: 4891–4894.

53 Nishiyama, H., Kino, T., and Tomita, I. (2012). *Macromol. Rapid Commun.* 33: 545–549.

54 Amaral, M.F.Z.J., Callejon, D.R., Riul, T.B. et al. (2014). *J. Braz. Chem. Soc.* 25: 1907–1913.

55 Agapie, T., Henling, L.M., DiPasquale, A.G. et al. (2008). *Organometallics* 27: 6245–6256.

56 Liu, K., Tang, S., Huang, P., and Lei, A. (2017). *Nat. Commun.* 8: 775.

57 Wiebe, A., Lips, S., Schollmeyer, D. et al. (2017). *Angew. Chem. Int. Ed.* 56: 14727–14731; *Angew. Chem.* 129: 14920–14925.

58 Lips, S., Schollmeyer, D., Franke, R., and Waldvogel, S.R. (2018). *Angew. Chem. Int. Ed.* 57: 13325–13329; *Angew. Chem.* 130: 13509–13513.

59 Lips, S., Frontana-Uribe, B.A., Dörr, M. et al. (2018). *Chem. Eur. J.* 24: 6057–6061.

60 Schulz, L., Enders, M., Elsler, B. et al. (2017). *Angew. Chem. Int. Ed.* 56: 4877–4881; *Angew. Chem.* 129: 4955–4959.

61 Schulz, L., Franke, R., and Waldvogel, S.R. (2018). *ChemElectroChem* 5: 2069–2072.

62 Dahms, B., Franke, R., and Waldvogel, S.R. (2018). *ChemElectroChem* 5: 1249–1252.

63 McPhail, K.L., Rivett, D.E.A., Lack, D.E., and Davies-Coleman, M.T. (2000). *Tetrahedron* 56: 9391–9396.

64 Long, Y.-Q., Jiang, X.-H., Dayam, R. et al. (2004). *J. Med. Chem.* 47: 2561–2573.

65 Ahmad, M., Luo, J.K., Purnawali, H. et al. (2012). *J. Mater. Chem.* 22: 8192.

66 Wang, S., Zhang, C., Shu, Y. et al. (2017). *Sci. Adv.* 3: e1602610.

67 Imada, Y., Röckl, J.L., Wiebe, A. et al. (2018). *Angew. Chem. Int. Ed.* 57: 12136–12140; *Angew. Chem.* 130: 12312–12317.

68 Röckl, J., Imada, Y., Chiba, K. et al. (2018). *ChemElectroChem* 6: 4184–4187.

69 Tian, J. and Moeller, K.D. (2005). *Org. Lett.* 7: 5381–5383.

70 Amatore, C., Cammoun, C., and Jutand, A. (2007). *Adv. Synth. Catal.* 349: 292–296.

71 Shundo, R., Nishiguchi, I., Matsubara, Y., and Hirashima, T. (1991). *Chem. Lett.* 20: 235–236.

72 Shizuri, Y., Suyama, K., and Yamamura, S. (1986). *J. Chem. Soc., Chem. Commun.* 1986: 63–64.

73 Shizuri, Y., Nakamura, K., and Yamamura, S. (1985). *J. Chem. Soc., Chem. Commun.* 1985: 530.

74 Yamamura, S., Shizuri, Y., Shigemori, H. et al. (1991). *Tetrahedron* 47: 635–644.

75 Gates, B.D., Dalidowicz, P., Tebben, A. et al. (1992). *J. Org. Chem.* 57: 2135–2143.

76 Kerns, M.L., Conroy, S.M., and Swenton, J.S. (1994). *Tetrahedron Lett.* 35: 7529–7532.

77 Swenton, J.S., Carpenter, K., Chen, Y. et al. (1993). *J. Org. Chem.* 58: 3308–3316.

78 Quideau, S., Pouysegu, L., and Deffieux, D. (2004). *Curr. Org. Chem.* 8: 113–148.

79 Chiba, K., Fukuda, M., Kim, S. et al. (1999). *J. Org. Chem.* 64: 7654–7656.

80 Yamamura, S. and Niwa, M. (1981). *Chem. Lett.* 10: 625–626.

81 Shizuri, Y. and Yamamura, S. (1983). *Tetrahedron Lett.* 24: 5012–5013.

82 Maki, S., Kosemura, S., Yamamura, S. et al. (1992). *Chem. Lett.* 21: 651–654.

83 New, D.G., Tesfai, Z., and Moeller, K.D. (1996). *J. Org. Chem.* 61: 1578–1598.

84 Moeller, K.D. and New, D.G. (1994). *Tetrahedron Lett.* 35: 2857–2860.

85 Jiang, Y., Xu, K., and Zeng, C. (2018). *Chem. Rev.* 118: 4485–4540.

86 Hou, Z.-W., Mao, Z.-Y., Zhao, H.-B. et al. (2016). *Angew. Chem. Int. Ed.* 55: 9168–9172.

87 Wu, Z.-J. and Xu, H.-C. (2017). *Angew. Chem. Int. Ed.* 56: 4734–4738.

3

Fundamentals of Photochemical Redox Reactions

Daniel A. Corbin, Nicholas A. Swisher, and Garret M. Miyake

Colorado State University, Department of Chemistry, 1301 Center Ave, 1872 Campus Delivery, Fort Collins, CO, 80523-1872 USA

3.1 Introduction: A Brief History of Photochemistry

For centuries, scientists have sought to understand how light impacts chemical reactions. In 1790, Joseph Priestley reported the photochemical conversion of nitric acid to nitrogen dioxide, and around the same time he formed hypotheses on the basic nature of photosynthesis [1]. Serendipitously, in the late nineteenth century, the German chemist Heinrich Klinger observed photochemical reductions of various quinones to hydroquinones using sunlight (Scheme 3.1). In what are considered the first examples of synthetic organic photochemistry, Klinger established that sunlight drove the intermolecular reaction of quinones and aldehydes or ketones to yield interesting new products, although the mechanistic basis for these transformations was not understood at the time.

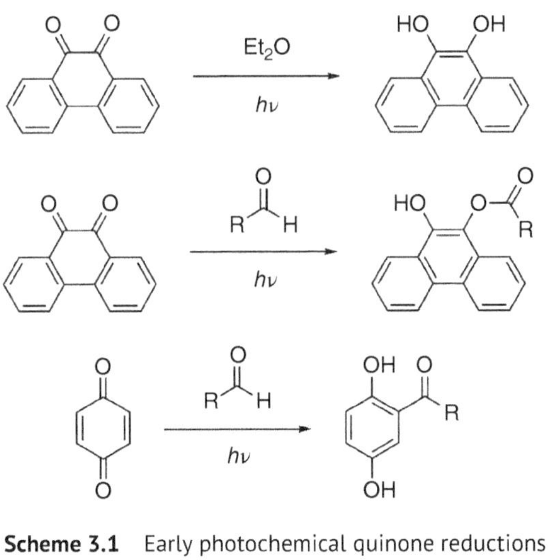

Scheme 3.1 Early photochemical quinone reductions studied by Klinger.

Organic Redox Chemistry: Chemical, Photochemical and Electrochemical Syntheses, First Edition.
Edited by Jun-Ichi Yoshida and Frédéric W. Patureau.

Many further developments in early solar photochemistry arose from the collaboration of Giacomo Ciamician and Paul Silber at the University of Bologna [2]. Over the course of many years, the pair discovered several fascinating reactions that were frequently conducted on their rooftop laboratory. Photochemical olefin isomerizations, pinacol-type couplings, and cycloadditions are just some of the reaction categories explored thoroughly. Both the reduction of nitrobenzene in ethanol to 2-methylquinoline and the coupling of acetone and methanol made apparent the potential of natural light in the synthesis of organic compounds (Scheme 3.2).

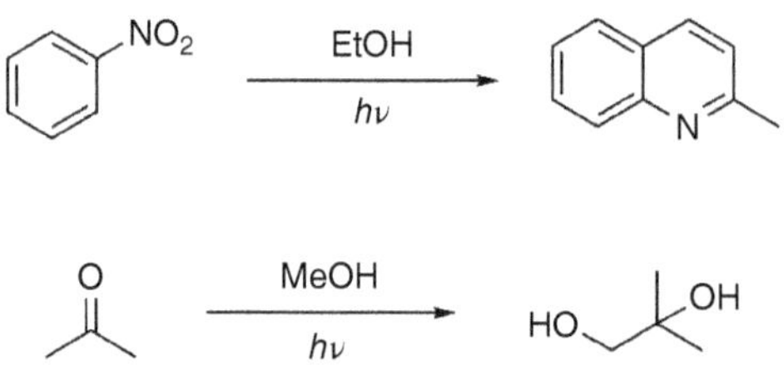

Scheme 3.2 Condensation of nitrobenzene and ethanol to form 2-methylquinoline and coupling reaction of acetone and methanol studied by Ciamician and Silber in Bologna.

Two key advances that spurred progress in synthetic photochemistry in the following decades were the development of modern light sources and transition metal photocatalysts, namely ruthenium and iridium polypyridyl complexes. The 1978 reduction of sulfonium salts with Hantzsch esters in the presence of visible light and catalytic amounts of the complex $[Ru(bpy)_3]Cl_2$ by Kellogg and coworkers highlights the power of these innovations (Scheme 3.3) [3]. This reaction is considered one of the earliest examples of **photoredox catalysis**, or catalysis using light energy to initiate single-electron transfer (SET) processes.

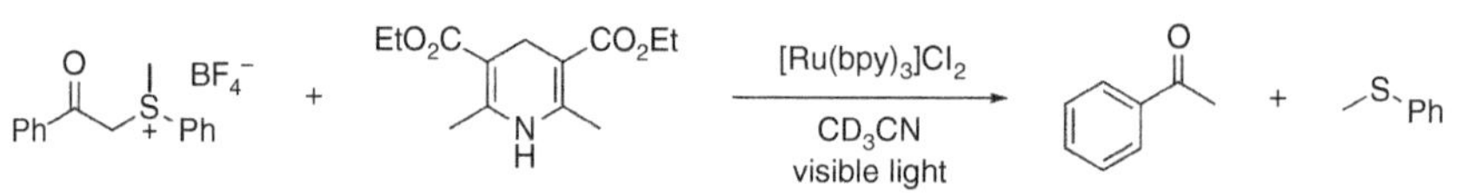

Scheme 3.3 First catalytic photoredox reaction for the reduction of sulfonium salts reported by Kellogg and coworkers in 1978. Source: Modified from Hedstrand et al. [3].

Other early instances of reductive photoredox catalysis also utilized $[Ru(bpy)_3]Cl_2$ as a photocatalyst and nitrogen-containing heterocyclic compounds as stoichiometric reductants. In 1981, Pac et al. disclosed an electron-poor olefin reduction with 1-benzyl-1,4-dihydronicotinamide as a reductant under visible light to model biological redox reactions (Scheme 3.4) [4]. Further, Fukuzumi et al. described the dehalogenation of phenacyl bromides with 10-methyl-9,10-dihydroacridine as an electron and H-atom donor in 1990 [5].

Net oxidative reactions in early photoredox catalysis include the conversion of benzyl alcohols to aldehydes mediated by aryl diazonium salts observed in 1984

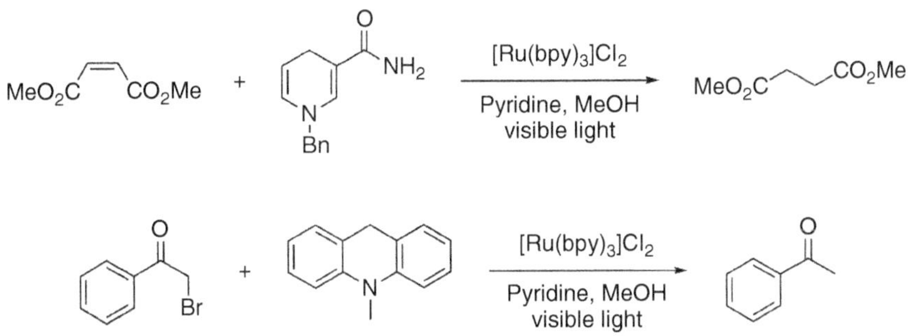

Scheme 3.4 Early photoredox reductions detailed by Pac and Fukuzumi.

by Cano-Yelo and Deronzier (Scheme 3.5) [6]. Similarly, an intramolecular Pschorr cyclization of an aryl diazonium salt from the same authors represents the first catalytic, redox-neutral organic photoredox reaction in the literature [7].

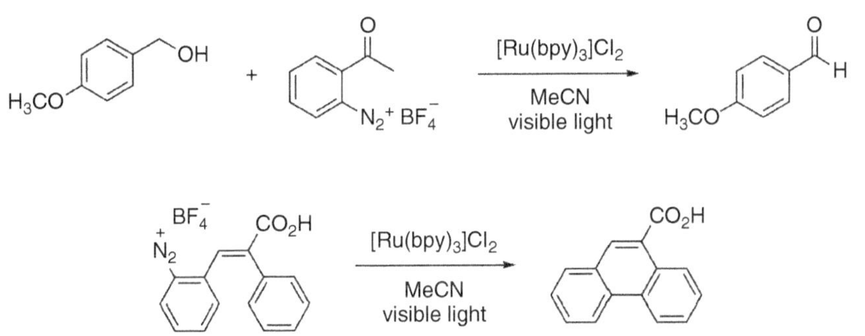

Scheme 3.5 Photoredox reactions of aryl diazonium salts developed by Cano-Yelo and Deronzier.

The modern resurgence of photoredox catalysis began in 2008 with reports by Yoon and MacMillan on stereoselective visible light–mediated reactions. Yoon's group communicated diastereoselective [2+2] photocycloadditions of bis(enones) to give bicyclic cyclobutane-containing diones, again using $[Ru(bpy)_3]Cl_2$ as a photocatalyst (Scheme 3.6) [8]. While the use of high-intensity flood lights enabled efficient cycloaddition reactions, sunlight from a laboratory window also promoted the reaction with high yield and stereoselectivity. By virtue of the milder reaction conditions employed as compared to conventional transition metal or electrochemical catalysis, Yoon's method enjoyed a wide substrate scope, including both electron-deficient and -rich enones.

The simultaneous publication by Nicewicz and MacMillan on the enantioselective α-alkylations of aldehydes using a combination of photoredox and organocatalysis was another seminal milestone in the field (Scheme 3.7) [9]. Using simple household 15-W compact fluorescent lightbulbs (CFLs), they found that alkyl bromides could serve as suitable coupling partners to aldehydes in the presence of chiral imidazolidinone catalysts and $[Ru(bpy)_3]Cl_2$ as the photocatalyst. Prior to this disclosure, the direct enantioselective functionalization of aldehydes with simple alkyl

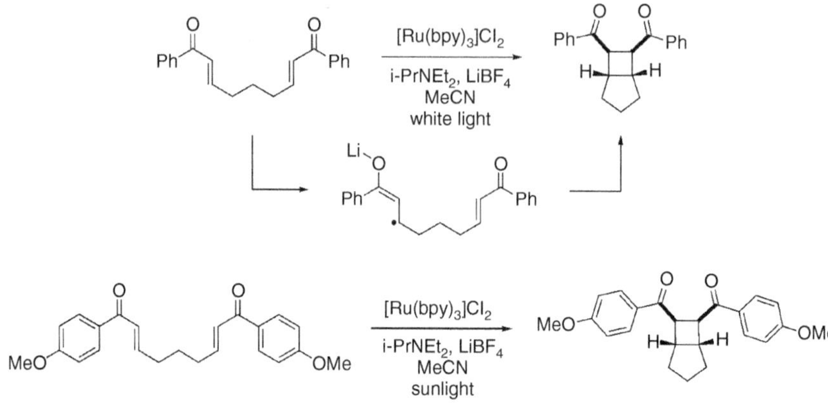

Scheme 3.6 Photoredox [2+2] enone cycloadditions reported by Yoon and coworkers.

groups was considered a fundamental challenge in organic synthesis, demonstrating the synergistic power of photoredox catalysis when merged with other forms of catalysis.

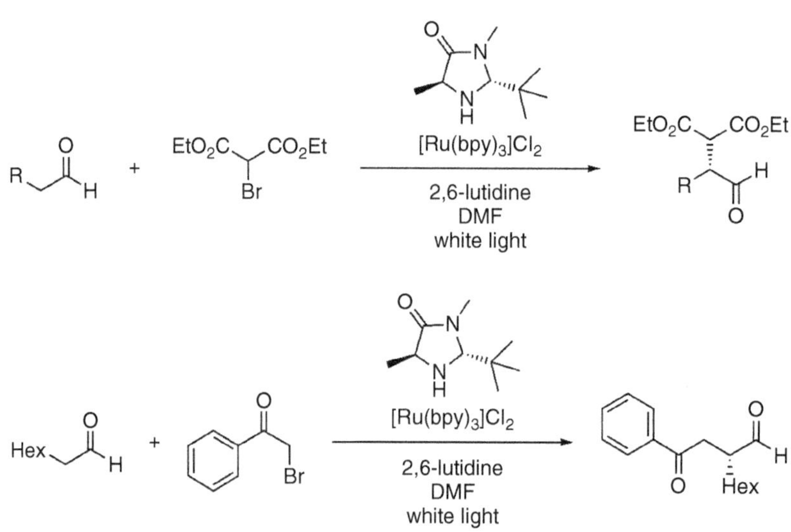

Scheme 3.7 MacMillan's report on the enantioselective alkylation of aldehydes with simple alkyl halide-coupling partners.

In 2009, Stephenson and coworkers further expanded the possibilities of photoredox catalysis in the dehalogenation of a diverse array of Csp^3—X (X=Br, Cl) bonds (Scheme 3.8) [10]. With a simple alkylamine base as a terminal reductant, this synthetic method provided an excellent alternative to the toxicity and product separation issues that plague classical tin-mediated dehalogenations. Thus, improvements in photoredox catalysis can be attractive from the perspective of substituting relatively benign reagents for inconvenient and hazardous compounds in interesting redox reactions.

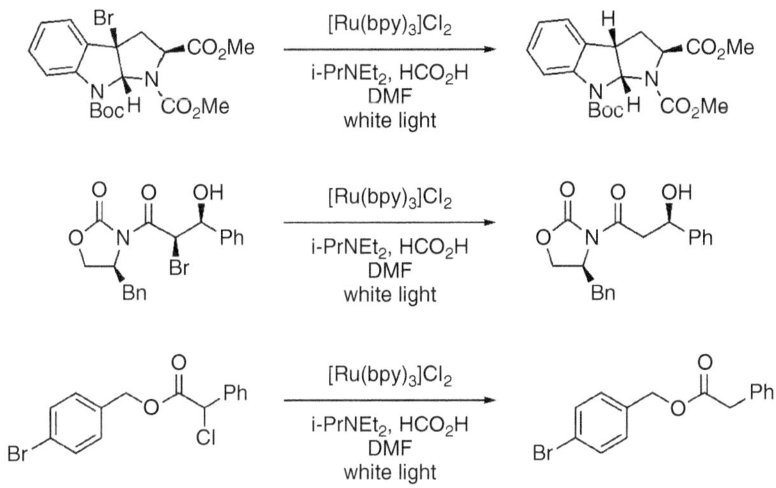

Scheme 3.8 Photoredox dehalogenation reactions developed by Stephenson and coworkers.

In the years following these landmark discoveries, applications of photoredox catalysis have expanded to a very broad array of transformations, including those once thought to be the domain of transition metal catalysis. Photoredox carbon–sulfur cross-coupling reactions between aryl halides and thiols have been performed by dual catalysis employing both transition metal and organic photoredox catalysts (PCs) [11]. Further, through control experiments, investigations have revealed some photoredox reactions can proceed in the absence of a photocatalyst (see Section 4.3.1) [12–14]. For example, Miyake and coworkers revealed that certain C–S cross-coupling reactions could be conducted under visible-light irradiation without any photocatalyst at all (Scheme 3.9) [15, 16]. Mechanistic studies supported that such reactions can proceed through the formation of electron donor–acceptor (EDA) complexes, which when irradiated undergo electron transfer (ET) to initiate

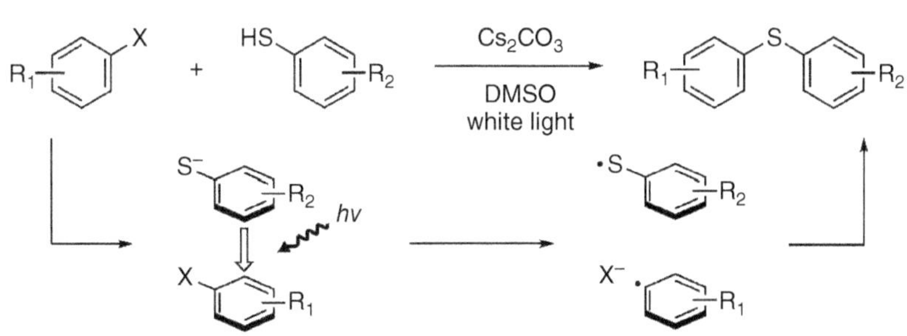

Scheme 3.9 Carbon–sulfur bond formation as a result of EDA complex formation and photoinduced intermolecular electron transfer.

redox reactions [13–15]. In the example reported by Miyake and coworkers, this intermolecular ET generates aryl and thiyl radicals, which subsequently combine to yield the C–S coupling products [15]. In photochemical transformations, the formation of such EDA complexes, also referred to as **charge-transfer (CT) complexes**, between electron-rich donor molecules and electron-poor acceptor molecules must always be considered a possibility for conducting ET without catalysts.

The remainder of this chapter begins with an overview of the fundamental concepts required to understand photoredox reactions. The theory behind common instrumental techniques and their application to the characterization of photoredox processes is also presented. With this foundation, the basic mechanisms of different classes of photoredox catalysis are then discussed, followed by a survey of the many different families of transition metal and organic PCs reported in the literature. Finally, the recent development and application of EDA complexes in synthetic organic transformations is briefly reviewed.

3.2 Photochemistry: Background and Theory

3.2.1 The Electromagnetic Spectrum

Light is all around us. Even when we cannot see it, light, or **electromagnetic radiation**, warms our food, allows us to communicate with each other over vast distances, and offers a glimpse into the human body through medical imaging. Just as light is at the center of numerous technologies we use in our everyday lives, it also plays a role in a variety of chemical processes.

Figure 3.1 shows a common depiction of the electromagnetic spectrum broken into seven categories: γ-ray, x-ray, ultra-violet (UV), visible, infrared, microwaves,

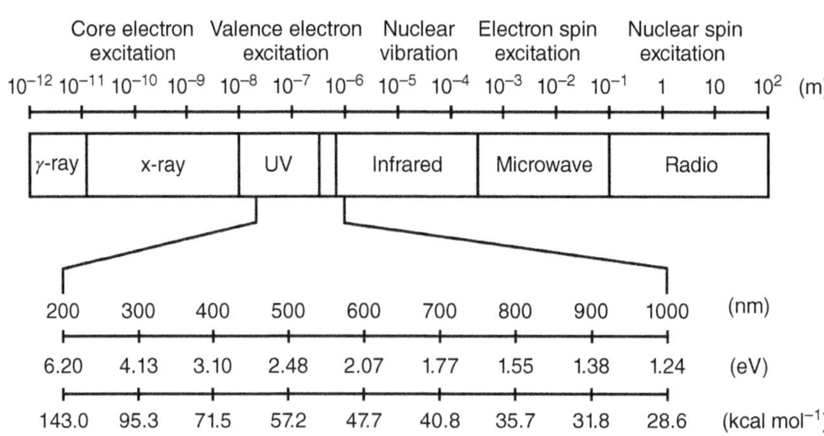

Figure 3.1 Electromagnetic spectrum, with a focus on regions associated with electronic transitions. On the top, labels indicate the approximate regions in which different transitions and excitations occur. On the bottom, wavelengths often associated with electronic transitions are converted into other useful units.

and radio waves. Each category is defined by a range of **wavelengths** or **frequencies**, which correspond to the energy of that light (Eq. (3.1)). Often, different units of energy are used for light, so some helpful conversions have been provided in Figure 3.1 and in Eqs. (3.2)–(3.4).

Importantly, the energy of light can sometimes correspond to chemical transitions (Figure 3.1), giving rise to a myriad of photochemical processes. For example, in nuclear magnetic resonance spectroscopy, radio waves are used to cause transitions between different nuclear spin states and probe the chemical environment around those nuclei. Since the difference in energy between nuclear spin states is quite small, very-low-energy light is necessary to probe these processes. On the other end of the spectrum, x-rays can be found at the heart of x-ray photoelectron spectroscopy, where they are used to promote high-energy transitions and ionize core electrons of various elements. One application of light in chemistry that has recently garnered significant attention is the use of light to promote valence electronic transitions that lead to interesting redox reactions. Since such reactions usually involve breaking bonds, a process that generally requires 20–100 kcal mol^{-1} of energy, it is unsurprising that UV and visible light have been explored extensively for this application.

$$E = h\nu = \frac{hc}{\lambda} \tag{3.1}$$

$$E = \frac{1240 \text{ eV nm}^{-1}}{\lambda} \tag{3.2}$$

$$E = \frac{2.86 \times 10^4 \text{ kcal nm mol}^{-1}}{\lambda} \tag{3.3}$$

$$1 \text{ eV} = 23.06 \text{ kcal mol}^{-1} \tag{3.4}$$

3.2.2 Allowed and Forbidden Transitions

Within an atom or molecule, one can imagine a number of different electronic transitions from one state to another. Broadly, we can classify transitions that are theoretically predicted to occur as quantum mechanically **allowed**, whereas those that are not predicted to occur are **forbidden**. It is important to note that the origin of allowed and forbidden transitions is quantum mechanical, meaning one can use theory to predict whether certain transitions will be allowed based on changes in quantum numbers and symmetry. These predictions regarding what transitions allowed or forbidden are often referred to as **selection rules**.

Curiously, although some transitions are considered forbidden, they can still occur to some degree through various processes. For example, in centrosymmetric metal complexes (those containing inversion center symmetry), transitions between two d-orbitals are considered forbidden based on the Laporte selection rule, yet d-to-d transitions are still observed for many metal complexes. This phenomenon arises because the Laporte selection rule applies only to symmetric molecules. Therefore, if a vibration within a molecule breaks its symmetry, this selection rule no longer applies and d-to-d transitions can become allowed. This specific process is referred to as **vibronic coupling**. Another process through which forbidden transitions might occur is called spin–orbit coupling and this phenomenon will be discussed later in the text (see Section 2.3.6). Importantly, while forbidden transitions can occur under the appropriate conditions, they typically occur to only a small degree, making them relatively weak or infrequent transitions.

3.2.3 Photophysical Processes

3.2.3.1 Jablonski Diagrams

When discussing photophysical processes, it is common to depict them on a **Jablonski diagram**. First proposed by Aleksander Jabłoński in 1933 [17], this diagram depicts different electronic states and the possible transitions between them (Figure 3.2). Sometimes, electronic states may be depicted as potential wells, but they are often simplified into lines representing the lowest-energy vibrational state for each electronic state. Within each electronic level, one can also find several vibrational states. Since vibrational transitions are usually lower in energy than electronic transitions, the spacing between different vibrational states is smaller than that between electronic states.

In addition, the electronic states can be labeled according to their multiplicity and relative energies. Multiplicity (M) refers to the number of unpaired electrons in a given state and is defined as $M = 2S + 1$, where S is the total spin angular momentum quantum number for a multielectron system. Put simply, each unpaired electron has a value of $S = ½$. Therefore, a state with no unpaired electrons has $M = 1$ and is termed a **singlet** state. Instead, a system with one unpaired electron has $M = 2$ and is termed a **doublet** state, whereas a system with two unpaired electrons has $M = 3$

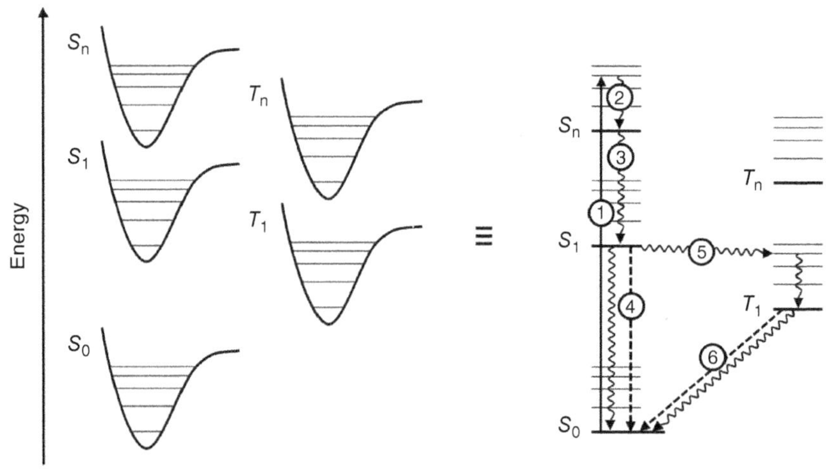

Figure 3.2 A Jablonski diagram depicted in two ways. On the left, electronic states are represented by potential wells, within which each energy level represents a different vibrational state. For ease of viewing, the diagram on the right is often used, which simplifies each electronic state to a line rather than a potential well. Transitions between different states can occur through absorption of light (1), vibrational relaxation (2), internal conversion (3), fluorescence (4), intersystem crossing (5), or phosphorescence (6). S, singlet state; T, triplet state.

and is termed a **triplet** state. States of even higher multiplicity are certainly possible and have been identified, especially in inorganic and organometallic systems where it is common to find molecules with a number of unpaired d-electrons. However, the three states outlined here are most common for organic molecules and their excited states and will be the focus of this discussion. Of course, multiple electronic states may exist with the same multiplicity, so it is common to label them based on their relative energies. For example, S_0 corresponds to the lowest-energy singlet state, often the **ground state** of many organic molecules, while S_1 corresponds to an electronic **excited state** with singlet character.

Finally, since we are interested in the processes for moving from one state to another on the Jablonski diagram, we can label these transitions with one of several different arrows. Typically, absorption (Figure 3.2(1)) is represented by a solid arrow, whereas emission (Figure 3.2(3,6)) is depicted using a dashed arrow. In addition, a number of nonradiative relaxation processes can occur (Figure 3.2(2,3,5)), which are commonly represented by wavy arrows. Importantly, transitions between states occur when their potential energy surfaces cross at one or more points, although these crossing points are often not depicted in Jablonski diagrams for simplicity. In the following sections, we will delve deeper into each of these photophysical processes and important related concepts.

3.2.3.2 Absorption

Absorption is the process by which energy from light is taken in by electrons to promote transitions from lower-energy electronic states to ones of higher energy. This process is quite fast, generally occurring on the time scale of 10^{-16} to 10^{-14} seconds. As a result, very few processes can compete with the absorption of light, since these electronic transitions tend to be much faster than anything else (Table 3.1). For example, even nuclear motion is relatively slow compared to this process, occurring over 10^{-13} to 10^{-12} seconds. In fact, one can consider nuclei to be effectively motionless during an electronic transition, a concept known as the **Frank–Condon principle**. As a consequence of this principle, the geometries of two interconverting states must be identical prior to an electronic transition, since the geometry of the molecule does not change during the transition.

Table 3.1 Typical time scales of photophysical processes for organic molecules.

Process	Time scale (s)
Absorption	$10^{-16}-10^{-14}$
Vibrational relaxation	$10^{-12}-10^{-10}$
Internal conversion	$10^{-11}-10^{-9}$
Fluorescence	$10^{-8}-10^{-5}$
Intersystem crossing	$10^{-8}-10^{-3}$
Phosphorescence	$10^{-3}-10^{2}$

For a photon of light to be absorbed by a molecule, the energy of the photon must exactly match the difference in energy between two states. Further, only one photon can be absorbed to promote a transition (although two photons can be absorbed consecutively to access higher-energy excited states). Together, these principles are known as the **Stark–Einstein Law**, and they have important implications in chemistry. For instance, since the energy of light absorbed must correspond to the energy of a transition, we can measure what light is absorbed or emitted by a molecule (or atom) to learn about the relative energies of its electronic states. Further, considering the use of light in chemical synthesis, if we desire to use certain wavelengths of light – such as those in the visible spectrum – we must design the system appropriately to absorb those wavelengths of light.

With these thoughts in mind, it can be useful when performing photochemistry to investigate what wavelengths of light a molecule absorbs when irradiated. Often this technique, called **absorption spectroscopy**, is performed using UV and visible light to probe electronic transitions, but it is certainly not limited to just these wavelength ranges. While a more detailed description of this technique will be provided in a later section (see Section 2.5.1), here it will suffice to understand that absorption spectroscopy quantifies how much light is absorbed at a specific wavelength. This quantity is often given as the unitless value of **absorbance** (A), often used interchangeably with **optical density** (OD), which corresponds to the logarithm of the intensity of light entering the sample (I_0) over the intensity of light exiting the sample and reaching the detector (I).

$$A = \mathrm{OD} = \log_{10}\left(\frac{I_0}{I}\right) \tag{3.5}$$

When the absorption of an atom or molecule is measured across a range of wavelengths, the resulting spectrum is called an **absorption spectrum**. In the simplest case, one can imagine a molecule with a single electronic transition, which would absorb a single wavelength of light. The absorption spectrum of this molecule would have a vertical line corresponding to the wavelength of light absorbed. In reality, molecules often have numerous possible transitions between different electronic and vibrational states that are close in energy, resulting in absorption spectra with broad features rather than lines corresponding to single transitions.

Of course, not all electronic transitions are equally probable, and some occur to a greater degree than others. As a result, one often sees that the intensities of peaks in an absorption spectrum vary from one to the next, corresponding to different amounts of light being absorbed to promote certain transitions. The propensity of a molecule to absorb a certain wavelength of light can be described using **molar absorptivity** (ε), which is also referred to as the **extinction coefficient**. However, other factors can also impact the amount of light absorbed, such as the concentration of the absorbing molecule (c) as well as the path length (b) of the sample through which light must travel. These concepts are summarized as **Beer's law** (Eq. 3.6), which relates the amount of light absorbed at a particular wavelength to each factor above:

$$A = \in bc \tag{3.6}$$

Table 3.2 Absorption characteristics of common organic functional groups.

Functional group	Approximate λ_{max} (nm)	Approximate ε_{max} (L mol^{-1} cm^{-1})	Transition
Alkane	<180	10^3	$\sigma \rightarrow \sigma^*$
Alcohol	180	10^2	$n \rightarrow \sigma^*$
Alkene	180	10^4	$\pi \rightarrow \pi^*$
Alkyne	180	10^4	$\pi \rightarrow \pi^*$
Carboxylic acid	200	10^1	$n \rightarrow \pi^*$
Conjugated alkene	>200	10^4	$\pi \rightarrow \pi^*$
Amide	210	10^1	$n \rightarrow \pi^*$
Bromo	210	10^2	$n \rightarrow \sigma^*$
Iodo	260	10^2	$n \rightarrow \sigma^*$
Benzene	280	10^2	$\pi \rightarrow \pi^*$
Carbonyl	280	10^1	$n \rightarrow \pi^*$
Naphthalene	310	10^2	$\pi \rightarrow \pi^*$
Anthracene	380	10^4	$\pi \rightarrow \pi^*$

Finally, when employing light as a reagent in chemical synthesis, it is important to consider how that light will interact with different compounds in the reaction. Table 3.2 lists the approximate wavelength of maximum absorption (λ_{max}), approximate molar absorptivity at λ_{max} (ε_{max}), and the type of transition observed for common organic functional groups. As can be seen, many organic functionalities tend to absorb light in the UV range, meaning that irradiation with light in this range of the spectrum could give rise to a number of different reactions. While this reactivity can be useful in certain situations, it is often undesirable as it can be difficult to control. Therefore, synthetic strategies employing visible light are often desirable, since under these conditions, reactions can be targeted without unwanted excitation of other molecules in solution.

3.2.3.3 Vibrational Relaxation

When light is absorbed by a molecule, it promotes the formation of a number of different electronic and vibrational excited states. For simplicity, this discussion will focus on a generic organic system, in which excitation occurs from a singlet ground state (S_0) to a generic singlet excited state (S_n, $n > 0$). From these excited states, a number of possible relaxation processes can occur, including **radiative** – ones involving the emission of light – and **nonradiative** processes. The first that will be considered is **vibrational relaxation**, by which a molecule relaxes from a vibrational excited state to the vibrational ground state within a given S_n (Figure 3.2(2)). During this nonradiative process, excess energy is converted to kinetic energy and distributed throughout the molecule and surrounding environment through vibrations. As a result, vibrational relaxation is generally quite fast, occurring on the time scale of 10^{-12} to 10^{-10} seconds.

3.2.3.4 Internal Conversion

Once a molecule has relaxed to the lowest vibrational state within an electronic excited state, it can then relax to lower-energy electronic excited states through a nonradiative process known as **internal conversion** (Figure 3.2(3)). Practically, this process is very similar to vibrational relaxation, although it can be slower due to the greater energy difference between electronic states relative to vibrational states. A typical time scale for internal conversion is 10^{-11} to 10^{-9} seconds, though this process can be slower for transitions between states of significantly different energies (e.g. $S_1 \rightarrow S_0$).

In most cases, a molecule in the condensed phase will rapidly relax through internal conversion and vibrational relaxation to the lowest-energy excited state, in this case S_1 (**Kasha's rule**) [18]. Practically, this means that most photochemistry in condensed phases occurs from the lowest-energy excited state [19], a phenomenon that has been extensively observed for a variety of different systems. As a result, many of the processes that follow, as well as most photochemical processes that are of interest to the synthetic chemist, occur from either S_1 or T_1, the latter of which will be discussed further in subsequent sections (see Sections 2.3.6 and 2.3.7). Of course, several exceptions to Kasha's rule have been observed [20] with one notable example being azulene, in which fluorescence occurs primarily from S_2 rather than S_1. In addition, Kasha's rule is less applicable in the gas phase, where internal conversion and vibrational relaxation are much slower due to decreased intermolecular interactions.

3.2.3.5 Fluorescence

Once a molecule has relaxed to S_1 (with a few exceptions as discussed in Section 3.2.3.4), it can undergo a radiative relaxation process to S_0 called **fluorescence** (Figure 3.2(4)), typically within 10^{-8} to 10^{-5} seconds. Importantly, while fluorescence is often observed from S_1 to S_0, it is relevant for any two states of the same multiplicity. Sometimes, this process may also be referred to as **emission** or **luminescence**, but one should note that these latter terms are more general and can also refer to emission from a triplet state (see Section 2.3.7). Fluorescence may occur to any vibrational state within S_0, leading to emission over a range of wavelengths.

Much useful information can be obtained by monitoring the fluorescence of a compound, some of which will be discussed further in a later section (see Section 2.5.2). One example is the S_1 excited-state energy, since the energy of the photon emitted during relaxation is equal to the energy difference between S_0 and S_1. Although the same relationship also holds true for absorption, fluorescence has the advantage of operating primarily from a single excited state due to Kasha's rule. By contrast, absorption may promote formation of a number of different excited states, making it difficult to measure the energy of just one.

In addition, since some energy is lost to various relaxation processes prior to fluorescence, it is common that the λ_{max} of fluorescence is **red-shifted** (shifted to lower energies of light, also known as a **bathochromic shift**) relative to the λ_{max} of absorption. This shift in the λ_{max} is known as a **Stokes shift**. On rare occasions, a molecule may exhibit an **anti-Stokes shift**, where the λ_{max} of fluorescence is **blue-shifted** (shifted to higher energies of light, also known as a **hypsochromic shift**) relative to the λ_{max} of absorption. This phenomenon is made possible when absorption occurs from a vibrational excited state of S_0, such that the energy absorbed is less than the energy emitted during the transition from S_1 to the vibrational ground state of S_0.

Finally, the fluorescence of a molecule can sometimes be sensitive to environmental factors, providing another opportunity to investigate the properties of the excited state. One example of such a property is **solvatochromism**, where the fluorescence of the molecule changes as a function of solvent polarity. By measuring this solvatochromic effect, one can gain insight into the relative stability of the excited state, as well as charge-transfer effects that might give rise to a polar excited state. Solvatochromism is not unique to emission and can even be observed in absorption, although this phenomenon is less common.

3.2.3.6 Intersystem Crossing

An alternative process that may occur is intersystem crossing (ISC), although this process is generally slower ($10^{-8}$to 10^{-3}) and cannot always compete with faster relaxation processes. Specifically, **ISC** is the process by which the spin – an intrinsic form of angular momentum exhibited by elementary particles – of the excited electron is inverted, converting a singlet excited state to a triplet excited state. For ISC to occur, the corresponding singlet and triplet states must share a common geometry (the Frank–Condon principle, see Section 2.3.2), which corresponds to a crossing point of their potential energy surfaces. Since the inversion of spin is considered a forbidden transition, it is unsurprising that this process is generally slower than other relaxation events. Regardless, ISC from S_n produces a triplet excited state (T_n), which then rapidly relaxes to T_1 in accordance with Kasha's rule (see Section 2.3.4). Importantly, the reader should remember that this discussion focuses on singlet and triplet states for simplicity, but ISC is certainly not limited to these states. For example, it is also feasible for ISC to convert a doublet to a quartet state [21].

When considering ISC, the question arises: how does a spin-flip occur if it is quantum mechanically forbidden? The answer is through spin–orbit coupling. To understand this phenomenon, it is important to first understand why a spin-flip is forbidden. Typically, spin-flips are considered forbidden because it is thought that the spin angular momentum of an electron must remain constant. However, a more accurate statement is that the total angular momentum of the electron must remain constant, which is a sum of the spin angular momentum and the orbital angular momentum. As a consequence, while a spin-flip does change spin angular momentum, this change can be compensated by an equal but opposite change in orbital angular momentum. Hence, **spin–orbit coupling** combines a spin-flip with an orbital change to conserve the total angular momentum of the electron, making ISC allowed under the appropriate conditions.

As a result of spin–orbit coupling, atoms or molecules with a greater number of orbitals are more likely to undergo ISC. This concept is known as the **heavy-atom effect**, since heavier atoms tend to also have a greater number of atomic orbitals (e.g. iodine vs. fluorine), leading to a greater probability that an orbital change can be coupled to a spin-flip. Therefore, ISC is generally more common for inorganic or organometallic molecules than for organic molecules, although several strategies exist for increasing the triplet yield of organic molecules. For example, incorporating heavy atoms such as bromine or iodine can increase ISC in a molecule. In addition, the presence of carbonyls can be advantageous, as spin–orbit coupling can be achieved by converting between an (n,π*) S_1 state to a (π,π*) T_1 state (**El Sayed's rules**), providing the necessary orbital change. Finally, **twisted intramolecular charge-transfer (TICT)** states can also be advantageous for promoting ISC. TICT refers to charge transfer between an electron donor and acceptor that are connected by a single bond, which gives rise to a twisted charge-separated state. The product of TICT is a molecule that closely resembles a radical ion pair, where an electron from a donor-centered orbital is transferred to an acceptor-centered orbital. As such, this orbital change can also serve to promote ISC.

3.2.3.7 Phosphorescence

After ISC produces a triplet excited state, several possible relaxation pathways analogous to those previously discussed for S_1 exist. For instance, the molecule may relax to S_0 via nonradiative decay processes, such as vibrational relaxation or internal conversion. For organic molecules, these nonradiative processes are generally predominant, leading to limited **phosphorescence** (Figure 3.2(6)), or emission from T_1. Similar to fluorescence, phosphorescence can occur between any two states that involve and inversion of spin, but T_1 to S_0 is most common in organic molecules. Often, organic molecules must be cooled to low temperatures in a glassing solvent, such as 2-methyl tetrahydrofuran, to reduce nonradiative relaxation, allowing phosphorescence to be observed. Further, compared to other relaxation processes, the time scale of phosphorescence is generally quite slow – on the order of 10^{-3} to 10 seconds or even longer. As a consequence, triplet excited states tend to be long lived – on the order of tens of nanoseconds up to microseconds, milliseconds, seconds, and even minutes – making them well suited to engage in bimolecular reactions.

3.2.4 Electron Transfers

Up to this point, we have mainly discussed photophysical processes that occur when a molecule interacts with light. In this section, we will briefly discuss theories of ET and how they can be applied to photoredox reactions.

3.2.4.1 Photoinduced Electron Transfer

When applying photoredox chemistry to chemical synthesis, we are interested in harnessing energy from light to enable interesting transformations under mild conditions that might not otherwise be possible. As such, it is important to understand how that light is converted to chemical energy, and how this strategy is unique to others.

When light is absorbed by a compatible molecule, that light promotes an electron into an excited state with different properties relative to the molecule's electronic ground state. In some cases, the excited state may have drastically different redox properties than the ground state, leading to ET events that would otherwise be challenging. When ET is promoted by absorption of light, it is termed **photoinduced electron transfer (PET)**. Such ETs will be the basis of much of the chemistry discussed in the coming sections (see Sections 3.3 and 3.4).

The Gibbs free energy of PET ($\Delta G^{\circ}_{\mathrm{ET}}$) in units of ($\mathrm{J\,mol^{-1}}$) can be calculated according to Eq. (3.7):

$$\Delta G^{o}_{\mathrm{ET}} = N_A \; \{e[E^{o}(\mathrm{D^+/D}) - E^{o}(\mathrm{A/A^-})] + w(\mathrm{D^+A^-}) - w(\mathrm{DA})\} - \Delta E_{0-0} \quad (3.7)$$

where N_{A} is Avogadro's number ($6.022 \times 10^{23}\ \mathrm{mol^{-1}}$), e is the elementary charge (1.602×10^{-19} C), D is a generic donor, A is a generic acceptor, E° is a standard potential (V), w is an electrostatic work function (J), and ΔE_{0-0} is the energy difference between the lowest vibrational level of the ground state and relevant excited state ($\mathrm{J\,mol^{-1}}$). The electrostatic work function is given by Eq. (3.8):

$$w(XY) = \frac{z_x z_y e^2}{4\pi\varepsilon_0 \varepsilon_r a} \quad (3.8)$$

where x and y are two generic components (e.g. D and A), z is the signed magnitude of the charge (unitless), ε_0 is the vacuum permittivity ($8.854 \times 10^{-12}\ \mathrm{C^2\,J^{-1}\,m^{-1}}$), ε_{r} is the relative medium static permittivity (also referred to as the solvent dielectric constant) (unitless), and a is the distance between the charged species after ET (m). Equation (3.7) is sometimes referred to as the Rehm–Weller equation, although IUPAC has recommended against doing so as this name is inaccurate [22].

Data for use in Eq. (3.7) can be obtained using two techniques. The standard potentials can be approximated by the $E_{1/2}$ for the relevant redox couple obtained using cyclic voltammetry (CV) (see Section 2.5.4). ΔE_{0-0} corresponds to the excited-state energy, which can be measured using emission spectroscopy (see Section 2.5.2). Further, while electrostatic work can be more challenging to determine, it can often be omitted during photoredox reaction development, as the magnitude of w is generally negligible [23].

3.2.4.2 Mechanisms of Electron Transfer

Broadly, ETs can be divided into two categories: **inner-sphere electron transfer (ISET)** and **outer-sphere electron transfer (OSET)**. In each case, ET occurs between an electron donor (D) and an electron acceptor (A), but the mechanism of ET is different. In **ISET**, the D and A are connected via a covalent bond through which the electron is transferred. During this process, it is common that bonds may be broken and new ones formed to facilitate the ET. In addition, ISET is often associated with adiabatic ET, where **electronic coupling (V_{el})** between the D and A is significant. Schematically, this property can be visualized by considering a reaction coordinate diagram where a smooth transition connects the reactant (R) and product (P) energy surfaces (Figure 3.3b).

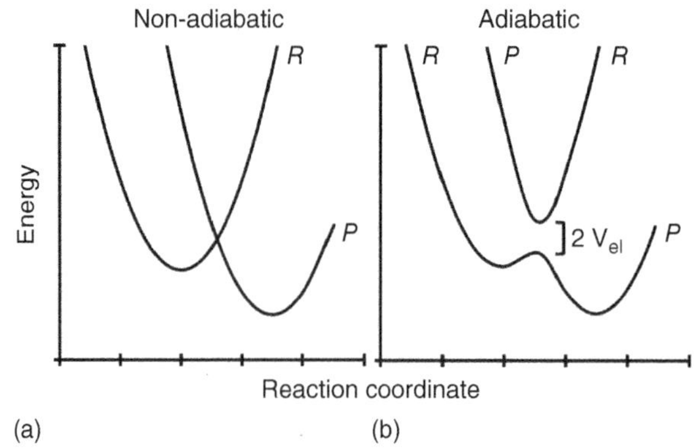

Figure 3.3 Potential energy surfaces for electron transfer in the nonadiabatic (a) and adiabatic (b) regimes. R and P refer to the reactant and product energy surfaces, respectively.

By contrast, in **OSET** the D and A do not necessarily need to be covalently bound. ET occurs through space, so it is sufficient for the D and A to associate through intermolecular interactions prior to ET and form an **encounter complex**. Due to the Frank–Condon principle (see Section 2.3.2), preassociation to form the encounter complex must occur *prior* to OSET. Further, electronic coupling of the D and A is minimal, meaning the electron must "jump" from one energy surface to the other at a crossing point (Figure 3.3a). This crossing point is a key feature of the **nonadiabatic** regime.

3.2.4.3 Marcus Theory

PET often occurs through an OSET mechanism. This section will provide a brief introduction to **Marcus theory**, which is the predominant theory used to understand OSET. Marcus theory begins by considering the ET between D and A. If one restricts this process to an outer-sphere mechanism, the D and A must associate to form the encounter complex [D–A]. Considering the ET as a chemical reaction, [D–A] is the reactant. After the ET occurs, the product is the complex $[D^+A^-]$, which can then dissociate or undergo back-electron transfer (BET) to regenerate [D–A]. To visualize this process, Marcus theory considers a simplified reaction coordinate diagram, where the R and P energy surfaces are represented by simple parabolas. Figure 3.4 shows reaction coordinates for the three most common cases.

In the **normal region** (Figure 3.4a), ET occurs generally as one might expect. Considering an exergonic ET, where the free-energy change of the reaction ($\Delta G°$) is negative, one can anticipate finding a barrier to the reaction ($\Delta G^{\ddagger}$) associated with the formation of a transition state. However, once this barrier is overcome, ET occurs to form the desired product $[D^+A^-]$. One term that is often referred to in Marcus theory is the **reorganization energy (λ_r)**. This term represents the energy that would be required to achieve the product ground-state geometry within the reactant, and it is a sum of inner-sphere (within the [D–A] complex) and outer-sphere (solvent molecules around the encounter complex) components (i.e. $\lambda_r = \lambda_i + \lambda_o$).

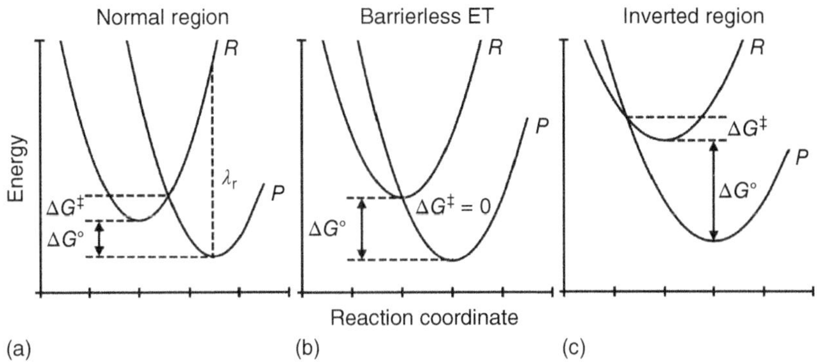

Figure 3.4 Simplified reaction coordinate diagrams used to describe Marcus theory in the Marcus normal region (a), for barrierless electron transfer (b), and in the Marcus-inverted region (c). $\Delta G^{\ddagger}$, free energy of the transition; ΔG°, free energy of the reaction; λ_r, reorganization energy.

If the magnitude of ΔG° is increased (i.e. making it more negative), this process can be visualized on a reaction coordinate diagram as a gradual lowering of the product energy surface in the y-direction. As ΔG° becomes more negative, one notices that $\Delta G^{\ddagger}$ becomes smaller and smaller until eventually $\Delta G^{\ddagger} = 0$. Practically, this statement has several important consequences. The first is that in Marcus theory, thermodynamics and kinetics are intertwined. That is, as the reaction becomes more thermodynamically favorable (i.e. more negative ΔG°), the rate of the reaction increases due to a lowering of the barrier to the reaction. Second, when $\Delta G^{\ddagger} = 0$, the reaction becomes barrierless such that ET occurs almost instantaneously.

Interestingly, Marcus theory predicts a limit to which this phenomenon holds true, as one will eventually enter the Marcus **inverted region** (Figure 3.4c). To understand what this means, consider the case where the reaction becomes more and more thermodynamically favorable. As one continues to lower ΔG° past the point where the reaction becomes barrierless, the crossing point between the R and P energy surfaces begins to rise again, resulting in an increase in $\Delta G^{\ddagger}$. In other words, as the reaction becomes more exergonic, the barrier to ET increases, resulting in a decrease in the rate of ET. While this phenomenon may seem counterintuitive, it has been observed experimentally [24].

3.2.5 Laboratory Techniques for Studying Photoredox Processes

The following section will provide a general overview of techniques commonly used to study photochemical and photophysical processes. While this list is certainly not exhaustive, it serves an introduction to relevant techniques and aims to point the reader to further useful reading where possible.

3.2.5.1 sUV–Visible Spectroscopy

UV–Visible spectroscopy (UV–vis) is a form of absorption spectroscopy focused on the UV and visible regions of the electromagnetic spectrum. In its most basic

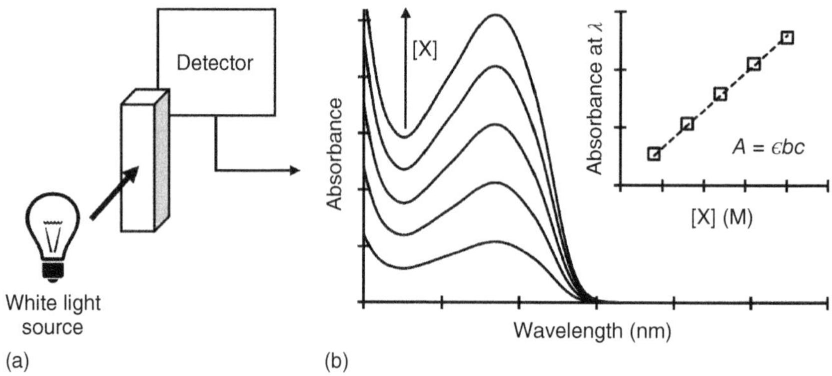

Figure 3.5 General instrument layout for absorption spectroscopy (a) and examples of data obtained by this technique (b). Once an absorption spectrum has been obtained, the molar absorptivity of a compound at a wavelength can be determined by measuring absorbance as a function of concentration (inset).

form, instrument design for this measurement is quite simple, involving a light source that shines through a sample and a detector on the other side to measure the light transmitted through the sample (Figure 3.5a). Of course, modern instruments can be more complicated as well, allowing for more accurate and precise measurements on a range of different samples (e.g. solutions, solids, films, reactions *in situ*, etc.).

When the measurement is performed, the instrument collects transmittance data as it measures light that passes through the sample without being absorbed. However, a more useful quantity for spectroscopy is absorbance (used interchangeably with OD, see Section 2.3.2), as absorbance can be related to other important quantities using Beer's law. Equation (3.9) shows the conversion of transmittance (T) to absorbance.

$$\mathrm{A} = \mathrm{OD} = \log_{10}\left(\frac{1}{T}\right) \tag{3.9}$$

It's worth noting that scattering, reflection, and refraction of light that results in loss of detected transmittance is treated as absorption when applying Eq. (3.9), as is commonly done automatically in modern spectrometers. Typically, the use of optically transparent cuvettes positioned at a right angle to the incoming light minimizes these effects as long as the sample consists of a homogeneous solution, such that absorption and OD can be used interchangeably.

Perhaps the simplest use of UV–vis is to examine the absorbance of different reaction components. In photoredox chemistry, such a study can be an important control experiment to ensure irradiation does not cause background reactivity through direct excitation of a substrate. In addition to measuring the absorbance of each reaction component, it can also be useful to measure the absorbance of different substrates in combination. As will be discussed later in the chapter (see Section 3.4.1.2), EDA complexes can form unique absorption bands relative to their individual components. As such, measurement of the components alone would yield little insight into the possibility of such a reactive complex.

In addition, it is often useful to know the molar absorptivity of a molecule, as this property can yield insight into the efficiency with which it absorbs a certain wavelength of light. In particular, this value can be important in photoredox catalysis, where the molar absorptivity of the catalyst directly impacts the penetration depth of the light into the reaction vessel. To measure this quantity, one can take advantage of Beer's law (Eq. (3.6)), which states that absorbance is directly proportional to the concentration of the absorbing species. Since the molar absorptivity of a molecule is constant at a certain wavelength, and one can maintain a constant path length, plotting absorbance at a wavelength vs. concentration for a molecule of interest gives a straight line with the slope corresponding to the molar absorptivity at that wavelength (Figure 3.5, inset). Importantly, since the percentage of light transmitted, and therefore detected, above $A = 1$ is quite small, molar absorptivity measurements should be conducted on solutions with $A < 1$ when possible to ensure the highest accuracy.

3.2.5.2 Emission Spectroscopy

Using **emission spectroscopy**, one can probe the nature of an excited state as well as its reactivity. The instrument used to measure emission is called a fluorimeter, although most fluorimeters can also measure phosphorescence. In its simplest form, a fluorimeter has many of the same components as an absorption spectrometer. However, since the intensity of emission is generally quite weak compared to the light source, the detector is placed at a 90° angle to the light source, allowing most of the source light to go undetected (Figure 3.6a).

In a simple emission measurement, one chooses an excitation wavelength – determined from the absorption spectrum of the molecule being probed – and monitors for emission at longer wavelengths of light (Figure 3.6b). Since most molecules exhibit a Stokes shift (see Section 2.3.5), it is often not necessary to monitor areas blue-shifted to the excitation wavelength. Once an emission spectrum has been collected, it can be used to estimate the energy of the excited state ($E_{0,0}$) based on the energy of light emitted. In theory, since the highest-energy

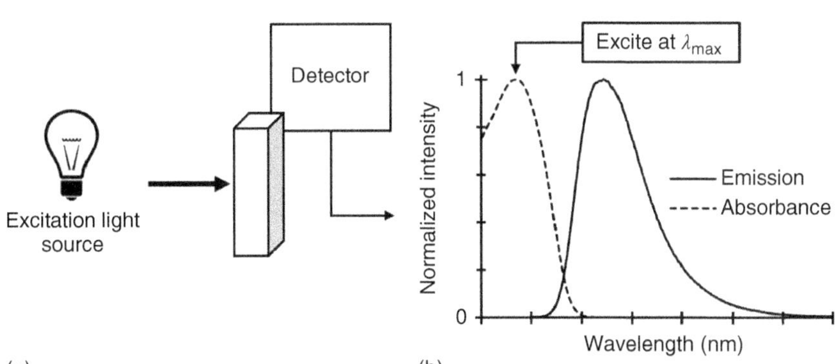

Figure 3.6 General instrument layout for emission spectroscopy (a) and an example of data obtained by this technique (b). Once an absorption spectrum has been obtained, a sample is typically excited at its λ_{max} to obtain an emission spectrum.

transition should be between the vibrational ground state of S_1 and that of S_0, the maximum wavelength of emission should correspond to this transition. However, several conventions exist for estimating $E_{0,0}$, including the use of the emission maximum, the onset of emission, and the crossing point between the emission and absorption spectra.

Of course, which excited state one investigates depends on the emission being measured. By monitoring fluorescence, one can determine the energy of the S_1 state, whereas monitoring phosphorescence will give the energy of the T_1 state. For the latter measurement, it might be necessary to cool the sample in a glassing solvent to slow down nonradiative decay pathways from T_1. In addition, to isolate phosphorescence from fluorescence, one may need to use a time delay between excitation and measurement of the sample, since phosphorescence is generally longer lived than fluorescence.

Closely related to an emission spectrum is an **excitation spectrum**, in which emission intensity at a single wavelength is monitored as the excitation wavelength is varied. Since the amount of light emitted by a **fluorophore**, or a fluorescent molecule, is proportional to the amount of light absorbed (i.e. Beer's law), a fluorophore's excitation spectrum should be identical to its absorption spectrum as long as Kasha's rule is obeyed. When this relationship is not true, the excitation spectrum can yield insight into the presence of anti-Kasha behavior (emission from a higher state or vibrational mode), or more commonly the presence of an impurity or aggregate species in solution.

Emission spectroscopy can also be used to probe the quantum yield of different states. **Quantum yield (Φ)** is the ratio of the quantity of emission (or states formed) relative to the quantity of light absorbed by a system. Typically, quantum yield is reported as a value between 0 and 1 or as a percentage, where 1 (or 100%) corresponds to every photon absorbed leading to the desired outcome. Quantum yield can be used to describe various processes, such as fluorescence, phosphorescence, ISC, or nonradiative decay. In addition, Φ is commonly used to report the degree to which a certain state forms, such as T_1. However, since both the light absorbed and the light emitted must be quantified precisely, the measurement of Φ is more complicated than a standard emission measurement, requiring comparison to a standard or use of an integrating sphere.

The final technique employing emission spectroscopy that will be discussed is **Stern–Volmer quenching**. In this technique, the emission of a molecule is compared in the absence and in the presence of a potential quencher to determine if electron or energy transfer (see Section 3.1) can occur from the excited state to a substrate. In theory, if no electron or energy transfer occurs, the emission of the molecule should be unchanged in the presence of another molecule. By contrast, if one of these processes is operative, it will lower the concentration of the excited state, thereby reducing the intensity of emission. Equation (3.10) relates the intensity of emission in the absence ($I_f°$) and presence (I_f) of a quencher (Q) to the rate constant for quenching (k_q) and the lifetime of the excited state in the absence of quencher (τ_0). Importantly, since emission quenching would be observed both in the case of ET and energy transfer, this technique cannot provide insight into which mechanism is operative in a reaction.

$$\frac{I_f^0}{I_f} = 1 + k_q \tau_0 \cdot (Q) \tag{3.10}$$

While this technique can be quite powerful, one must be wary of other processes that might reduce emission as well, such as the ground-state (or excited state) association of the quencher with the emitting molecule to form a nonemissive complex. In addition, one must be careful to avoid the **inner-filter effect**, which is the absorption of light used for excitation or the emission of the fluorophore by either the fluorophore or the quencher. For this reason, emission spectroscopy is often performed at low concentrations of the fluorophore ($A < 0.1$), even though mathematical and experimental corrections for the inner-filter effect have been devised [25].

3.2.5.3 Transient Absorption Spectroscopy

While many common spectroscopic techniques perform measurements under steady-state conditions on relatively long time scales, transient techniques make measurements on very short time scales to gain insights into chemical dynamics. In this section, we will focus our discussion on transient absorption spectroscopy, as this is perhaps the most common transient technique used in the investigation of photochemical systems.

Broadly, **transient absorption spectroscopy (TAS)**, also called **pump-probe spectroscopy**, is very similar to standard absorption spectroscopy. Commonly, the UV and visible regions of the electromagnetic spectrum are probed, so this discussion will focus on those regions, although the technique can certainly be applied to other regions as well. In the case of UV–visible absorption, a white light source is used to irradiate a cuvette containing the sample, and the light transmitted is monitored by a detector (Figure 3.7a). Under such conditions, one would observe a spectrum corresponding to the ground-state molecule, as one would observe in a standard UV–visible absorption measurement. To probe the characteristics of an excited state, a laser pulse (or pump) is applied to the sample orthogonal to the probe pulse – so as to not interfere with the detector – generating a small region of excited-state molecules for observation. Since the cross-section of the laser is often quite small, the probe light will still interact with a significant amount of ground-state molecules, making it difficult to observe the excited-state species. As such, data are often reported with units of ΔA (or ΔOD, Figure 3.7b), such that absorption from the ground state can be subtracted from the overall signal and the absorption of excited-state species can be isolated. Related to this technique is **transient emission spectroscopy** or **time-resolved emission spectroscopy**, where a sample is simply excited with a laser pulse and its emission monitored in the absence of a probe light.

Using modern instruments, measurements on the picosecond and femtosecond time scale can be achieved, providing information about even the fastest photophysical processes. As such, these techniques have been used to monitor the fast relaxation processes that occur after the absorption of light by a molecule, allowing chemists to probe the formation and characteristics of excited states within photoactive molecules. In particular, two measurements are common in such studies.

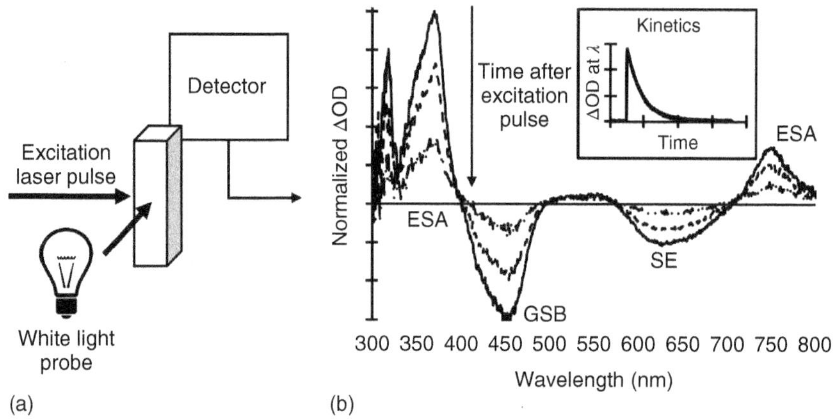

Figure 3.7 General instrument layout for transient absorption (a) and examples of data obtained by this technique (b). Excited-state spectra can be observed using absorption spectroscopy at different time delays after an excitation pulse with a laser. By then selecting a wavelength corresponding to an excited state and monitoring absorbance at that wavelength as a function of time (inset), excited-state lifetimes can be determined. Abbreviations: ESA, excited-state absorption; GSB, ground-state bleach; SE, stimulated emission.

In the first, the absorption spectrum of an excited sample is collected at different time delays after the excitation pulse (Figure 3.7b), allowing one to monitor changes in the excited-state identity after absorption. Several common features may be observed in such a measurement, including a **ground-state bleach (GSB)**, corresponding to a negative ΔOD where ground-state molecules have been converted to an excited state in a region of the spectrum where the excited state does not absorb; an **excited-state absorption (ESA)**, where newly formed excited-state species absorb the probe light and are promoted to higher-energy excited states; and **stimulated emission (SE)**, where excited-state molecules are forced to emit light through an interaction with the probe light. Second, once a feature has been identified, the lifetime of this species can be determined by monitoring the change in absorbance at a relevant wavelength over time. This measurement is sometimes called **single-wavelength kinetics** (Figure 3.7b inset) and requires fitting the data to an exponential function to determine the excited-state lifetime. If multiple features are identified that belong to the same excited state, monitoring any feature should give the same single-wavelength kinetics (within error). If significant differences are observed, it may indicate the presence of another species.

In addition, these measurements can be employed in a number of other important studies probing the reactivity of certain excited states. For instance, in photocatalysis, kinetic measurements are often used to probe whether a reaction occurs through an ET or energy-transfer mechanism (see Section 3.1), as the products of these mechanisms can be differentiated spectroscopically. Further, Stern–Volmer quenching can be performed by monitoring how different quenchers impact the lifetimes (or decay rate constants k_1 and k_2) of different excited-state species (Eq. (3.11)), allowing one to determine which excited state is responsible for an observed reaction.

$$\frac{k_2}{k_1} = 1 + \frac{k_q}{k_1}(Q) \tag{3.11}$$

3.2.5.4 Cyclic Voltammetry

To investigate the reduction and oxidation of molecules, cyclic voltammetry is commonly used. Broadly, **cyclic voltammetry (CV)** involves applying an electrochemical potential to a sample, which is changed (or swept) at a constant rate, and then measuring the current that flows in and out of the sample at each applied potential (Figure 3.8). In essence, by changing the applied potential, one is altering the thermodynamic driving force for ET, and by monitoring the current response of the sample, one is measuring the kinetics of ET between the sample and the electrode. By doing so, one can determine the ground-state reduction and oxidation potentials of a molecule, which correspond to the average of the peak potentials ($E_{1/2}$) as observed in Figure 3.8 (solid line).

In addition, one can determine from the shape of the current response whether an ET is reversible, where a perfectly **reversible** one-ET has an oxidation and reduction peak of equal magnitude that are separated by 57 mV (Eq. 3.12), where ΔE_p is the peak separation (V), R is the ideal gas constant (8.314 J mol^{-1} K^{-1}), T is the absolute temperature (K), n is the number of electrons being transferred, and F is Faraday's constant (96 485 C mol^{-1}).

$$\Delta E_p = 2.22\frac{RT}{nF} \tag{3.12}$$

By contrast, a completely **irreversible** ET would only exhibit an oxidation or reduction peak (Figure 3.8, dashed line), as the oxidized or reduced compound disappears prior to the return scan and is not regenerated in the original oxidation state. A system can appear irreversible for both chemical and electrochemical reasons, although it is not always simple to determine which is the case. Sometimes, a system may appear to exist somewhere between totally reversible or irreversible (Figure 3.8, dotted line) and can be referred to as **quasi-reversible**.

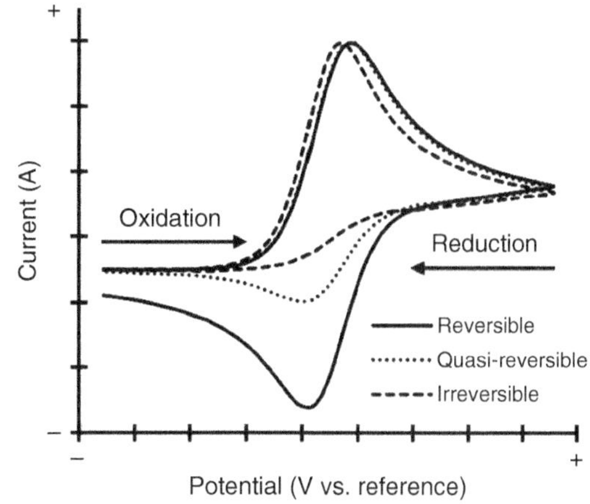

Figure 3.8 Examples of cyclic voltammograms for reversible, quasireversible, and irreversible systems.

With regard to photoredox chemistry, several useful pieces of information can be obtained using CV. In photoredox catalysis, determination of a PC's ground-state oxidation and reduction potentials can provide insight into possible catalytic mechanisms, as well as compatibility with certain substrates when the redox properties of the substrates are also known or measured. In addition, by combining CV with emission spectroscopy, it is possible to estimate the redox properties of a photocatalyst's excited state, providing insight into the reactivity of this key catalytic species (see Section 3.2.4). Finally, CV can also be used to guide the development of reactions employing EDA complexes, as the reduction and oxidation potentials of molecules can serve as estimates of their electron-donating and -accepting abilities.

3.2.6 Practical Considerations for Performing Photochemical Reactions

3.2.6.1 Factors Influencing Bimolecular Reactions

The rate of a bimolecular reaction in solution is limited by the rate of diffusion. While the rate of a reaction may be slower than diffusion, the fastest two molecules can react is only as quickly as they can diffuse to each other. As such, it is important to keep this consideration in mind when designing a photochemical reaction, as the relevant photoexcited species must be long enough lived to interact with the other molecules in solution. For a bimolecular system, the lower limit for this excited-state lifetime is generally on the order of a few nanoseconds, although some exceptions certainly exist. Sometimes, a short excited-state lifetime can be overcome by increasing the concentration of the reagents in solution, although this can come at the cost of increased catalyst loadings or waste generation.

3.2.6.2 Photoreactor Design

Another important consideration that is often overlooked is the design of the photoreactor. In fact, reactor design in photochemistry has drawn significant criticism over the years, as few standardized designs exist, leading to significant variability among different research groups, though several different commercially available photoreactors are currently available. As such, careful reactor design and reporting are necessary to ensure the reproducibility of photochemical reactions.

Broadly, two different reactor designs can be considered, each with its own set of design parameters that can be tuned. A batch reactor (Figure 3.9a) is perhaps the simplest and involves surrounding a reaction flask with light source. In its most basic form, this reactor design can be achieved by placing a light source next to a reaction flask, although this design can introduce significant variability from one reaction to the next if care is not taken. Another simple but more reproducible design involves wrapping LED strips on the inside of a beaker or recrystallization dish, such that the average distance of the light source from the reaction vessel can be easily controlled. In addition, the beaker itself can be wrapped with a reflective coating, directing more of the light into the reaction vessel rather than the surroundings. Regardless of the design employed, it is important to remember that the intensity of light entering the reaction is inversely proportional to the distance squared ($1/d^2$) between the reaction

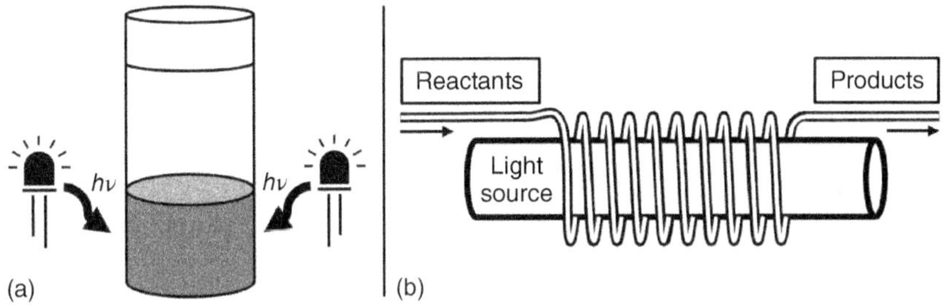

Figure 3.9 General design of a batch reactor (a) and a flow reactor (b).

and the light source. Therefore, it is advantageous to keep the reaction as close as possible to the light source.

Toward this end, continuous-flow reactors (Figure 3.9b) can be advantageous, as they achieve very uniform and consistent irradiation conditions by passing the reagents through a transparent tube and around a light source. In addition, flow reactors have the advantage of being scalable, as they are not limited by reaction volume, whereas batch photochemistry is typically challenging to scale-up. The primary disadvantage of a flow-reactor design is that it introduces added complexity and new factors such as fluid dynamics and mixing that must be considered, requiring some degree of experience to optimize reaction conditions. In addition, reaction screening can be slow with this reactor design, although it is not impossible. However, a number of resources exist to introduce the interested reader to this apparatus [26, 27].

Regardless of the reactor design, temperature control of the reactor is an important consideration as that heat produced by the light source could give rise to thermal background reactivity. In most cases, cooling can easily be achieved through the use of cooling fans.

3.2.6.3 Choice of Light Source

Of course, one of the most important reactor components is the light source. In modern photochemistry, a number of different light sources are available for use, including multicolor and white light emitting diodes (LEDs), monochromatic LEDs, fluorescent lights, incandescent bulbs, arc lamps, lasers, sunlight, and more. While each may have different advantages and disadvantages, white and monochromatic LEDs are among the most popular due to their ease of operation, generally low cost, compact size, and easy incorporation into a variety of reactor designs.

3.3 Photoredox Catalysis

3.3.1 General Mechanisms of Photocatalysis

Photoredox catalysis is generally divided into two mechanisms based on whether the PC behaves as an excited-state oxidant or reductant. When the PC is an excited-state oxidant, it accepts an electron from the substrate or a sacrificial electron donor and

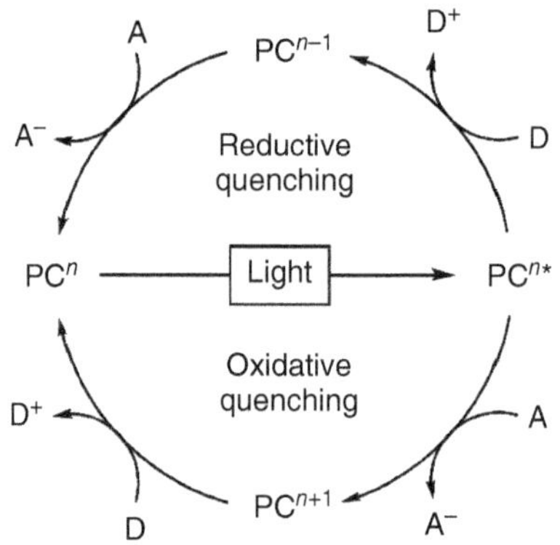

Figure 3.10 Generalized photoredox catalytic cycles proceeding through oxidative (bottom) and reductive (top) quenching mechanisms. D, electron donor, A, electron acceptor.

is itself reduced (Figure 3.10). As such, this mechanism is referred to as **reductive quenching**, since the excited state is quenched by reduction of the PC. To regenerate the ground-state catalyst, the reduced PC must then donate an electron to a substrate or sacrificial electron acceptor. By contrast, an **oxidative quenching** mechanism involves reduction of a substrate or sacrificial electron acceptor to generate the oxidized form of the PC, which then accepts an electron to regenerate the ground state. Typically, the mechanism that dominates depends on the excited-state redox properties of the PC, as well as the redox properties of the other molecules in solution.

In addition, some photocatalytic transformations can proceed through an **energy-transfer** pathway, where energy from the excited-state PC is transferred to a substrate, generating the PC ground state and the substrate excited state. The substrate excited state can then participate in ET reactions that are inaccessible to the substrate ground state. Energy transfer is quite common in photocatalysis and can proceed through different mechanisms. However, since this pathway is not a redox process and has been discussed by others [23, 28], it will not be covered in this chapter.

3.3.2 Design Principles for Effective Photoredox Catalysts

The following section outlines several important properties commonly exhibited by effective PCs. While this list is not universal and each reaction might have its own additional requirements, the properties listed below are generally considered desirable and can serve as a useful starting point for developing a photoredox-catalyzed reaction.

3.3.2.1 Effective Absorption of Light

Since the energy to generate the excited-state PC is derived from light, effective absorption of light is necessary; this statement means two things. First, the PC must absorb light in a desirable region of the electromagnetic spectrum. For organic synthesis, near-IR, visible, and low-energy UV light ($\lambda > 350$ nm) are most often used, since these wavelengths of light have sufficient energy to promote reactivity but are not absorbed by most organic molecules (see Section 2.3.2). Second, the PC should have a sufficient molar absorptivity in the spectral region corresponding to the light source. Importantly, while a high molar absorptivity could allow for more efficient absorption of light and therefore lower catalyst loadings, it also decreases the penetration depth of the incident light. As such, one has to consider the requirements of a given reaction system to determine what value of molar absorptivity is most desirable. For example, a reaction performed in a flow reactor might benefit from a strongly absorbing catalyst, whereas one performed in a batch reactor might benefit from a weaker absorbing PC due to the greater depth of the reactor.

3.3.2.2 High Quantum Yield of Desired Excited State

Quantum yield refers to the ratio of a state formed relative to the number of photons absorbed by the system. For a PC to maximize the utility of each photon, it is desirable for the quantum yield of the catalytic state to be as close to one as possible. For example, if a reaction is mediated by a PC triplet excited state, then a PC with a high triplet yield would be most desirable. Generally, the specific state desired may vary depending on the reaction, the reaction conditions, and the specific PC being employed.

3.3.2.3 Long-Lived Excited State

In photoredox catalysis, reactions commonly proceed through a bimolecular mechanism, where the excited-state PC must collide with a substrate in solution prior to ET taking place. As such, the excited state of the PC must be long enough lived to undergo bimolecular collisions – a process that generally occurs on the order of a few nanoseconds. In many cases, triplet excited states can exhibit sufficient lifetimes to participate in bimolecular reactivity because of the forbidden nature of a spin flip, which makes relaxation to the ground state a relatively slow process. However, they also have some drawbacks. Due to photophysical relaxation processes, triplet states tend to be less energetic than singlet states. In addition, triplet excited states can be quenched by oxygen, often necessitating that reactions be performed under an inert atmosphere. This reactivity arises because oxygen has a triplet ground state, which can undergo energy transfer with many triplet excited-state PCs to produce the ground-state PC and singlet oxygen.

Singlet excited states can be advantageous because they are not as easily quenched by oxygen. However, the lifetimes of these states are generally much shorter than those for triplet excited states. As a consequence, many singlet excited states may not exhibit long enough lifetimes to participate in bimolecular ET, sometimes limiting their utility in synthesis. With regard to doublet excited states, which have only recently begun to be employed in synthesis through photoexcitation of radical species (see Section 3.3.6), the lifetimes of these states are generally quite short, often in the picosecond timescale [29, 30].

3.3.2.4 Favorable Thermodynamics

Like all redox reactions, one must consider whether a reaction of interest is thermodynamically feasible based on the reduction and oxidation potentials of the substrates and the catalyst. For a PC, the excited-state redox properties can be estimated using modified Rehm–Weller equations (Eqs. (3.13) and (3.14)):

$$E^*(\mathrm{PC}^{n+1}/\mathrm{PC}^{n*}) = E^o(\mathrm{PC}^{n+1}/\mathrm{PC}^{n}) - E_{0,0} \tag{3.13}$$

$$E^*(\mathrm{PC}^{n*}/\mathrm{PC}^{n-1}) = E^o(\mathrm{PC}^{n}/\mathrm{PC}^{n-1}) + E_{0,0} \tag{3.14}$$

where E° is approximated using the $E_{1/2}$ obtained from cyclic voltammetry, and $E_{0,0}$ is the excited-state energy in eV obtained from the emission spectrum of the molecule.

The redox properties of the oxidized or reduced catalyst (i.e. PC^{n+1} or PC^{n-1}) are also important to consider, since ET to or from these species is necessary to regenerate the ground-state catalyst. These redox properties can be determined by cyclic voltammetry and are also necessary for calculation of the excited-state redox properties above.

3.3.2.5 Redox Reversibility

Finally, to ensure catalyst turnover, redox reversibility of the PC is important. Put simply, it is crucial that PC^{n+1} or PC^{n-1} does not undergo side reactivity; otherwise, catalyst degradation can occur. Using cyclic voltammetry, the reversibility of a catalyst can be evaluated based on the reversibility of the relevant redox couple.

3.3.3 Inorganic Photocatalysts

Transition-metal-catalyzed photoredox synthesis has been historically dominated by the use of d^6 polypyridyl Ru(II) and Ir(III) catalysts. The complex $[\mathrm{Ru(bpy)_3}]\mathrm{Cl_2}$ (Figure 3.11, bpy = 2,2′-bipyridine) in particular has been employed across a wide range of catalytic reactions, aided by thorough study of its photophysical properties [31]. The complex strongly absorbs visible light (λ_{max} = 452 nm, ε = 14 600 $\mathrm{M^{-1}\,cm^{-1}}$ in H_2O) to access long-lived ($\tau \sim 1\,\mu s$) triplet excited states, which can be assigned as **metal-to-ligand charge transfer (MLCT)** [32]. As a result of relatively low-lying π^* orbitals of the bipyridyl ligands, excitation of a d electron from the Ru(II) center results in a rich excited-state topology for $[\mathrm{Ru(bpy)_3}]^{2+*}$. The widespread use of $[\mathrm{Ru(bpy)_3}]\mathrm{Cl_2}$ in photoredox catalysis also stems from its versatile redox properties. While the ground-state reduction potential of the catalyst is modest [$E^\circ(\mathrm{Ru^{II}/Ru^{III}}) \sim 1.26$ V vs. saturated calomel electrode [SCE]], in the excited state the PC becomes significantly more reducing [$E^\circ(\mathrm{Ru^{II*}/Ru^{III}}) \sim -0.81$ V vs. SCE] and oxidizing [$E^\circ(\mathrm{Ru^{I*}/Ru^{II}}) \sim 0.77$ V vs. SCE] [33]. Furthermore, reductive quenching of the PC excited state with sacrificial donors such as amines to form $[\mathrm{Ru(bpy)_3}]^+$ results in significantly more reducing species [$E^\circ(\mathrm{Ru^{I}/Ru^{II}}) \sim -1.33$ V vs. SCE]. Modifications to the complex's ancillary ligand are also useful for tuning PC redox properties, as in the case of substituting 2,2′-bipyrazine (bpz) for bipyridine where [$E^\circ(\mathrm{Ru^{II}/Ru^{III}}) \sim 1.86$ V vs. SCE] for $[\mathrm{Ru(bpz)_3}]\mathrm{Cl_2}$. As the π-accepting character of the ancillary ligand increases, the decreased electron density at the metal makes oxidation less favorable.

In modern photoredox catalysis, Ir(III) PCs with 2-phenylpyridine (ppy) ligands are ubiquitous (Figure 3.11). As compared to 2,2-bipyridine, use of the strongly

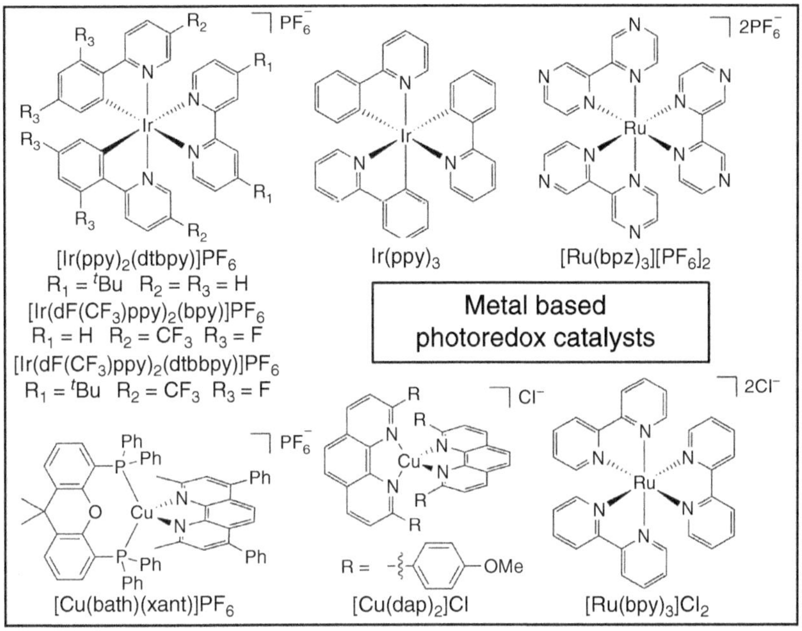

Figure 3.11 Common photoredox catalysts based on Ru, Ir, and Cu complexes.

donating 2-phenylpyridine ligand such as in *fac*-[Ir(ppy)$_3$] helps to stabilize the resulting Ir(IV) complex after oxidative quenching, as evidenced by the very negative excited-state reduction potential observed [E°(Ir^{III*}/Ir^{IV}) ~ −1.73 V vs. SCE] [33]. Stephenson showed that *fac*-[Ir(ppy)$_3$] catalyzes the reduction of unactivated alkyl and aryl iodides, while dehalogenation by [Ru(bpy)$_3$]Cl$_2$ is mainly limited to activated bromides (Scheme 3.10) [34].

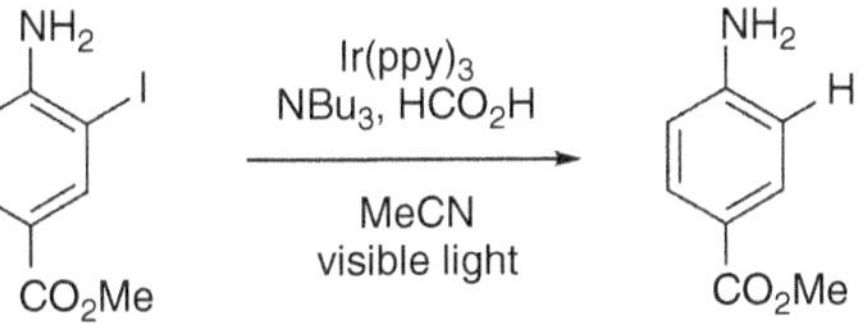

Scheme 3.10 Ir(III) photocatalyzed hydrodehalogenation of an aryl iodide. Source: Modified from Nguyen et al. [34].

Heteroleptic Ir(III) PCs with combinations of substituted 2-phenylpyridine and bipyridine ligands are able to access a wide range of redox potentials and thus are very attractive in terms of broad applicability in photoredox catalysis for organic synthesis. This versatility is evident in their ability to affect tertiary amine functionalization through reductive quenching pathways. Stephenson reported that under visible-light irradiation, [Ir(ppy)$_2$(dtbbpy)]PF$_6$ catalyzes an aza-Henry reaction between *N*-aryl tetrahydroisoquinolines and nitromethane, likely proceeding through iminium ion intermediates (Scheme 3.11) [35].

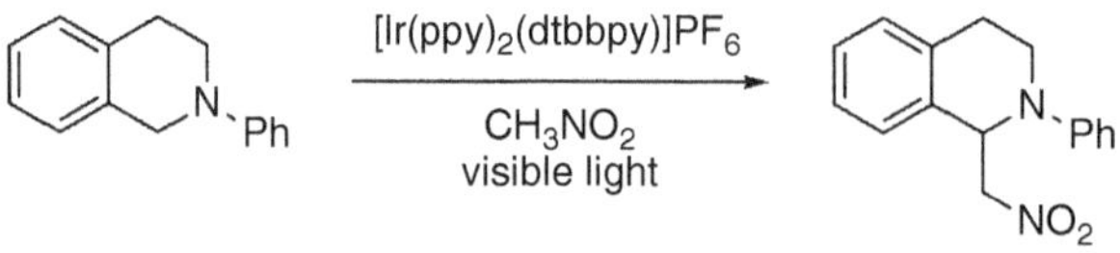

Scheme 3.11 Visible light–mediated aza-Henry reaction of an *N*-aryl tetrahydroisoquinoline. Source: Modified from Condie et al. [35].

In an example of an overall redox-neutral reaction, Koike and Akita demonstrated the *fac*-Ir(ppy)$_3$ catalyzed atom transfer radical addition (ATRA) of trifluoromethyl radicals derived from Umemoto's reagent to styrenes [36]. The authors hypothesized that oxidation of intermediate benzylic radicals followed by trapping with oxygen nucleophiles led to the observed oxytrifluoromethylation products (Scheme 3.12).

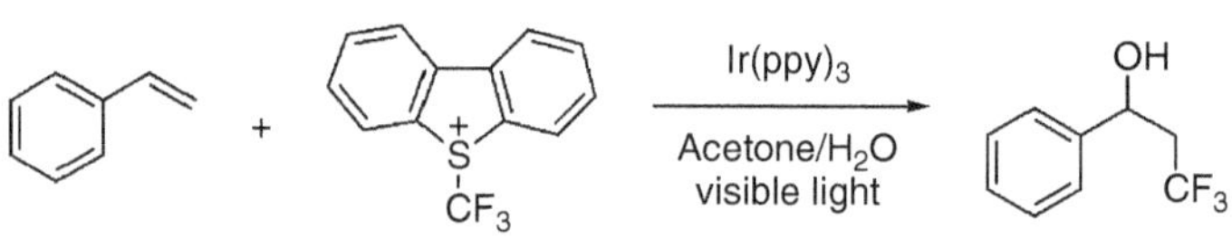

Scheme 3.12 Oxytrifluoromethylation of styrene using an Ir(III) photocatalyst.

A particular strength of photoredox catalysis has been the mild generation of carbon radicals from carboxylic acid precursors. MacMillan disclosed a method for decarboxylation of a wide scope of aliphatic carboxylic acids and subsequent trapping of the formed carbon radicals with Selectfluor (Scheme 3.13) [37]. The reaction could be catalyzed by either Ru or Ir PCs.

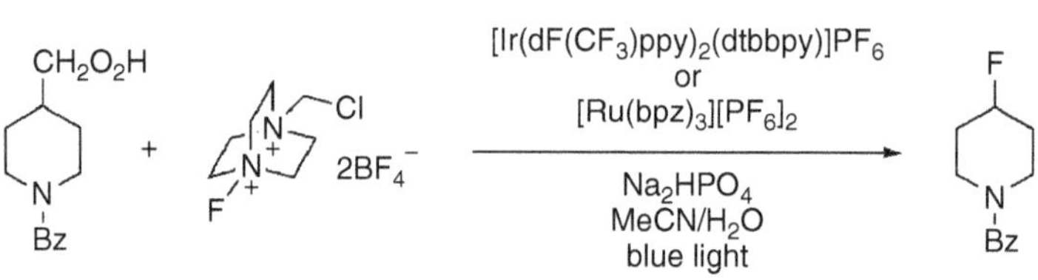

Scheme 3.13 Decarboxylative fluorination with metal photocatalysts. Source: Modified from Ventre et al. [37].

Unfortunately, noble metal PCs suffer from fundamental drawbacks such as their low natural abundance and potential toxicity. First-row transition metal complexes offer attractive alternatives as more sustainable PCs, but the photophysical properties of first-row metal complexes often differ greatly from their second- and third-row analogs. For example, photoexcited Fe(II) polypyridyl complexes are typically unable to operate via MLCT excited states because of deactivation from relatively low-lying metal-centered states, a consequence of smaller ligand field splitting for 3d metal complexes [38]. Ultimately, metal-centered excited states in potential first-row PCs provide undesirable nonradiative decay pathways, typically resulting in short (ps) excited-state lifetimes. However, increasing ligand field strength by replacing pyridyl ligands with strongly donating N-heterocyclic carbene ligands has shown promise to mitigate the issue of short excited-state lifetimes in Fe(II) PCs [39].

A more successful family of non-noble metal PCs is that of Cu(I) photocatalysts. Because of their d^{10} configuration, Cu(I) complexes do not possess metal-centered excited states and do exhibit MLCT. Cu(I) complexes like $[Cu(dap)_2]Cl$ (Figure 3.11) have been shown to be highly reducing in their excited state [$E°(Cu^{I*}/Cu^{II}) = -1.43$ V vs. SCE] and have found use in selective ATRA reactions under green light irradiation (Scheme 3.14) [40]. Thus, PCs based on earth abundant metals are promising candidates to rival the redox power and synthetic utility of their more venerable precious metal counterparts.

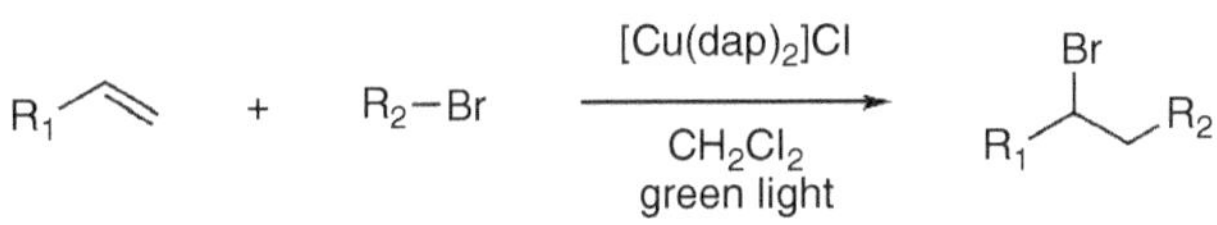

Scheme 3.14 ATRA catalyzed by a Cu(I) photocatalyst. Source: Pirtsch et al. [40].

3.3.4 Organic Excited-State Oxidants

Fully organic PCs are inherently attractive from the perspective of sustainability as compared to rare noble metal complexes. Xanthene dyes are well established in photoredox catalysis, with PCs such as Rose Bengal (Figure 3.12) capable of promoting the same transformations as precious metal photocatalysts such as the oxidative functionalization of tertiary amines under visible-light irradiation (Scheme 3.15) [50].

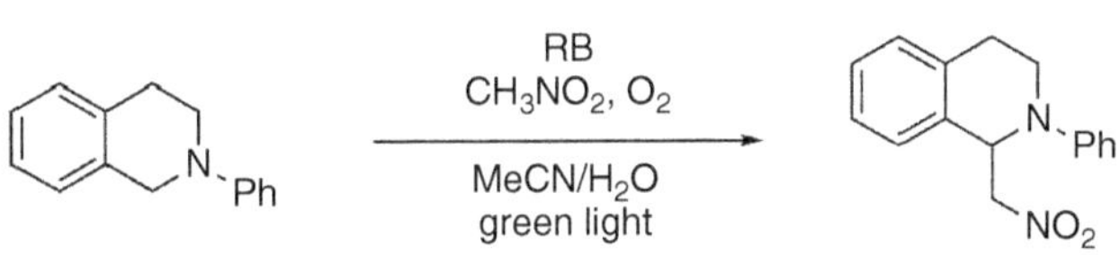

Scheme 3.15 Oxidative amine functionalization by an organic photocatalyst. Source: Pan et al. [50].

Additionally, certain organic PCs have been found to exhibit excited-state redox potentials that exceed typical ranges found in transition metal catalysts. Acridinium salts (Figure 3.12) have been fascinating subjects of study since early controversy over the precise dynamics of their photoexcited states [47, 48, 51]. The existence of highly oxidizing [$E°(PC\cdot^{-}/PC^{*}) > 2.00$ V vs. SCE] excited singlet states in 9-mesitylacridinium dyes, however, is widely agreed upon and key to their exceptional oxidative reactivity with organic compounds. In-depth mechanistic studies by Nicewicz have revealed the importance of these states in anti-Markovnikov alkene functionalization reactions [52]. In combination with redox-active hydrogen donors, acridinium dyes have been reported to catalyze the anti-Markovnikov addition of a range of nucleophiles such as amines to putative alkene cation radical intermediates (Scheme 3.16) [53].

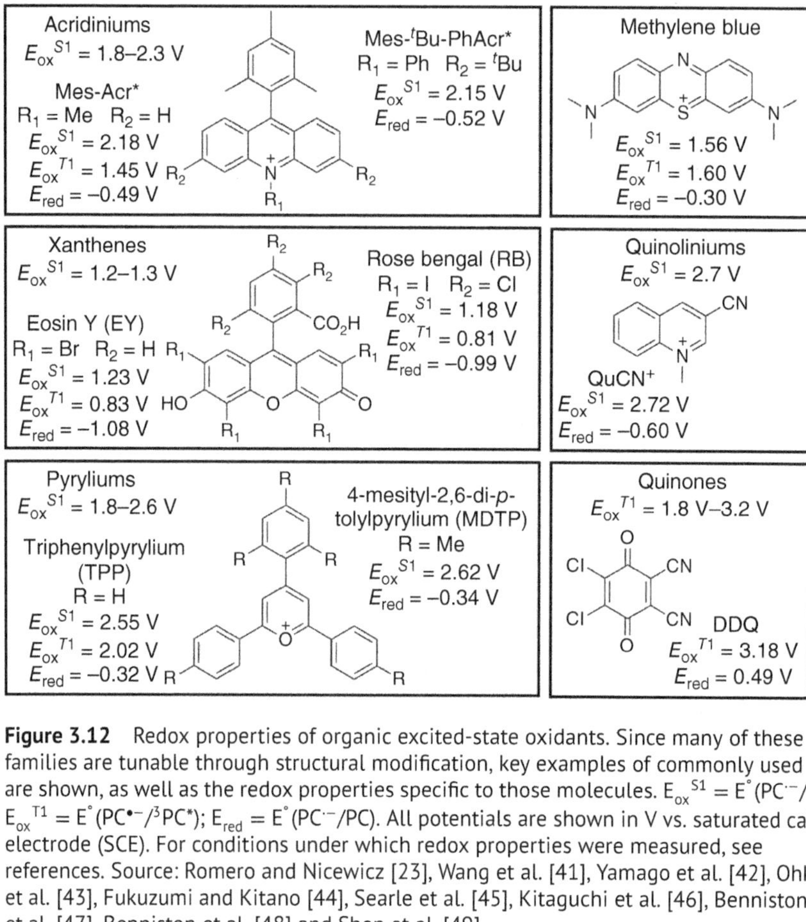

Figure 3.12 Redox properties of organic excited-state oxidants. Since many of these families are tunable through structural modification, key examples of commonly used PCs are shown, as well as the redox properties specific to those molecules. $E_{ox}^{S1} = E^{\circ}(PC^{\cdot-}/^{1}PC^{*})$; $E_{ox}^{T1} = E^{\circ}(PC^{\bullet-}/^{3}PC^{*})$; $E_{red} = E^{\circ}(PC^{\cdot-}/PC)$. All potentials are shown in V vs. saturated calomel electrode (SCE). For conditions under which redox properties were measured, see references. Source: Romero and Nicewicz [23], Wang et al. [41], Yamago et al. [42], Ohkubo et al. [43], Fukuzumi and Kitano [44], Searle et al. [45], Kitaguchi et al. [46], Benniston et al. [47], Benniston et al. [48] and Shen et al. [49].

Acridinium PCs have also been employed for the C–H functionalization of aromatic substrates. Under aerobic conditions, Nicewicz and coworkers reported the C–H amination of substituted arenes with nitrogen heterocycles such as pyrazole, again via likely cation radical intermediates (Scheme 3.17) [54]. One drawback to this approach to aromatic functionalization with acridinium catalysts is its limited utility in the oxidation of unactivated arenes, whose oxidation potentials exceed the power of visible light–excited acridinium PCs.

In contrast, Fukuzumi and coworkers have exploited high-energy UV light to generate excited 3-cyanoquinolinium ions [$QuCN^+$, $E^{\circ}(PC/PC^{+*}) = 2.72$ V vs. SCE] for the direct aerobic oxidation of benzene to phenol with water [55]. As an alternative method, the same group later disclosed the use of stoichiometric, visible-light excited DDQ (Figure 3.12) [$E^{\circ}(PC^{\cdot-}/PC^{*}) = 3.18$ V vs. SCE] to affect the same transformation [43]. Catalytic amounts of *tert*-butyl nitrite ($^{t}BuNO_2$) were used to turn over reduced DDQ, but oxygen remained the terminal oxidant for the overall reaction (Scheme 3.18).

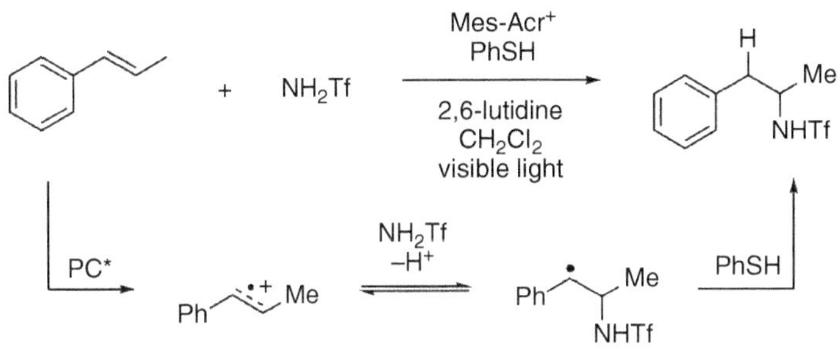

Scheme 3.16 Anti-Markovnikov hydroamination of olefins via radical cation intermediates.

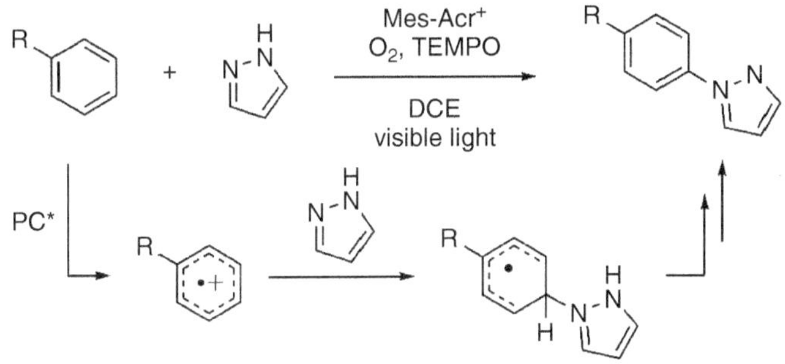

Scheme 3.17 Photoredox-catalyzed C–H amination reported by Nicewicz. Source: Modified from Romero et al. [54].

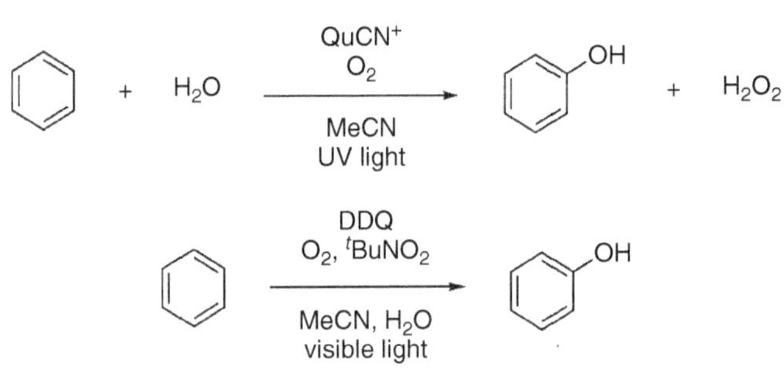

Scheme 3.18 Photocatalytic oxygenation of benzene to form phenol.

Triarylpyrylium ions are another class of visible-light-absorbing PCs, which possess highly oxidizing excited states. To overcome issues of PC degradation by nucleophilic species in catalysis, Beeler and coworkers designed a relatively sterically hindered catalyst 4-mesityl-2,6-di-*p*-tolylpyrylium (MDTP, Figure 3.12) capable of promoting oxidation of benzylic epoxides to carbonyl ylides [56]. The generated carbonyl ylides were subsequently trapped by dimethyl acetylenedicarboxylate (DMAD) in [3+2] dipolar cycloadditions to give diverse dihydrofuran products (Scheme 3.19). Introduction of a mesityl group to increase catalyst durability was inspired by similar effects seen for acridinium compounds, illustrating the importance of catalyst development across multiple families of PCs [57].

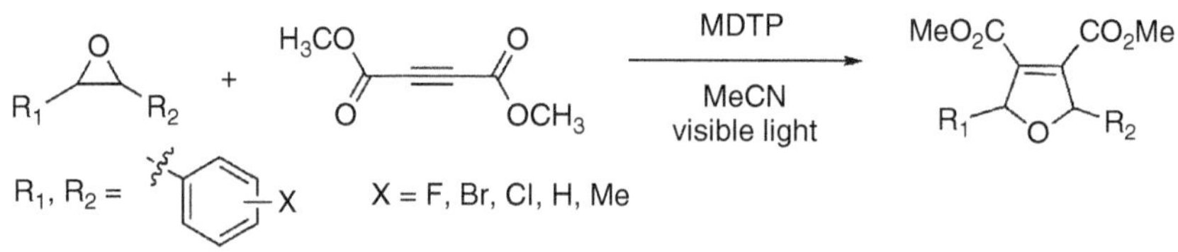

Scheme 3.19 Oxidation of benzylic epoxides and cycloaddition with DMAD to form dihydrofurans.

3.3.5 Organic Excited-State Reductants

Until about 2016, the range of redox properties accessible by excited-state reductants was limited compared to excited-state oxidants. Early examples of common excited-state reductants include α-sexithiophene [58, 59] and benzophenones such as Michler's ketone (MK) [23, 49, 60–62] (Figure 3.13). However, both systems have associated disadvantages, such as the limited tunability of α-sexithiophene and the necessity of a UV light source for MK.

In 2014, motivation for the development of strongly reducing organic PCs grew after two concurrent reports from the Miyake [78] and Hawker [68] groups detailing the first organocatalyzed atom transfer radical polymerization (O-ATRP). Analogous to atom transfer radical addition, ATRP involves reversibly breaking and forming carbon–halide bonds at the end of polymer chains to synthesize polymers with well-defined structures from vinyl monomers (Scheme 3.20). In O-ATRP, a strongly reducing organic PC mediates this process, where PC* reduces polymer C—Br bonds to "activate" the polymer growth, after which the oxidized catalyst species reforms the C—Br bond to "deactivate" the polymer and prevent radical-based side reactions.

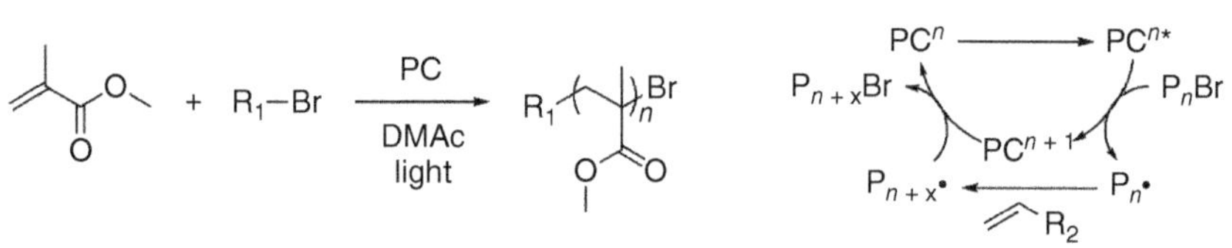

Scheme 3.20 General scheme (left) and mechanism (right) of O-ATRP.

In early reports, both perylene [78] and 10-phenyl phenothiazine (PTH) [68] were demonstrated as PCs for O-ATRP, although each had associated advantages and disadvantages. Using perylene, O-ATRP could be performed under visible-light irradiation with moderate control over polymer structure [78]. By contrast, the use of PTH provided better control over polymer structure while requiring UV light [68], increasing the risk of photoinduced side reactions. Since the reduction potential [$E°$(C–Br/C–Br$\cdot^-$)] of a typical C—Br bond in ATRP is about −0.8 to −0.6 V vs. SCE [73], the superior ability of PTH to mediate a controlled polymerization was attributed to its stronger excited-state reduction potential (Figure 3.13), motivating

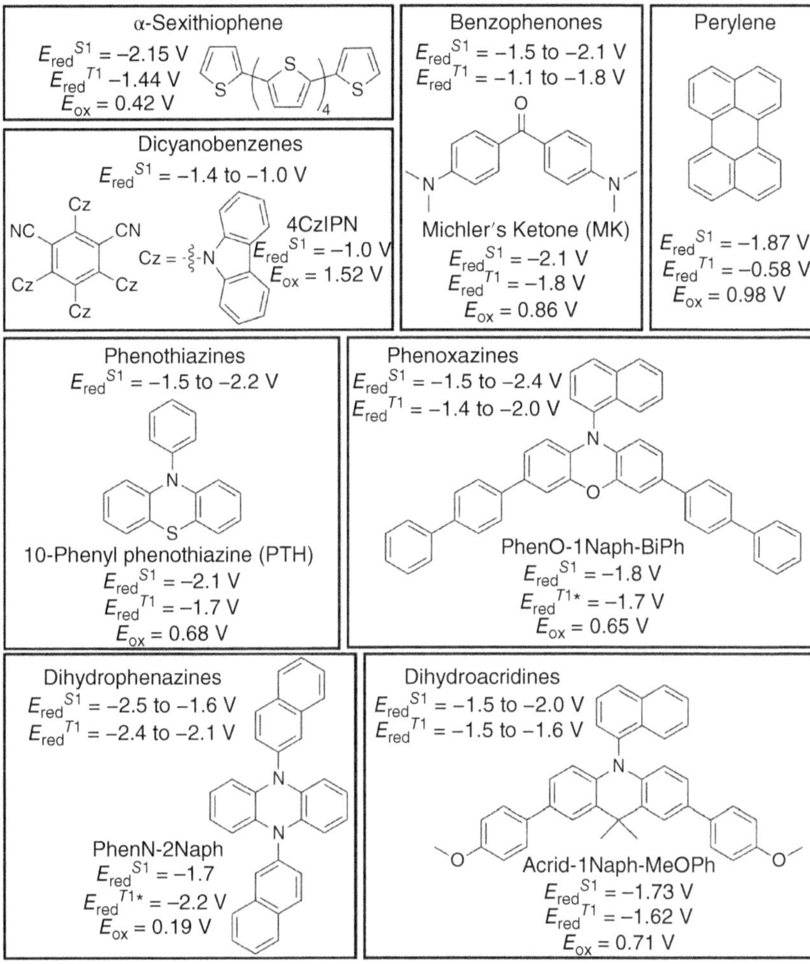

Figure 3.13 Redox properties of organic excited-state reductants. Since many of these families are tunable through structural modification, key examples of commonly used PCs are shown, as well as the redox properties specific to those molecules. E_{red}^{S1} = $E^{\circ}(PC^{\cdot+}/{}^{1}PC^{*})$; E_{red}^{T1} = $E^{\circ}(PC^{\cdot+}/{}^{3}PC^{*})$; E_{ox} = $E^{\circ}(PC^{\cdot+}/PC)$. All potentials are shown in V vs. saturated calomel electrode (SCE). For conditions under which redox properties were measured, see references [23, 49, 59–75]. Computed redox potentials indicated by * (e.g. E_{red}^{T1*}), see references for details [71, 72]. Source: Luo and Zhang [67], Singh et al. [76], Korobov et al. [69], Ishimatsu et al. [77].

the development of more powerful excited-state reductants that could operate under visible light.

Early advances in catalyst design came through the development of dihydrophenazine and phenoxazine PCs. Dihydrophenazines (Figure 3.13) were the first organic catalyst family capable of mediating the synthesis of well-defined polymers by O-ATRP under visible light. Through deeper investigation of catalysts in this family, researchers discovered important design principles for effective catalysis. For example, using density functional theory (DFT), it was found that the best catalysts for O-ATRP exhibited intramolecular CT in the triplet excited state from the phenazine core to the *N*-aryl substituent (Figure 3.13) [71]. In later studies, it was shown that this improvement in catalysis is likely due to increased ISC enabled by the charge-transfer state [79], which produces a long-lived triplet excited state.

Formation of the CT state could be encouraged through installation of electron-withdrawing groups (EWGs) or substituents bearing extended conjugation at the *N*-aryl position. In addition, through both computational and experimental studies, it was found that EWGs at the *N*-aryl position decreased the excited-state reduction potential (i.e. made it less reducing) and increased the oxidation potential of the PC radical cation, whereas electron-donating groups (EDGs) had the opposite effect [71]. In turn, these discoveries formed the basis for tuning the photophysical and redox properties of these and similar PCs in future investigations (*vide infra*).

Shortly thereafter, phenoxazines (Figure 3.13) were introduced as organic PCs that could also mediate O-ATRP under visible light. Computational and crystallographic investigations compared the phenoxazine and phenothiazine cores to understand the impact of the PC core on catalysis. Ultimately, this work showed that although the redox properties of analogous phenoxazines and phenothiazines are similar, the core geometries of key catalytic intermediates varied significantly. In the case of phenoxazines, the core is predicted to remain relatively planar transitioning from the triplet excited state ($^3PC^*$) to the radical cation ($PC^{\cdot+}$) to the PC ground state. By contrast, the phenothiazine core transitions from a twisted geometry in $^3PC^*$ to planar in $PC^{\bullet+}$ and finally to a bent geometry in the ground state. As a result, phenothiazines are predicted to have higher reorganization energies during ET, resulting in a lower rate of ET (see Section 2.4.3) [72].

In the studies that followed, the scope and tunability of these and similar PCs was greatly widened through several strategies. For example, a number of researchers have reported on methods to tune phenoxazines [73, 80] and phenothiazines [81–85] by installation of different substituents at the *N*-aryl position and on the catalyst core – usually at the three and seven positions for these families. Similarly, a number of new dihydrophenazine PCs have been reported through variation of the *N*-aryl substituent [76, 86] as well as substitution of the phenazine core, the latter of which produced PCs that can operate at extremely low catalyst loadings in O-ATRP [74]. More recently, dihydroacridines were introduced as organic PCs for O-ATRP, featuring strong excited-state reduction potentials [$E°(PC^{\cdot+}/^3PC^*) < -1.5$ V vs. SCE] and even more oxidizing radical cations than phenoxazines or phenothiazines (Figure 3.13) [75].

Given the impressive photophysical and redox properties of these molecules, it is not surprising that they have found numerous applications in small-molecule and polymer synthesis [87]. For example, the Hawker group demonstrated the utility of phenothiazines in aryl–halide dehalogenation reactions, first in hydrogenation reactions [88] and later in C–C coupling reactions [81]. Notably, the later report demonstrated that selectivity for different carbon halide bonds could be achieved through modulation of the PC's excited-state reduction potential (Scheme 3.21)

[81], emphasizing motivation for the development of highly tunable catalyst systems.

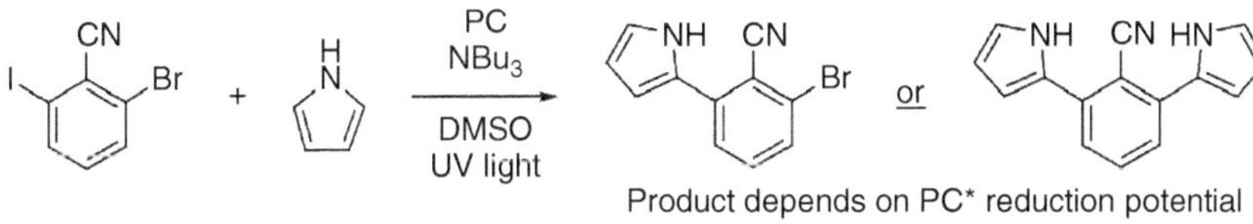

Scheme 3.21 Selective functionalization of aryl-halides using phenothiazine PCs. Source: Modified from Poelma et al. [81].

In another example highlighting the capabilities of these PCs, the Jui group reported on the ability of PTH to activate aryl trifluoromethyl C—F bonds for catalytic defluoroalkylations in the presence of cyclohexanethiol as a cocatalyst (Scheme 3.22). While direct reduction of the C—F bond was not proposed, the strong excited-state reduction potentials of these PCs were nonetheless critical to the success of this transformation, which began through reduction of the aromatic substrate [E°(sub/sub·$^-$) ~ −2.0 V vs. SCE] followed by mesolytic cleavage of a C—F bond to generate the desired radical intermediate [89].

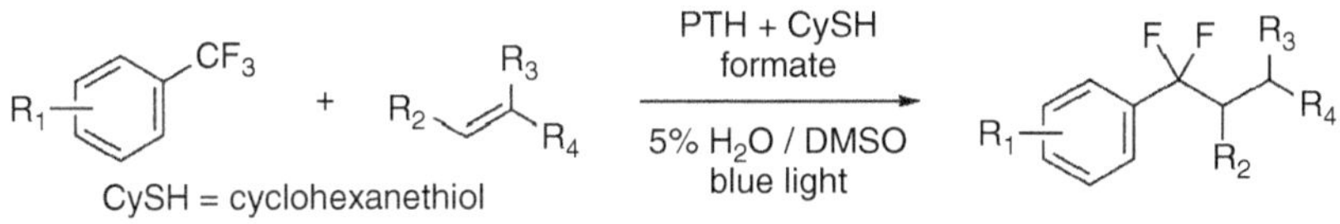

Scheme 3.22 Alkylation of trifluoromethyl arenes through defluorination using an organic excited-state reductant.

Finally, the ability of phenoxazines and dihydrophenazines to act as precious-metal-free alternatives to Ru and Ir based PCs was demonstrated through several transformations. Using a 2-naphthyl-substituted dihydrophenazine (Phen-*N*-2-Naphth, Figure 3.13), trifluoromethylations were carried out under visible light to afford a variety of substituted alkenes (Scheme 3.23). In addition, through the use of phenoxazine and dihydrophenazine PCs with a Ni cocatalyst, both aryl C–N (Scheme 3.24) and aryl C–S (Scheme 3.25) couplings were successful, providing a more sustainable approach to these transformations that reduces dependence on precious metals [11].

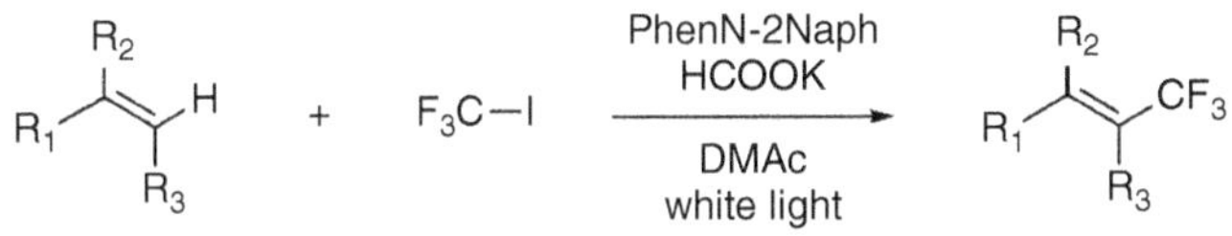

Scheme 3.23 Trifluoromethylation of alkenes using a dihydrophenazine PC.

While these PCs have been used in a number of other transformations and continue to find new applications, they are too numerous to discuss in great

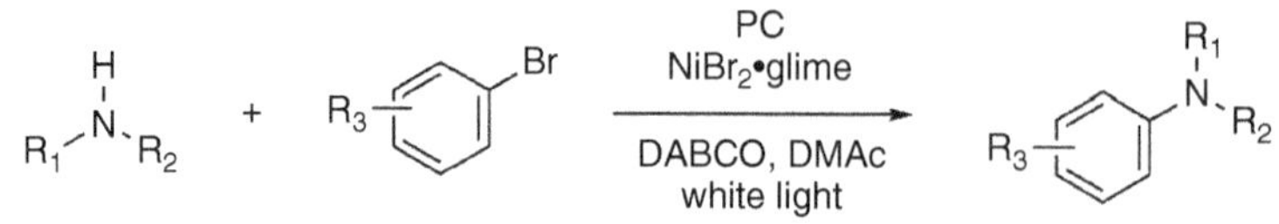

Scheme 3.24 Aryl C–N coupling reactions employing dihydrophenazine and phenoxazine PCs.

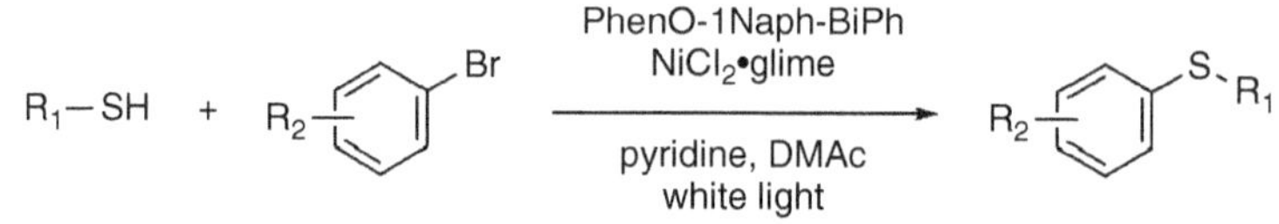

Scheme 3.25 Aryl C–S couplings using a phenoxazine PC.

detail here and have been discussed elsewhere [87]. In addition, a number of other strong excited-state reductants have been explored, such as dicyanobenzenes (Figure 3.13) [67, 76, 77, 90], naphthochromenones [91], coumarin dyes [92], diketopyrrolopyrroles [93, 94], dihydropyridines [95], anthracenes [96], naphthalenes [97], carbazoles [98, 99], and more.

3.3.6 Open-Shell Photoredox Catalysts

To expand the energies accessible to excited PCs beyond those of visible-light photons, one strategy described in the literature involves **consecutive photoinduced electron transfer (ConPET)**. A single photon of blue light (440 nm) possesses 2.8 eV of energy, which could theoretically be employed by an excited PC, but ISC and nonradiative decay pathways diminish the energy available for bimolecular PET with substrate compounds. Wasielewski studied the photophysical properties of electrochemically generated radical anions of a series of aromatic diimides and found that they were highly reducing upon photoexcitation [100]. In 2014, König and coworkers reported multiphoton visible-light excitation of one such perylene diimide (PDI) photocatalyst capable of reducing aryl chlorides (Scheme 3.26), a substrate class for which typical reduction potentials [$E°(ArCl^{•-}/ArCl) < -2.0$ V vs. SCE] exceed the reducing power of conventional PCs excited by a single photon of visible light [101].

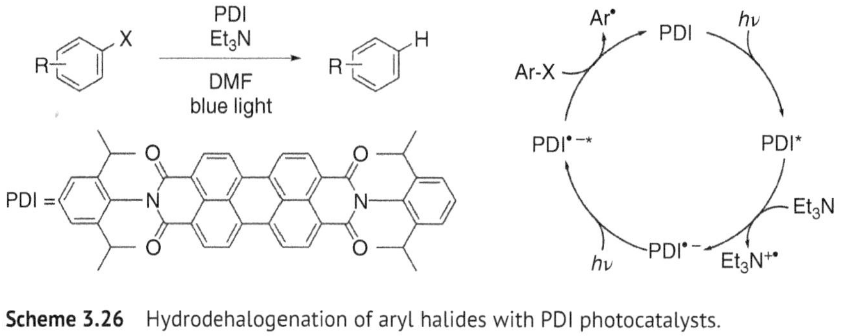

Scheme 3.26 Hydrodehalogenation of aryl halides with PDI photocatalysts.

The reaction likely proceeds first through initial visible-light excitation of the PDI to generate the excited-state PDI*. Reductive quenching with Et_3N generates the radical anion $PDI^{\cdot-}$, which after excitation by a second photon yields an excited radical anion $PDI^{\cdot-*}$. This species is a potent reductant capable of ET to an aryl halide acceptor and subsequent regeneration of the ground-state catalyst [102]. In addition to organic radical anion photoexcitation, Nicewicz has shown that the neutral acridine radicals produced by reductive quenching of excited acridinium PCs with simple amines can also undergo excitation to generate a powerful reductant [$E°(PC^+/PC^{\cdot *}) \sim -3.36$ V vs. SCE] [30]. Although these examples demonstrate the power of photoredox catalysis via ConPET, drawbacks include possible side reactions from radical cation byproducts of the sacrificial reductant.

Electrochemistry provides an alternative approach to generating open-shell intermediates for subsequent photoexcitation. The groups of Wickens and coworkers and Lin and coworkers have disclosed photoexcitation of cathodically generated organic radical anions followed by selective catalytic reductive couplings of aryl halides [103, 104]. In the latter case, electrochemical reduction of 1,9-dicyanoanthracene (DCA) to form a radical anion $DCA^{\cdot-}$ and subsequent irradiation with visible light was proposed to form the excited species $DCA^{\cdot-*}$ with an extreme estimated reduction potential [$E°(PC/PC^{\cdot-*}) \sim -3.20$ V vs. SCE]. Importantly, under these conditions, aryl halides could be selectively reduced and borylated even in the presence of potentially sensitive substrate functional groups (Scheme 3.27). The authors propose that the observed chemoselectivity is a result of the controlled generation of low concentrations of the highly reactive excited-state species $DCA^{\cdot-*}$.

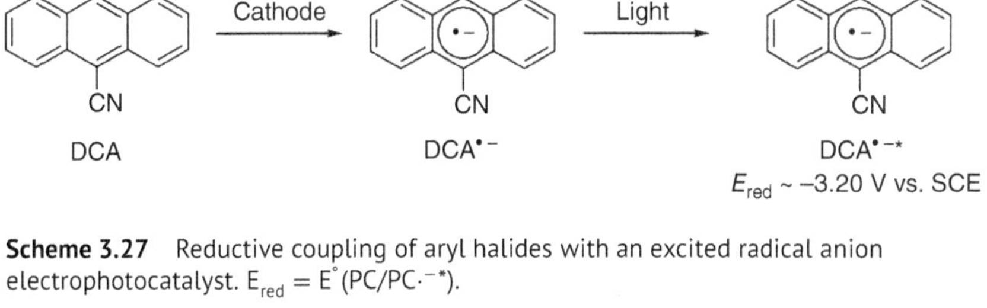

Scheme 3.27 Reductive coupling of aryl halides with an excited radical anion electrophotocatalyst. $E_{red} = E°(PC/PC^{\cdot-*})$.

As an example of an aromatic reduction not involving aryl halides, Miyake and coworkers have developed a visible light–mediated photoredox Birch reduction using benzo[ghi]perylene imide PCs (Scheme 3.28). In this system, basic reductants like OH^- are proposed to form an anionic adduct with the PC, which after photolytic fragmentation yields a radical anion, which can absorb a second photon of visible light. It is proposed that the subsequent excited radical anion species can ionize to form solvated electrons, which might be responsible for the observed conversion of arene substrates to unconjugated cyclohexadienes [105].

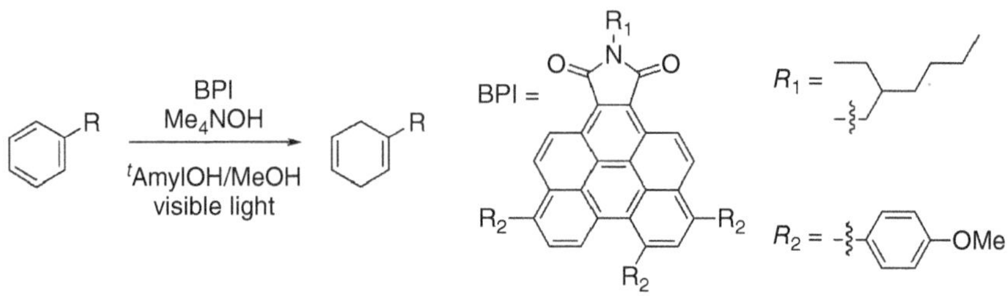

Scheme 3.28 Organocatalyzed visible light-mediated photoredox Birch reduction.

In contrast to reports of photoredox catalysis via reduced open-shell species, photoexcitation of oxidized open-shell species is less common. Wasilewski and coworkers discovered that phenothiazine radical cations can be excited by visible light to access doublet excited states with oxidation potentials upwards of E°(PC/PC^{+*}) ~ 2.1 V vs. SCE (Figure 3.14) [106]. Lambert and coworkers found that visible-light irradiation of an anodically generated trisaminocyclopropenium radical dication (Figure 3.14) yielded a highly oxidizing species [E°(PC$^+$/PC^{2+*}) ~ 3.33 V vs. SCE], which could be applied to the oxidation of benzene and subsequent C–N coupling [107].

Finally, Kerzig et al. demonstrated how multiple excitation events can be combined to produce higher-energy reactants, which in their case led to the formation of solvated electrons that could perform otherwise challenging reductions. In this example, a collimating lens was used to focus light from a 1-W laser into a small reaction volume, creating a region of very high light intensity. As a result, the PC employed in these reactions was able to undergo two consecutive photoexcitation events, ultimately leading to ionization of the PC to form a solvated electron [108]. In later work, the synthetic utility of this approach was demonstrated in comparison to traditional photoexcitation of the same PC. While irradiation with lower intensity light only allowed for reduction of aryl C—Br bonds (due to the lower excited-state reduction potential of PC*), the use of high-intensity light to produce solvated electrons allowed for the reduction of aryl C—Cl bonds that would otherwise be inaccessible to the PC [109].

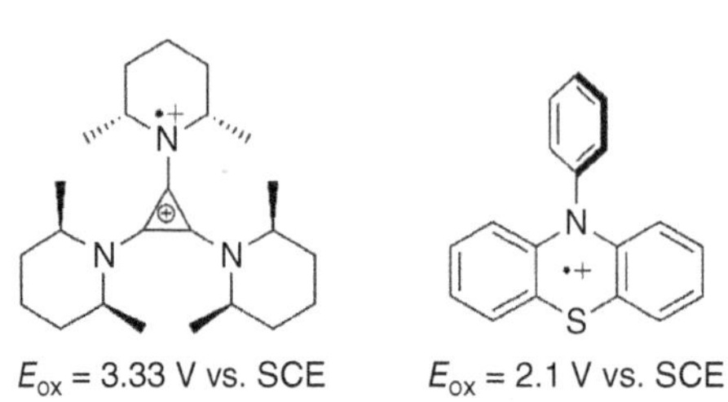

Figure 3.14 Highly oxidizing excited organic radical cations. E_{ox} = E°(PC$^+$/PC^{2+*}) or E°(PC/PC$^{\cdot+*}$). Source: Modified from Christensen et al. [106].

3.4 Photochemistry of Electron Donor–Acceptor Complexes

3.4.1 Background and Theory

3.4.1.1 What Is an EDA Complex?

An **EDA complex**, also referred to as a **charge-transfer complex**, is composed of an electron-rich molecule (donor) and an electron-poor molecule (acceptor) that reversibly associate in the ground state through intermolecular forces. Due to the generally weak nature of these interactions (e.g. Van der Waals forces), EDA complexes are quite sensitive to environmental factors, such as temperature, solvent, and D/A concentrations. These complexes have been studied extensively since the 1950s and were observed even earlier, such as in the interaction of iodine with various solvents to give rise to different colored solutions [110].

Early work in this area was carried out by Mulliken, who first proposed the existence of EDA complexes in his charge-transfer theory in 1952 [111–113]. It was here that Mulliken first defined an EDA complex and proposed that the donor should be electron rich and have a low ionization potential, whereas the electron acceptor should be electron poor and have a high electron affinity. Based on these definitions, one can estimate the electron-donating or -accepting ability of a molecule according to its redox properties, as the ionization potential of a donor can be approximated by its oxidation potential in solution, and the electron affinity of an acceptor can be approximated by its reduction potential. Importantly, using these data in combination with Eq. 3.7 (see Section 2.4.1), the likelihood of reactivity between an electron donor and acceptor pair can be predicted. Even when the excited-state energy (ΔE_{0-0}) of the EDA complex is unavailable, comparison of the D and A redox properties can serve as a guide for reaction design. For examples of common donors and acceptors that form EDA complexes, the reader can refer to reviews by Rosokha and Kochi [114] and Paixaõ snd coworkers [115]. In addition, work by Nicewicz and coworkers has tabulated the reduction and oxidation potentials of numerous common organic molecules that could be relevant to this chemistry [116].

3.4.1.2 How do EDA Complexes Interact with Light?

Upon formation of an EDA complex, new molecular orbitals (MOs) form through mixing of the D and A highest-occupied and lowest-unoccupied molecular orbitals (highest-occupied molecular orbital [HOMO] and lowest-unoccupied molecular orbital [LUMO], respectively). The formation of this new MO gives rise to a charge-transfer band in the absorption spectrum of the EDA complex, which is typically red-shifted relative to the absorption bands of the D and A alone. Often, this CT band can appear in the visible spectrum, allowing the EDA complex to absorb visible light even when the D and A cannot. As a result, irradiation of the EDA complex can give rise to PET (Figure 3.15), creating the opportunity for reactivity in these complexes.

Broadly, EDA complexes can also be considered an **absorption complex**, which refers to two molecules that cooperatively absorb a single photon of light. Closely

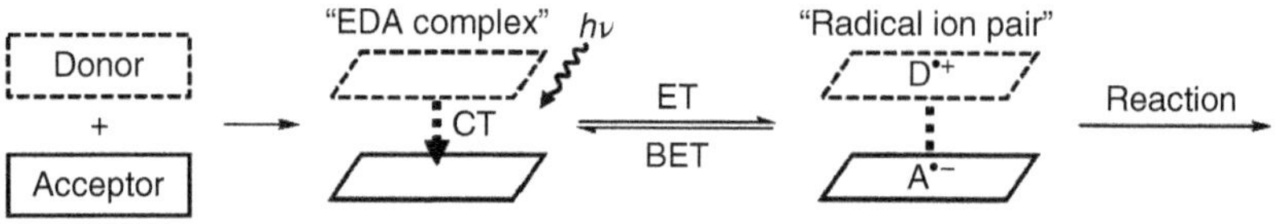

Figure 3.15 General diagram showing the association of a donor and acceptor to form an EDA complex, which through irradiation undergoes electron transfer and subsequent reactions.

Table 3.3 Typical properties of EDA complexes that undergo ISET and OSET.

	ISET	OSET
D–A distance (Å)	3.0–3.3	5–6
V_{el} (cm^{-1})	1000–3000	100–300
K_{EDA} (M^{-1})	0.1–1	n/a

Source: Roth [1], Tobisu et al. [14] and Lima et al. [115].

related to this concept is an **exciplex**, or an excited-state complex, as well as an **excimer**, which is an excited-state dimer, both of which form from excited states associating with other molecules.

3.4.1.3 Electron Transfer in EDA Complexes

ET in an EDA complex can be considered as either ISET (adiabatic) or OSET (nonadiabatic, see Section 2.4). While ISET is most often observed [115], the specific mechanism of ET depends on both the structural and electronic characteristics of the EDA complex. For example, EDA complexes that operate through ISET typically have a D–A distance of about 3 Å, strong electronic coupling (V_{el} ~ 1000–3000 cm^{-1}), and moderate equilibrium association constants (K_{EDA} ~ 0.1–1 M^{-1}). By contrast, EDA complexes that undergo OSET have larger D–A distances (5–6 Å), weaker electronic coupling (V_{el} ~ 100–300 cm^{-1}), and small association constants that are often too small to measure (Table 3.3) [114, 115].

An important consideration in the chemistry of EDA complexes is **BET**, where the radical anion of the acceptor formed after ET can donate an electron back to the radical cation of the donor and regenerate the ground-state EDA complex (Figure 3.15). Practically, this process can lead to limited product formation if BET is not minimized. A strategy to address this issue includes designing the acceptor molecule to contain a leaving group, which rapidly and irreversibly cleaves after ET to prevent BET.

3.4.1.4 Environmental Factors Affecting EDA Complexes

As mentioned previously, the weak intermolecular interactions that lead to D–A association also make an EDA complex sensitive to several environmental factors. For example, since the D, A, and EDA complex are in equilibrium, the EDA complex will be sensitive to typical equilibrium perturbations such as temperature and the concentration of the reactants (D and A). Further, given the charged nature of the products (a radical anion and cation), solvents can have a significant impact on EDA complex reactivity. Generally, polar solvents stabilize the radical anion and cation formed after ET, favoring dissociation of the ion pair over BET [114].

3.4.2 Early Examples of EDA Photochemistry

Photoredox chemistry in EDA complexes was observed as early as the 1970s and 1980s by several researchers, although examples of such reactivity were limited for a number of reasons. Melchiorre and coworkers have proposed that early examples of EDA complex photochemistry might have been limited by challenges in overcoming BET [117]. In addition, until about 2008, photochemistry in organic synthesis remained relatively underexplored. Therefore, the number of researchers investigating EDA complex photochemistry prior to this time was likely limited, further slowing the development of this chemistry.

Cantacuzene reported an early example of this chemistry in 1977, which involved the condensation of enamines with perfluoroalkyl iodides to yield α-substituted ketones [118]. Similar products were obtained by Bunnett and coworker in the same year through the reaction of ketones with potassium alkoxides, which upon irradiation gave α-substituted ketones [119]. In 1983, Fox et al. showed that this reaction likely proceeds through formation and photoexcitation of an EDA complex, ultimately leading to ET [120].

Another researcher who contributed several early examples of EDA complex photochemistry is Kochi. For example, in 1979, Kochi's group studied the addition of a tetraalkyltin compounds to tetracyanoethylene, which was found to proceed through photoexcitation of an EDA complex [121]. In addition, his group also reported aromatic nitration reactions proceeding through irradiation of an EDA complex in 1987 [122].

In the same year, Kornblum and coworkers disclosed a reaction between p-nitrobenzylchloride and sodium azide, which he proposed proceeded through EDA complex photochemistry [123]. Similar reactivity was observed in 1991 by Russell and Wang with other nitrogen-containing donors [124]. Finally, in 1991, Hall's group reported on the cycloaddition of vinyl carbazole to a substituted dicyanoethylene, which were proposed to form an EDA complex prior to the cycloaddition [125].

3.4.3 Recent Examples of EDA Photochemistry

3.4.3.1 Rediscovering EDA Complexes through Photoredox Catalysis

The renaissance of EDA complex photochemistry arguably began in 2013 as a result of two concurrent reports from the Chatani [14] and Melchiorre [13] groups. An even earlier example of this chemistry might have been observed in 2011 by the MacMillan group during work on photoredox-catalyzed trifluoromethylations, which in some cases was observed to proceed to high yields in the absence of a PC.

While photoexcitation of an EDA complex was proposed, this mechanism was not confirmed at the time [12].

The example by Chatani and coworkers in 2013 was also discovered serendipitously through investigations focused on photoredox catalysis, which revealed that arylations of pyrrole using diaryliodonium salts could proceed efficiently in the absence of a photocatalyst (Scheme 3.29) [14]. In the same year, Melchiorre reported a stereoselective approach to synthesizing α-alkylated aldehydes, which was found to proceed through photoexcitation of an EDA complex. This interesting example, which will be referred to several times throughout this section, combines a number of strategies to catalytically generate an EDA complex and then stereoselectively substitute aldehydes to produce the desired products (Scheme 3.30) [13].

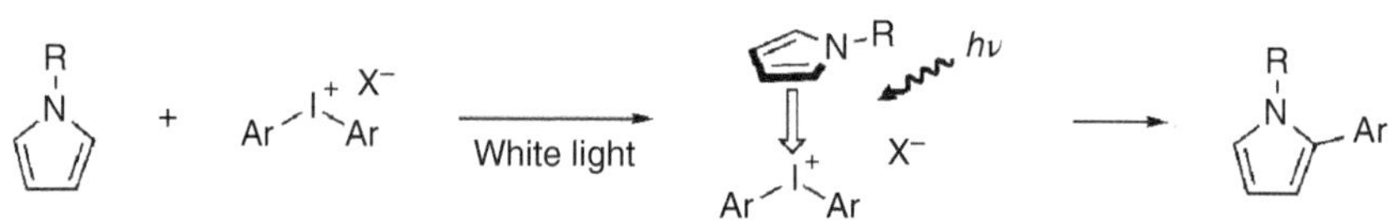

Scheme 3.29 An example of EDA complex reactivity for coupling pyrroles and aryl-iodonium salts reported by Chatani and coworkers in 2013. Source: Modified from Tobisu et al. [14].

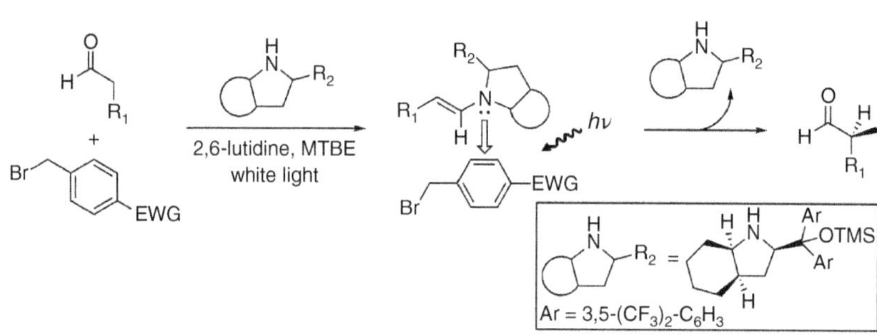

Scheme 3.30 An example of EDA complex reactivity for enantioselective-coupling reactions reported by Melchiorre and coworkers in 2013. Source: Modified from Arceo et al. [13].

3.4.3.2 Stoichiometric EDA Reactions

The simplest reactions involving EDA complexes are stoichiometric, where the donor and acceptor ultimately couple to each other to generate the product. An excellent example of a stoichiometric EDA reaction was reported by Melchiorre and coworkers in 2015 for coupling electron-deficient benzyl bromides to indoles (Scheme 3.31). In this reaction, indole acts as an electron donor with the electron-deficient arene to generate the EDA complex, which upon absorption of visible light generates a radical ion pair. The C—Br bond of the radical anion then rapidly cleaves to form a radical at the benzylic position, which is trapped to yield various substituted indoles. Excitingly, Melchiorre and coworkers were able to obtain a crystal structure of the EDA complex formed during this reaction, which is typically challenging given the weak association inherent to these complexes [126].

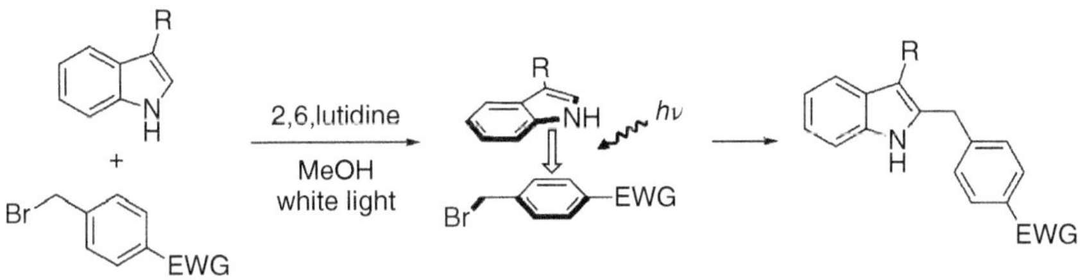

Scheme 3.31 A reaction reported by Melchiorre and coworkers in 2015 that proceeds through photoexcitation of an EDA complex to couple benzyl-bromides to indoles.

A similar strategy is seen in the aryl–thiol couplings reported by Miyake and coworkers in 2017, which generates an EDA complex from aryl halides and aryl thiolates (Scheme 3.32). Although neither compound alone absorbs light in the visible region, the EDA complex is colored, allowing these C–S couplings to be performed selectively under mild conditions [15, 16]. Closely related to this reaction, Wang and coworkers reported a strategy for coupling aryl halides and phenols, which was proposed to undergo a similar mechanism to the aryl halide–thiol coupling [127].

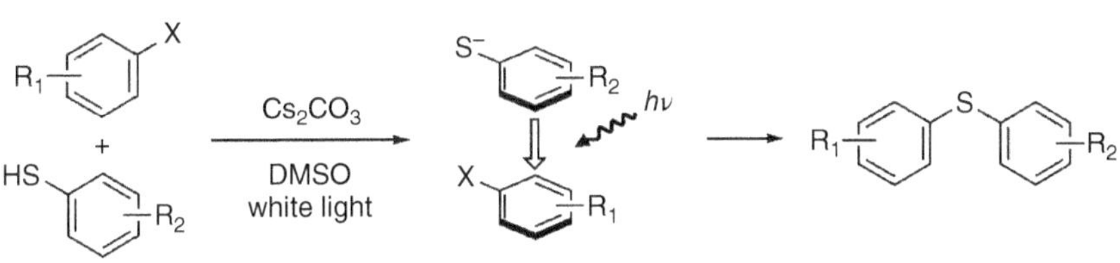

Scheme 3.32 A C–S coupling reaction enabled by EDA complex reactivity reported in 2017 by Miyake and coworkers.

3.4.3.3 Use of Sacrificial Donors and Acceptors

In some cases, couplings between moieties that are poor electron donors or acceptors have been achieved through the use of sacrificial donors or acceptors, which enable formation of the EDA complex but are not incorporated into the final product. An example from Paixão and coworkers in 2015 exhibits this strategy, where intramolecular cyclizations to form indoles were performed utilizing tris(TMS)silane (TMS = trimethylsilyl) as a sacrificial electron donor to generate the EDA complex and perform ET to the substrate (Scheme 3.33). Upon generation of the acceptor radical anion, the aryl C—X bond (X = halide) cleaves to generate an aryl radical, enabling an intramolecular cyclization with the alkyne to yield substituted indoles. The same approach was also demonstrated to synthesize oxindoles when the appropriate amide was employed as the acceptor rather than an amine [128].

Along these lines, in 2017 Chen and coworkers reported the alkylation of alkenes using photoexcited EDA complexes that employed both a sacrificial electron donor

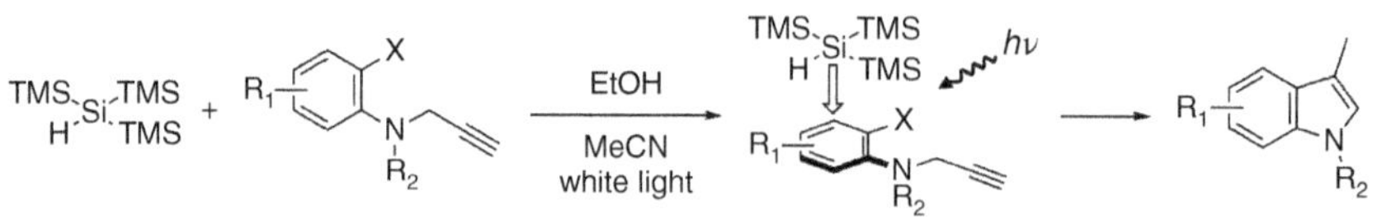

Scheme 3.33 An example from Paixão and coworkers in 2015 employing a sacrificial donor to form an EDA complex, which upon photoexcitation generates substituted indoles.

and a redox auxiliary electron acceptor. The redox auxiliary moiety can be thought of as a sacrificial electron acceptor, which upon ET generates a reactive species for use in the reaction. In Chen's example, an N-substituted phthalimide is used as the redox auxiliary, whereas Hantzsch ester is used as the sacrificial donor. Upon irradiation of the EDA complex with blue light, an alkyl radical is generated and then trapped to yield the product (Scheme 3.34) [129].

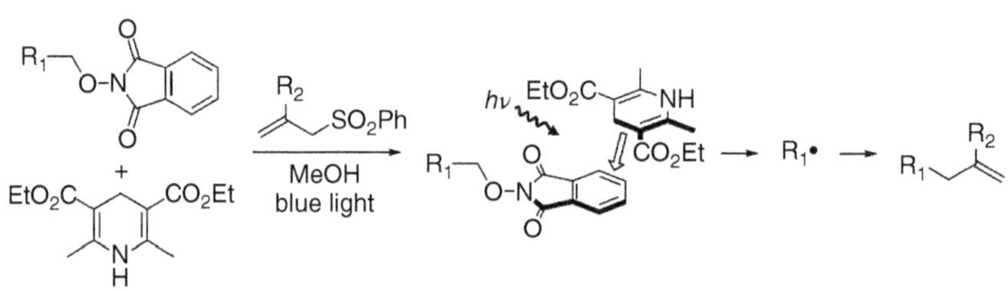

Scheme 3.34 Reaction reported in 2017 by Chen and coworkers that employs both a sacrificial donor and a redox auxiliary to yield substituted alkenes. Source: Modified from Zhang et al. [129].

3.4.3.4 Redox Auxiliaries to Expand Donor and Acceptor Scope

Expanding on the use of redox auxiliaries, Leonori and coworkers disclosed a reaction employing an electron-deficient aryl moiety as an acceptor to yield substituted pyrrolines through intramolecular cyclizations. Interestingly, hydrogenation or hydroxylation could be selectively achieved in the last step of the reaction through addition or exclusion of 1,4-cyclohexadiene, respectively (Scheme 3.35) [130].

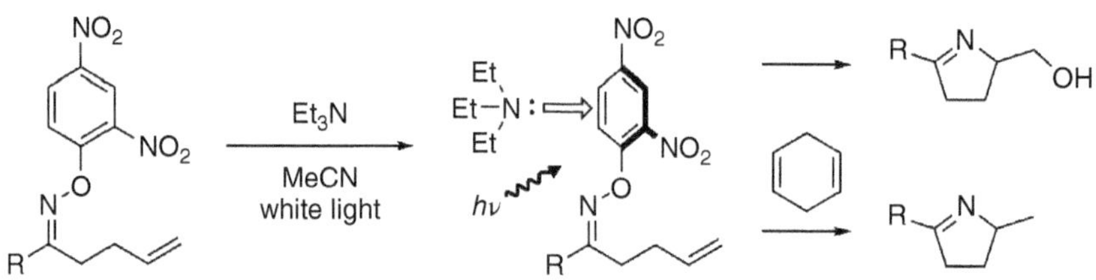

Scheme 3.35 Reported by Leonori and coworkers in 2015, this reaction employs an auxiliary redox moiety to generate substituted pyrrolines. Source: Modified from Davies et al. [130].

Further, Aggarwal and coworkers developed a series of borylation reactions employing redox auxiliaries [131–133]. Notably, his report in 2017 demonstrates how a substituted phthalimide can undergo PET with bis(catecholato)diboron to generate an alkyl radical and a catecholboryl radical. While this reaction was originally proposed to proceed through formation of a ternary complex [131], it has since been suggested this reaction might proceed through the formation of an EDA complex (Scheme 3.36) [117]. Regardless, upon coupling of the radicals produced and addition of pinacol, the desired pinacolborane can be obtained under mild conditions (Scheme 3.36) [131].

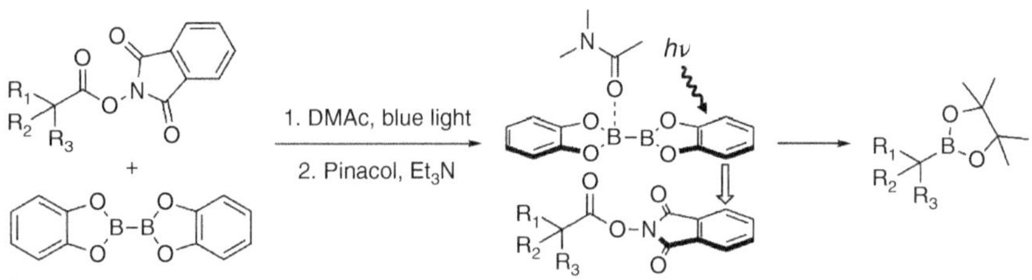

Scheme 3.36 This reaction, reported by Aggarwal and coworkers in 2017, employs an auxiliary redox moiety to enable C–B couplings through photoexcitation of an EDA complex. Source: Roth [1], Cano-Yelo and Deronzier [7] and Fawcett et al. [131].

3.4.3.5 Catalytic EDA Reactions

In contrast to stoichiometric EDA reactions, catalytic reactions employ acceptors or donors that are regenerated during the course of the reaction and then subsequently reused. An early example of such reactivity is evident in Melchiorre's seminal 2013 report, where an amine reacts with an aldehyde to form the donor species and is later regenerated upon product formation (Scheme 3.30) [13]. Similarly, Bosque and Bach reported a reaction employing a catalytic donor in 2019, where a substituted quinuclidine served as a sacrificial donor and was regenerated later in the reaction to perform a series of decarboxylations (Scheme 3.37) [134].

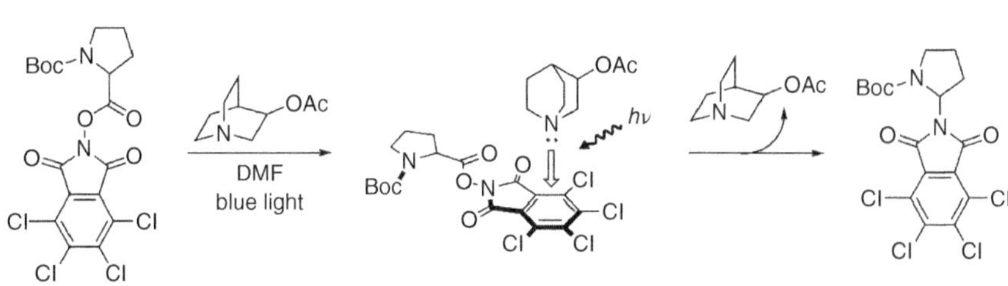

Scheme 3.37 An example from Bosque and Bach in 2019 of an EDA complex reaction in which a catalytic donor is employed. Source: Modified from Bosque and Bach [134].

3.4.3.6 Enantioselective Reactions of EDA Complexes

Several strategies have been reported for performing enantioselective reactions using EDA complexes. Again, the seminal example by Melchiorre and coworkers demonstrates this reactivity, where a chiral amine catalyst was used to direct the formation of the final product (Scheme 3.30) [13].

In 2016, Hyster and coworkers reported the use of ketoreductase enzymes to perform enantioselective hydrogenations of halolactones, where the enantiomer produced could be precisely controlled through selection of the appropriate ketoreductase (Scheme 3.38). Further, through investigation of the reaction mechanism, evidence was found suggesting formation of an EDA complex between the substrate and dihydronicotinamide-adenine dinucleotide phosphate (NADPH), which only occurred in the presence of a ketoreductase enzyme [135].

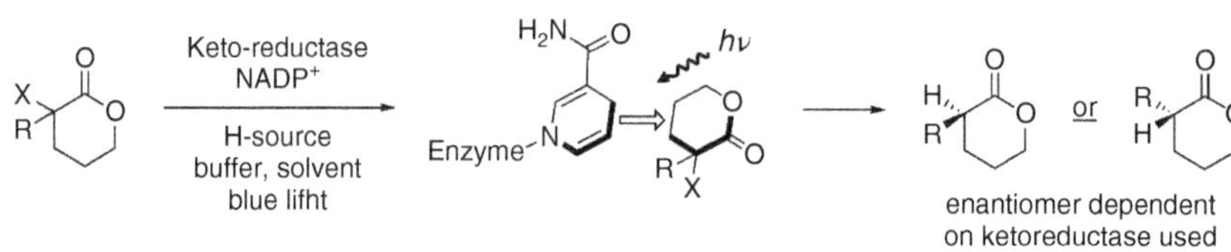

Scheme 3.38 Enantioselective reductive dehalogenation reported by Hyster and coworkers in 2016.

3.5 Concluding Thoughts

In this chapter, the fundamentals of photoredox chemistry were discussed, including photophysical processes and theories of ET relevant to photochemistry. While much of this fundamental knowledge was uncovered over the last century, it has provided the foundation for the recent renaissance of photochemistry for organic synthesis that began around 2008. Enabled by these previous discoveries and advances in lighting technology, recent work in photochemistry has led to the widespread implementation of photoredox catalysis in organic synthesis. As a result, countless reactions can now be performed under mild reaction conditions that were previously inaccessible or required harsh reagents or forcing conditions. In addition, studies focused on photocatalysis have revealed that some reactions can proceed in the absence of a catalyst, reviving interest in EDA complex photochemistry. Combined, these photoredox strategies offer many unique and exciting opportunities to perform challenging reactions in a more sustainable manner that are becoming a pillar of synthetic chemistry.

Suggested Additional Reading

Photochemistry and Photophysical Processes

Anslyn, E.V. and Dougherty, D.A. (2006). Photochemistry. In: *Modern Physical Organic Chemistry*, Chapter 16 (ed. J. Murdzek), 935–1000. University Science Books.

Lakowicz, J.R. (2006). *Principles of Fluorescence Spectroscopy*. Springer.

Turro, N.J. (1991). *Modern Molecular Photochemistry*. University Science Books.

Turro, N.J. (2010). *Modern Molecular Photochemistry of Organic Molecules*. University Science Books.

Electrochemical Methods

Bard, A.J. and Faulkner, L.R. (2001). *Electrochemical Methods: Fundamentals and Applications*. Wiley.

Elgrishi, N., Rountree, K.J., McCarthy, B.D. et al. (2018). A practical beginner's guide to cyclic voltammetry. *J. Chem. Educ.* 95: 197.

Photoredox Catalysis

Arias-Rotondo, D.M. and McCusker, J.K. (2016). The photophysics of photoredox catalysis: a roadmap for catalyst design. *Chem. Soc. Rev.* 45: 5803.

Corrigan, N., Shanmugam, S., Xu, J., and Boyer, C. (2016). Photocatalysis in organic and polymer synthesis. *Chem. Soc. Rev.* 45: 6165.

Gesmundo, N.J., Shaw, M.H., Twilton, J. et al. (2019). *Photoredox Catalysis Desk Reference and User's Guide*. Sigma-Aldrich.

Prier, C.K., Rankic, D.A., and MacMillan, D.W.C. (2013). Visible light photoredox catalysis with transition metal complexes: applications in organic synthesis. *Chem. Rev.* 113: 5322.

Romero, N.A. and Nicewicz, D.A. (2016). Organic photoredox catalysis. *Chem. Rev.* 116: 10075.

Earth Abundant Metal Photoredox Catalysis

Hockin, B.M., Li, C., Robertson, N., and Zysman-Colman, E. (2019). Photoredox catalysts based on earth abundant metal complexes. *Catal. Sci. Technol.* 9: 889.

McCusker, J.K. (2019). Electronic structure in the transition metal block and its implications for light harvesting. *Science* 363: 484.

Wenger, O.S. (2018). Photoactive complexes with earth-abundant metals. *J. Am. Chem. Soc.* 140: 13522.

EDA Complexes

Crisenza, G.E.M., Mazzarella, D., and Melchiorre, P. (2020). Synthetic methods driven by photoactivity of electron donor-acceptor complexes. *J. Am. Chem. Soc.* 142: 5461.

Lima, C.G.S., Lima, T.d.M., Duarte, M. et al. (2016). Organic synthesis enabled by light-irradiation of EDA complexes: theoretical background and synthetic applications. *ACS Catal.* 6: 1389.

Rosokha, S.V. and Kochi, J.K. (2008). Fresh look at electron-transfer mechanisms via the donor/acceptor bindings in the critical encounter complex. *Acc. Chem. Res.* 41: 641.

References

1 Roth, H.D. (1989). The beginnings of organic photochemistry. *Angew. Chem. Int. Ed. Engl.* 28 (9): 1193–1207.

2 Heindel, N.D. and Pfau, M.A. (1965). A profitable partnership: Giacomo Ciamician and Paul Silber. *J. Chem. Educ.* 42 (7): 383.

3 Hedstrand, D.M., Kruizinga, W.H., and Kellogg, R.M. (1978). Light induced and dye accelerated reductions of phenacyl onium salts by 1,4-dihydropyridines. *Tetrahedron Lett.* 19 (14): 1255–1258.

4 Pac, C., Ihama, M., Yasuda, M. et al. (1981). $Ru(bpy)_3^{2+}$-mediated photoreduction of olefins with 1-benzyl-1,4-dihydronicotinamide: a mechanistic probe for electron-transfer reactions of NAD(P)H-model compounds. *J. Am. Chem. Soc.* 103 (21): 6495–6497.

5 Fukuzumi, S., Mochizuki, S., and Tanaka, T. (1990). Photocatalytic reduction of phenacyl halides by 9,10-dihydro-10-methylacridine: control between the reductive and oxidative quenching pathways of tris(bipyridine)ruthenium complex utilizing an acid catalysis. *J. Phys. Chem.* 94 (2): 722–726.

6 Cano-Yelo, H. and Deronzier, A. (1984). Photo-oxidation of some carbinols by the Ru(II) polypyridyl complex-aryl diazonium salt system. *Tetrahedron Lett.* 25 (48): 5517–5520.

7 Cano-Yelo, H. and Deronzier, A. (1987). Photocatalysis of the pschorr reaction by $Ru(bpy)_3^{2+}$. *J. Photochem.* 37 (2): 315–321.

8 Ischay, M.A., Anzovino, M.E., Du, J., and Yoon, T.P. (2008). Efficient visible light photocatalysis of [2+2] enone cycloadditions. *J. Am. Chem. Soc.* 130 (39): 12886–12887.

9 Nicewicz, D.A. and MacMillan, D.W.C. (2008). Merging photoredox catalysis with organocatalysis: the direct asymmetric alkylation of aldehydes. *Science* 322 (5898): 77–80.

10 Narayanam, J.M.R., Tucker, J.W., and Stephenson, C.R.J. (2009). Electron-transfer photoredox catalysis: development of a tin-free reductive dehalogenation reaction. *J. Am. Chem. Soc.* 131 (25): 8756–8757.

11 Du, Y., Pearson, R.M., Lim, C.-H. et al. (2017). Strongly reducing, visible-light organic photoredox catalysts as sustainable alternatives to precious metals. *Chem. Eur. J.* 23 (46): 10962–10968.

12 Pham, P.V., Nagib, D.A., and MacMillan, D.W.C. (2011). Photoredox catalysis: a mild, operationally simple approach to the synthesis of α-trifluoromethyl carbonyl compounds. *Angew. Chem. Int. Ed.* 50 (27): 6119–6122.

13 Arceo, E., Jurberg, I.D., Álvarez-Fernández, A., and Melchiorre, P. (2013). Photochemical activity of a key donor–acceptor complex can drive stereoselective catalytic α-alkylation of aldehydes. *Nat. Chem.* 5 (9): 750–756.

14 Tobisu, M., Furukawa, T., and Chatani, N. (2013). Visible light-mediated direct arylation of arenes and heteroarenes using diaryliodonium salts in the presence and absence of a photocatalyst. *Chem. Lett.* 42 (10): 1203–1205.

15 Liu, B., Lim, C.-H., and Miyake, G.M. (2017). Visible-light-promoted C–S cross-coupling via intermolecular charge transfer. *J. Am. Chem. Soc.* 139 (39): 13616–13619.

16 Liu, B., Lim, C.-H., and Miyake, G. (2018). Transition-metal-free, visible-light-promoted C–S cross-coupling through intermolecular charge transfer. *Synlett* 29 (19): 2449–2455.

17 Jablonski, A. (1933). Efficiency of anti-stokes fluorescence in dyes. *Nature* 131: 839.

18 Kasha, M. (1950). Characterization of electronic transitions in complex molecules. *Discuss. Faraday Soc.* 9: 14–19.

19 Demchenko, A.P., Heldt, J., Waluk, J. et al. (2014). Michael Kasha: from photochemistry and flowers to spectroscopy and music. *Angew. Chem. Int. Ed.* 53 (52): 14316–14324.

20 Demchenko, A.P., Tomin, V.I., and Chou, P.-T. (2017). Breaking the Kasha rule for more efficient photochemistry. *Chem. Rev.* 117 (21): 13353–13381.

21 Okamura, T., Sancar, A., Heelis, P.F. et al. (1989). Doublet-quartet intersystem crossing of flavin radical in DNA photolyase. *J. Am. Chem. Soc.* 111 (15): 5967–5969.

22 International Union of Pure and Applied Chemistry (2020). The Gold book: gibbs energy of photoinduced electron transfer. https://goldbook.iupac.org/terms/view/GT07388 (accessed 13 July 2020).

23 Romero, N.A. and Nicewicz, D.A. (2016). Organic photoredox catalysis. *Chem. Rev.* 116 (17): 10075–10166.

24 Miller, J.R., Calcaterra, L.T., and Closs, G.L. (1984). Intramolecular long-distance electron transfer in radical anions. The effects of free energy and solvent on the reaction rates. *J. Am. Chem. Soc.* 106 (10): 3047–3049.

25 Lakowicz, J.R. (2006). *Principles of Fluorescence Spectroscopy*. New York: Springer.

26 Garlets, Z.J., Nguyen, J.D., and Stephenson, C.R.J. (2014). The development of visible-light photoredox catalysis in flow. *Isr. J. Chem.* 54 (4): 351–360.

27 Buss, B.L. and Miyake, G.M. (2018). Photoinduced controlled radical polymerizations performed in flow: methods, products, and opportunities. *Chem. Mater.* 30 (12): 3931–3942.

28 Arias-Rotondo, D.M. and McCusker, J.K. (2016). The photophysics of photoredox catalysis: a roadmap for catalyst design. *Chem. Soc. Rev.* 45 (21): 5803–5820.

29 Weir, D. and Scaiano, J.C. (1986). Substituent effects on the lifetime and fluorescence of excited diphenylmethyl radicals in solution. *Chem. Phys. Lett.* 128 (2): 156–159.

30 MacKenzie, I.A., Wang, L., Onuska, N.P.R. et al. (2020). Discovery and characterization of an acridine radical photoreductant. *Nature* 580 (7801): 76–80.

31 Kalyanasundaram, K. (1982). Photophysics, photochemistry and solar energy conversion with tris(bipyridyl)ruthenium(II) and its analogues. *Coord. Chem. Rev.* 46: 159–244.

32 Thompson, D.W., Ito, A., and Meyer, T.J. (2013). $[Ru(bpy)_3]^{2+*}$ and other remarkable metal-to-ligand charge transfer (MLCT) excited states. *Pure Appl. Chem.* 85 (7): 1257–1305.

33 Tucker, J.W. and Stephenson, C.R.J. (2012). Shining light on photoredox catalysis: theory and synthetic applications. *J. Org. Chem.* 77 (4): 1617–1622.

34 Nguyen, J.D., D'Amato, E.M., Narayanam, J.M.R., and Stephenson, C.R.J. (2012). Engaging unactivated alkyl, alkenyl and aryl iodides in visible-light-mediated free radical reactions. *Nat. Chem.* 4 (10): 854–859.

35 Condie, A.G., González-Gómez, J.C., and Stephenson, C.R.J. (2010). Visible-light photoredox catalysis: aza-Henry reactions via C–H functionalization. *J. Am. Chem. Soc.* 132 (5): 1464–1465.

36 Yasu, Y., Koike, T., and Akita, M. (2012). Three-component oxytrifluoromethylation of alkenes: highly efficient and regioselective difunctionalization of C—C bonds mediated by photoredox catalysts. *Angew. Chem. Int. Ed.* 51 (38): 9567–9571.

37 Ventre, S., Petronijevic, F.R., and MacMillan, D.W.C. (2015). Decarboxylative fluorination of aliphatic carboxylic acids via photoredox catalysis. *J. Am. Chem. Soc.* 137 (17): 5654–5657.

38 Liu, Y., Persson, P., Sundström, V., and Wärnmark, K. (2016). Fe N-heterocyclic carbene complexes as promising photosensitizers. *Acc. Chem. Res.* 49 (8): 1477–1485.

39 Kjær, K.S., Kaul, N., Prakash, O. et al. (2019). Luminescence and reactivity of a charge-transfer excited iron complex with nanosecond lifetime. *Science* 363 (6424): 249–253.

40 Pirtsch, M., Paria, S., Matsuno, T. et al. (2012). [Cu(dap)$_2$Cl] as an efficient visible-light-driven photoredox catalyst in carbon-carbon bond-forming reactions. *Chem. Eur. J.* 18 (24): 7336–7340.

41 Wang, Y., Haze, O., Dinnocenzo, J.P. et al. (2007). Bonded exciplexes. A new concept in photochemical reactions. *J. Org. Chem.* 72 (18): 6970–6981.

42 Yamago, S., Miyazoe, H., Iida, K., and Yoshida, J. (2000). Highly efficient and chemoselective reductive bis-silylation of quinones by silyltellurides. *Org. Lett.* 2 (23): 3671–3673.

43 Ohkubo, K., Fujimoto, A., and Fukuzumi, S. (2013). Visible-light-induced oxygenation of benzene by the triplet excited state of 2,3-dichloro-5,6-dicyano-p-benzoquinone. *J. Am. Chem. Soc.* 135 (14): 5368–5371.

44 Fukuzumi, S. and Kitano, T. (1991). Mechanisms of reductive methylation of NAD^+ analogues by a *trans*-dimethylcobalt(III) complex. *J. Chem. Soc., Perkin Trans. 2* (1): 41.

45 Searle, R., Williams, J.L.R., DeMeyer, D.E., and Doty, J.C. (1967). The sensitization of stilbene isomerization. *Chem. Commun. Lond.* 22: 1165.

46 Kitaguchi, H., Ohkubo, K., Ogo, S., and Fukuzumi, S. (2006). Electron-transfer oxidation properties of unsaturated fatty acids and mechanistic insight into lipoxygenases. *J. Phys. Chem. A* 110 (5): 1718–1725.

47 Benniston, A.C., Harriman, A., Li, P. et al. (2005). Illumination of the 9-mesityl-10-methylacridinium ion does not give a long-lived photoredox state. *Chem. Commun.* (21): 2701.

48 Benniston, A.C., Harriman, A., Li, P. et al. (2005). Charge shift and triplet state formation in the 9-mesityl-10-methylacridinium cation. *J. Am. Chem. Soc.* 127 (46): 16054–16064.

49 Shen, T., Zhao, Z.-G., Yu, Q., and Xu, H.-J. (1989). Photosensitized reduction of benzil by heteroatom-containing anthracene dyes. *J. Photochem. Photobiol. Chem.* 47 (2): 203–212.

50 Pan, Y., Kee, C.W., Chen, L., and Tan, C.-H. (2011). Dehydrogenative coupling reactions catalysed by Rose Bengal using visible light irradiation. *Green Chem.* 13 (10): 2682.

51 Fukuzumi, S., Kotani, H., Ohkubo, K. et al. (2004). Electron-transfer state of 9-mesityl-10-methylacridinium ion with a much longer lifetime and higher energy than that of the natural photosynthetic reaction center. *J. Am. Chem. Soc.* 126 (6): 1600–1601.

52 Romero, N.A. and Nicewicz, D.A. (2014). Mechanistic insight into the photoredox catalysis of anti-Markovnikov alkene hydrofunctionalization reactions. *J. Am. Chem. Soc.* 136 (49): 17024–17035.

53 Nguyen, T.M., Manohar, N., and Nicewicz, D.A. (2014). Anti-Markovnikov hydroamination of alkenes catalyzed by a two-component organic photoredox system: direct access to phenethylamine derivatives. *Angew. Chem. Int. Ed.* 53 (24): 6198–6201.

54 Romero, N.A., Margrey, K.A., Tay, N.E., and Nicewicz, D.A. (2015). Site-selective arene C–H amination via photoredox catalysis. *Science* 349 (6254): 1326–1330.

55 Ohkubo, K., Kobayashi, T., and Fukuzumi, S. (2011). Direct oxygenation of benzene to phenol using quinolinium ions as homogeneous photocatalysts. *Angew. Chem. Int. Ed.* 50 (37): 8652–8655.

56 Alfonzo, E., Alfonso, F.S., and Beeler, A.B. (2017). Redesign of a pyrylium photoredox catalyst and its application to the generation of carbonyl ylides. *Org. Lett.* 19 (11): 2989–2992.

57 Lin, Y.-C. and Chen, C.-T. (2009). Acridinium salt-based fluoride and acetate chromofluorescent probes: molecular insights into anion selectivity switching. *Org. Lett.* 11 (21): 4858–4861.

58 McTiernan, C.D., Pitre, S.P., and Scaiano, J.C. (2014). Photocatalytic dehalogenation of vicinal dibromo compounds utilizing sexithiophene and visible-light irradiation. *ACS Catal.* 4 (11): 4034–4039.

59 Pitre, S.P., McTiernan, C.D., and Scaiano, J.C. (2016). Understanding the kinetics and spectroscopy of photoredox catalysis and transition-metal-free alternatives. *Acc. Chem. Res.* 49 (6): 1320–1330.

60 Suppan, P. (1975). Photoreactivity of Michler's ketone in solution. *J. Chem. Soc., Faraday Trans. 1 Phys. Chem. Condens. Phases* 71: 539–547.

61 Timpe, H.-J., Kronfeld, K.-P., Lammel, U. et al. (1990). Excited states of ketones as electron donors-ketone-iodonium salt systems as photoinitiators for radical polymerization. *J. Photochem. Photobiol. Chem.* 52 (1): 111–122.

62 Schweitzer, C., Mehrdad, Z., Noll, A. et al. (2001). Oxygen quenching of $n\pi^*$ triplet phenyl ketones: local excitation and local deactivation. *Helv. Chim. Acta* 84: 15.

63 Bachman, J.C., Kavian, R., Graham, D.J. et al. (2015). Electrochemical polymerization of pyrene derivatives on functionalized carbon nanotubes for pseudocapacitive electrodes. *Nat. Commun.* 6 (1): 7040.

64 Singh-Rachford, T.N. and Castellano, F.N. (2010). Triplet sensitized red-to-blue photon upconversion. *J. Phys. Chem. Lett.* 1 (1): 195–200.

65 Parac, M. and Grimme, S. (2003). A TDDFT study of the lowest excitation energies of polycyclic aromatic hydrocarbons. *Chem. Phys.* 292 (1): 11–21.

66 Kikuchi, K., Niwa, T., Takahashi, Y. et al. (1993). Quenching mechanism in a highly exothermic region of the Rehm-Weller relationship for electron-transfer fluorescence quenching. *J. Phys. Chem.* 97 (19): 5070–5073.

67 Luo, J. and Zhang, J. (2016). Donor–acceptor fluorophores for visible-light-promoted organic synthesis: photoredox/Ni dual catalytic $C(sp^3)$–$C(sp^2)$ cross-coupling. *ACS Catal.* 6 (2): 873–877.

68 Treat, N.J., Sprafke, H., Kramer, J.W. et al. (2014). Metal-free atom transfer radical polymerization. *J. Am. Chem. Soc.* 136 (45): 16096–16101.

69 Korobov, V.E., Shubin, V.V., and Chibisov, A.-K. (1977). Triplet state of rhodamine dyes and its role in production of intermediates. *Chem. Phys. Lett.* 45 (3): 4.

70 Yasui, S., Tsujimoto, M., Itoh, K., and Ohno, A. (2000). Quenching of a photosensitized dye through single-electron transfer from trivalent phosphorus compounds. *J. Org. Chem.* 65 (15): 4715–4720.

71 Theriot, J.C., Lim, C.-H., Yang, H. et al. (2016). Organocatalyzed atom transfer radical polymerization driven by visible light. *Science* 352 (6289): 1082–1086.

72 Pearson, R.M., Lim, C.-H., McCarthy, B.G. et al. (2016). Organocatalyzed atom transfer radical polymerization using N-aryl phenoxazines as photoredox catalysts. *J. Am. Chem. Soc.* 138 (35): 11399–11407.

73 McCarthy, B.G., Pearson, R.M., Lim, C.-H. et al. (2018). Structure–property relationships for tailoring phenoxazines as reducing photoredox catalysts. *J. Am. Chem. Soc.* 140 (15): 5088–5101.

74 Cole, J.P., Federico, C.R., Lim, C.-H., and Miyake, G.M. (2019). Photoinduced organocatalyzed atom transfer radical polymerization using low ppm catalyst loading. *Macromolecules* 52 (2): 747–754.

75 Buss, B.L., Lim, C.-H., and Miyake, G.M. (2020). Dimethyl dihydroacridines as photocatalysts in organocatalyzed atom transfer radical polymerization of acrylate monomers. *Angew. Chem. Int. Ed.* 59 (8): 3209–3217.

76 Singh, V.K., Yu, C., Badgujar, S. et al. (2018). Highly efficient organic photocatalysts discovered via a computer-aided-design strategy for visible-light-driven atom transfer radical polymerization. *Nat. Catal.* 1 (10): 794–804.

77 Ishimatsu, R., Matsunami, S., Kasahara, T. et al. (2014). Electrogenerated chemiluminescence of donor-acceptor molecules with thermally activated delayed fluorescence. *Angew. Chem. Int. Ed.* 53 (27): 6993–6996.

78 Miyake, G.M. and Theriot, J.C. (2014). Perylene as an organic photocatalyst for the radical polymerization of functionalized vinyl monomers through oxidative quenching with alkyl bromides and visible light. *Macromolecules* 47 (23): 8255–8261.

79 Sartor, S.M., McCarthy, B.G., Pearson, R.M. et al. (2018). Exploiting charge-transfer states for maximizing intersystem crossing yields in organic photoredox catalysts. *J. Am. Chem. Soc.* 140: 4778–4781.

80 Park, G.S., Back, J., Choi, E.M. et al. (2019). Visible light-mediated metal-free atom transfer radical polymerization with N-trifluoromethylphenyl phenoxazines. *Eur. Polym. J.* 117: 347–352.

81 Poelma, S.O., Burnett, G.L., Discekici, E.H. et al. (2016). Chemoselective radical dehalogenation and C—C bond formation on aryl halide substrates using organic photoredox catalysts. *J. Org. Chem.* 81 (16): 7155–7160.

82 Dadashi-Silab, S., Pan, X., and Matyjaszewski, K. (2017). Phenyl benzo[b]phenothiazine as a visible light photoredox catalyst for metal-free atom transfer radical polymerization. *Chem. Eur. J.* 23 (25): 5972–5977.

83 Gong, H., Zhao, Y., Shen, X. et al. (2018). Organocatalyzed photocontrolled radical polymerization of semifluorinated (meth)acrylates driven by visible light. *Angew. Chem. Int. Ed.* 57 (1): 333–337.

84 Zhao, Y., Gong, H., Jiang, K. et al. (2018). Organocatalyzed photoredox polymerization from aromatic sulfonyl halides: facilitating graft from aromatic C—H bonds. *Macromolecules* 51 (3): 938–946.

85 Sartor, S.M., Chrisman, C.H., Pearson, R.M. et al. (2020). Designing high-triplet-yield phenothiazine donor–acceptor complexes for photoredox catalysis. *J. Phys. Chem. A* 124 (5): 817–823.

86 Ryan, M.D., Theriot, J.C., Lim, C.-H. et al. (2017). Solvent effects on the intramolecular charge transfer character of N,N-diaryl dihydrophenazine catalysts for organocatalyzed atom transfer radical polymerization. *J. Polym. Sci., Part A: Polym. Chem.* 55 (18): 3017–3027.

87 Corbin, D.A., Lim, C.-H., and Miyake, G.M. (2019). Phenothiazines, dihydrophenazines, and phenoxazines: sustainable alternatives to precious-metal-based photoredox catalysts. *Aldrichim. Acta* 52 (1): 15.

88 Discekici, E.H., Treat, N.J., Poelma, S.O. et al. (2015). A highly reducing metal-free photoredox catalyst: design and application in radical dehalogenations. *Chem. Commun.* 51 (58): 11705–11708.

89 Wang, H. and Jui, N.T. (2018). Catalytic defluoroalkylation of trifluoromethylaromatics with unactivated alkenes. *J. Am. Chem. Soc.* 140 (1): 163–166.

90 Uoyama, H., Goushi, K., Shizu, K. et al. (2012). Highly efficient organic light-emitting diodes from delayed fluorescence. *Nature* 492 (7428): 234–238.

91 Mateos, J., Rigodanza, F., Vega-Peñaloza, A. et al. (2020). Naphthochromenones: organic bimodal photocatalysts engaging in both oxidative and reductive quenching processes. *Angew. Chem. Int. Ed.* 59 (3): 1302–1312.

92 Gualandi, A., Rodeghiero, G., Della Rocca, E. et al. (2018). Application of coumarin dyes for organic photoredox catalysis. *Chem. Commun.* 54 (72): 10044–10047.

93 Jia, T., Huang, S., Bohra, H., and Wang, M. (2019). Examining derivatives of quinacridone, diketopyrrolopyrrole and indigo as the visible-light organic photocatalysts for metal-free atom transfer radical polymerization. *Dyes Pigm.* 165: 223–230.

94 Yang, L., Huang, Y., Peng, Y. et al. (2020). Pyridine-diketopyrrolopyrrole-based novel metal-free visible-light organophotoredox catalyst for atom-transfer radical polymerization. *J. Phys. Chem. A* 124 (6): 1068–1075.

95 Buzzetti, L., Prieto, A., Roy, S.R., and Melchiorre, P. (2017). Radical-based C—C bond-forming processes enabled by the photoexcitation of 4-alkyl-1,4-dihydropyridines. *Angew. Chem. Int. Ed.* 56 (47): 15039–15043.

96 Noto, N., Tanaka, Y., Koike, T., and Akita, M. (2018). Strongly reducing (diarylamino)anthracene catalyst for metal-free visible-light photocatalytic fluoroalkylation. *ACS Catal.* 8 (10): 9408–9419.

97 Noto, N., Koike, T., and Akita, M. (2019). Visible-light-triggered monofluoromethylation of alkenes by strongly reducing 1,4-bis(diphenylamino)naphthalene photoredox catalysis. *ACS Catal.* 9 (5): 4382–4387.

98 Matsubara, R., Shimada, T., Kobori, Y. et al. (2016). Photoinduced charge-transfer state of 4-carbazolyl-3-(trifluoromethyl)benzoic acid: photophysical property and application to reduction of carbon–halogen bonds as a sensitizer. *Chem. Asian J.* 11 (14): 2006–2010.

99 Matsubara, R., Yabuta, T., Md Idros, U. et al. (2018). UVA- and visible-light-mediated generation of carbon radicals from organochlorides using nonmetal photocatalyst. *J. Org. Chem.* 83 (16): 9381–9390.

100 Gosztola, D., Niemczyk, M.P., Svec, W. et al. (2000). Excited doublet states of electrochemically generated aromatic imide and diimide radical anions. *J. Phys. Chem. A* 104 (28): 6545–6551.

101 Ghosh, I., Ghosh, T., Bardagi, J.I., and Konig, B. (2014). Reduction of aryl halides by consecutive visible light-induced electron transfer processes. *Science* 346 (6210): 725–728.

102 Zeman, C.J., Kim, S., Zhang, F., and Schanze, K.S. (2020). Direct observation of the reduction of aryl halides by a photoexcited perylene diimide radical anion. *J. Am. Chem. Soc.* 142 (5): 2204–2207.

103 Cowper, N.G.W., Chernowsky, C.P., Williams, O.P., and Wickens, Z.K. (2020). Potent reductants via electron-primed photoredox catalysis: unlocking aryl chlorides for radical coupling. *J. Am. Chem. Soc.* 142 (5): 2093–2099.

104 Kim, H., Kim, H., Lambert, T.H., and Lin, S. (2020). Reductive electrophotocatalysis: merging electricity and light to achieve extreme reduction potentials. *J. Am. Chem. Soc.* 142 (5): 2087–2092.

105 Cole, J.P., Chen, D.-F., Kudisch, M. et al. (2020). Organocatalyzed birch reduction driven by visible light. *J. Am. Chem. Soc.* 142 (31): 13573–13581. https://doi.org/10.1021/jacs.0c05899.

106 Christensen, J.A., Phelan, B.T., Chaudhuri, S. et al. (2018). Phenothiazine radical cation excited states as super-oxidants for energy-demanding reactions. *J. Am. Chem. Soc.* 140 (15): 5290–5299.

107 Huang, H., Strater, Z.M., Rauch, M. et al. (2019). Electrophotocatalysis with a trisaminocyclopropenium radical dication. *Angew. Chem. Int. Ed.* 58 (38): 13318–13322.

108 Kerzig, C., Guo, X., and Wenger, O.S. (2019). Unexpected hydrated electron source for preparative visible-light driven photoredox catalysis. *J. Am. Chem. Soc.* 141 (5): 2122–2127.

109 Kerzig, C. and Wenger, O.S. (2019). Reactivity control of a photocatalytic system by changing the light intensity. *Chem. Sci.* 10 (48): 11023–11029.

110 Hildebrand, J.H. and Glascock, B.L. (1909). The color of iodine solutions. *J. Am. Chem. Soc.* 31 (1): 26–31.

111 Mulliken, R.S. (1950). Structures of complexes formed by halogen molecules with aromatic and with oxygenated solvents. *J. Am. Chem. Soc.* 72 (1): 600–608.

112 Mulliken, R.S. (1952). Molecular compounds and their spectra. II. *J. Am. Chem. Soc.* 74 (3): 811–824.

113 Mulliken, R.S. (1952). Molecular compounds and their spectra. III. The interaction of electron donors and acceptors. *J. Phys. Chem.* 56 (7): 801–822.

114 Rosokha, S.V. and Kochi, J.K. (2008). Fresh look at electron-transfer mechanisms via the donor/acceptor bindings in the critical encounter complex. *Acc. Chem. Res.* 41 (5): 641–653.

115 Lima, C.G.S., Lima, T.d.M., Duarte, M. et al. (2016). Organic synthesis enabled by light-irradiation of EDA complexes: theoretical background and synthetic applications. *ACS Catal.* 6 (3): 1389–1407.

116 Roth, H., Romero, N., and Nicewicz, D. (2015). Experimental and calculated electrochemical potentials of common organic molecules for applications to single-electron redox chemistry. *Synlett* 27 (05): 714–723.

117 Crisenza, G.E.M., Mazzarella, D., and Melchiorre, P. (2020). Synthetic methods driven by the photoactivity of electron donor–acceptor complexes. *J. Am. Chem. Soc.* 142 (12): 5461–5476.

118 Cantacuzène, D., Wakselman, C., and Dorme, R. (1977). Condensation of perfluoroalkyl iodides with unsaturated nitrogen compounds. *J. Chem. Soc., Perkin Trans. 1* (12): 1365–1371.

119 Hoz, S. and Bunnett, J.F. (1977). A quantitative study of the photostimulated reaction of iodobenzene with diethyl phosphite ion. *J. Am. Chem. Soc.* 99 (14): 4690–4699.

120 Fox, M.A., Younathan, J., and Fryxell, G.E. (1983). Photoinitiation of the $S_{RN}1$ reaction by excitation of charge-transfer complexes. *J. Org. Chem.* 48 (18): 3109–3112.

121 Fukuzumi, S., Mochida, K., and Kochi, J.K. (1979). A unified mechanism for thermal and photochemical activation of charge-transfer processes with organometals. Steric effects in the insertion of tetracyanoethylene. *J. Am. Chem. Soc.* 101 (20): 5961–5972.

122 Sankararaman, S., Haney, W.A., and Kochi, J.K. (1987). Annihilation of aromatic cation radicals by ion-pair and radical pair collapse. Unusual solvent and salt effects in the competition for aromatic substitution. *J. Am. Chem. Soc.* 109 (25): 7824–7838.

123 Wade, P.A., Morrison, H.A., and Kornblum, N. (1987). The effect of light on electron transfer substitution at a saturated carbon atom. *J. Org. Chem.* 52 (14): 3102–3107.

124 Russell, G.A. and Wang, K. (1991). Homolytic alkylation of enamines by electrophilic radicals. *J. Org. Chem.* 56 (11): 3475–3479.

125 Gotoh, T., Padias, A.B., and Hall, H.K. (1991). An electron donor-acceptor complex and thermal triplex as intermediates in the cycloaddition reaction of N-vinylcarbazole with dimethyl 2,2-dicyanoethylene-1,1-dicarboxylate. *J. Am. Chem. Soc.* 113 (4): 1308–1312.

126 Kandukuri, S.R., Bahamonde, A., Chatterjee, I. et al. (2015). X-ray characterization of an electron donor-acceptor complex that drives the photochemical alkylation of indoles. *Angew. Chem. Int. Ed.* 54 (5): 1485–1489.

127 Yang, Q.-Q., Liu, N., Yan, J.-Y. et al. (2020). Visible light- and heat-promoted C–O coupling reaction of phenols and aryl halides. *Asian J. Org. Chem.* 9 (1): 116–120.

128 da Silva, G.P., Ali, A., da Silva, R.C. et al. (2015). Tris(trimethylsilyl)silane and visible-light irradiation: a new metal- and additive-free photochemical process for the synthesis of indoles and oxindoles. *Chem. Commun.* 51 (82): 15110–15113.

129 Zhang, J., Li, Y., Xu, R., and Chen, Y. (2017). Donor-acceptor complex enables alkoxyl radical generation for metal-free C(sp^3)-C(sp^3) cleavage and allylation/alkenylation. *Angew. Chem. Int. Ed.* 56 (41): 12619–12623.

130 Davies, J., Booth, S.G., Essafi, S. et al. (2015). Visible-light-mediated generation of nitrogen-centered radicals: metal-free hydroimination and iminohydroxylation cyclization reactions. *Angew. Chem. Int. Ed.* 54 (47): 14017–14021.

131 Fawcett, A., Pradeilles, J., Wang, Y. et al. (2017). Photoinduced decarboxylative borylation of carboxylic acids. *Science* 357 (6348): 283–286.

132 Wu, J., He, L., Noble, A., and Aggarwal, V.K. (2018). Photoinduced deaminative borylation of alkylamines. *J. Am. Chem. Soc.* 140 (34): 10700–10704.

133 Wu, J., Bär, R.M., Guo, L. et al. (2019). Photoinduced deoxygenative borylations of aliphatic alcohols. *Angew. Chem. Int. Ed.* 58 (52): 18830–18834.

134 Bosque, I. and Bach, T. (2019). 3-Acetoxyquinuclidine as catalyst in electron donor–acceptor complex-mediated reactions triggered by visible light. *ACS Catal.* 9 (10): 9103–9109.

135 Emmanuel, M.A., Greenberg, N.R., Oblinsky, D.G., and Hyster, T.K. (2016). Accessing non-natural reactivity by irradiating nicotinamide-dependent enzymes with light. *Nature* 540 (7633): 414–417.

4

C—H Bond Functionalization with Chemical Oxidants

Jia-Xiang Xiang, Pooja Vemuri, and Frédéric W. Patureau

RWTH Aachen University, Institute of Organic Chemistry, Landoltweg 1, 52074 Aachen, Germany

4.1 Introduction

4.1.1 A Shift in the Rate-Determining Step

The field of C—H bond activation and functionalization has considerably expanded over the last 20 years. The concept is particularly attractive because it shortcuts classical cross-coupling techniques by avoiding preactivation or preoxidation steps [1]. This is particularly true in the case of dehydrogenative couplings, in which case an oxidant is then typically required to overcome the often-unfavorable thermodynamic situation [2]. Those recent coupling methods changed the paradigm of the previously reputed "inert C—H bond" toward the idea of a reactive and versatile functional group. The new catalysts that enable those transformations have become so efficient at it, through careful ligand, solvent, and directing group (DG) design, that C—H bond activation is now possible even at room temperature [1e]. In some systems, it is even no longer the rate-determining step of the coupling reaction. Indeed, it has become quite common to find low kinetic isotope effects (KIEs = k_H/k_D) on the initial rate of the reaction, and/or high H/D scrambling at the functionalized C—H bond [3]. The latter scrambling allows to directly probe the *reversibility* of the C—H bond activation step under reaction conditions and thus its non-rate-limiting character, provided of course that it does not arise from background reactions [4]. Thus, in many recent reaction systems, the C—H bond activation step is no longer the difficult step of the reaction. This has naturally an important consequence: the rate-determining barrier is increasingly shifted toward other elementary steps in the catalytic cycle, such as redox events.

4.1.2 The Nature of the Oxidant

So what does a facile C—H bond activation process mean for method developers? In particular, in the case of cross-dehydrogenative couplings? A direct consequence

☆ Jia-Xiang Xiang and Pooja Vemuri contributed equally.

Organic Redox Chemistry: Chemical, Photochemical and Electrochemical Syntheses, First Edition.
Edited by Jun-Ichi Yoshida and Frédéric W. Patureau.

is that future breakthroughs in this field are likely to come from optimizing the new, often redox rate-determining steps. Typically, when the C—H bond activation step is no longer rate significant, it is the final reductive elimination step, which can assume that role. It is interesting to observe that in the latter cases, the nature of the oxidant seems to have an immense impact on the outcome of the reaction. Its oxidizing strength of course would play an important role, but so also would its shape, size, solubility, concentration, and interacting ability with the catalyst and/or the substrates. In the case of metal-catalyzed reactions, this can be understood through the decisive oxidation or reoxidation of the metal catalyst, depending on whether reductive elimination precedes, is concomitant to, or succeeds the reductive elimination step. For metal-free dehydrogenative C–H functionalization systems, it is the direct interaction of the oxidant with the substrates, which should be decisive. We therefore propose in this chapter an oxidant class-orientated analysis of the recent literature on the topic. This overview is by no means exhaustive and only covers a very short selection from the most recent and representative examples.

4.2 Metal-Based Oxidants and Other Inorganic Oxidants

The use of metal salts as terminal oxidants is one of the most utilized strategies in cross-dehydrogenative couplings. The most common metal salts of this type are probably copper(II) and silver(I) salts, due to their appropriate redox potentials and their ability to host basic ligands such as carboxylates. Those have been utilized extensively for years as privileged oxidants to construct C—C, C—N, and C—O bonds (Scheme 4.1) [1]. One of the earliest and most significant C—H bond activation coupling reactions for instance, the Fujiwara–Moritani reaction, utilizes a Cu(II) salt as reoxidant of the Pd(II) active catalyst [5]. Copper(II) and silver(I) salts are particularly suitable terminal oxidants in combination with Rh, Ir, Ru, Pd, Co, and other C—H bond activation catalysts [1]. Their main drawbacks are however their cost, which is not likely to improve in the coming decades and centuries, and the technical difficulties associated to their recovery from often complex reaction mixtures. Their ability to exchange ligands and electrons under relatively mild conditions with active C—H bond activation catalysts marks them as privileged oxidants for cross-dehydrogenative couplings.

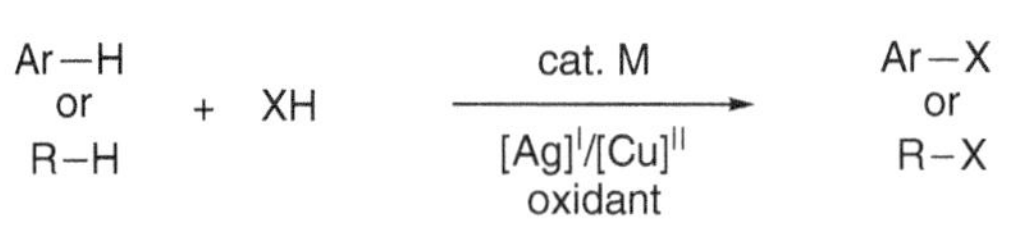

M = Co, Rh, Pd, Ru, etc
X = C, N, O.

Scheme 4.1 Ag(I) and Cu(II) oxidants. Source: Modified from Leitch and Frost [1a].

4.2.1 Silver Salt Oxidants

In this chapter, we wish to focus a bit more on the recent and promising cobalt-catalyzed C–H functionalizations, a considerably cheaper and arguably still underappreciated first-row transition metal. In 2015, for example, Wei, Niu and coworkers performed a remarkable theoretical and experimental study on a cobalt(II/III)-catalyzed radical C–H oxidative coupling [6]. They report a cobalt-catalyzed alkoxylation of benzamide derivatives with alcohols, involving a single-electron-transfer (hereafter: SET)-initiated radical mechanism (Scheme 4.2). The Ag(I) salt is therein able to (re-)oxidize the Co(II) into the Co(III) active catalyst.

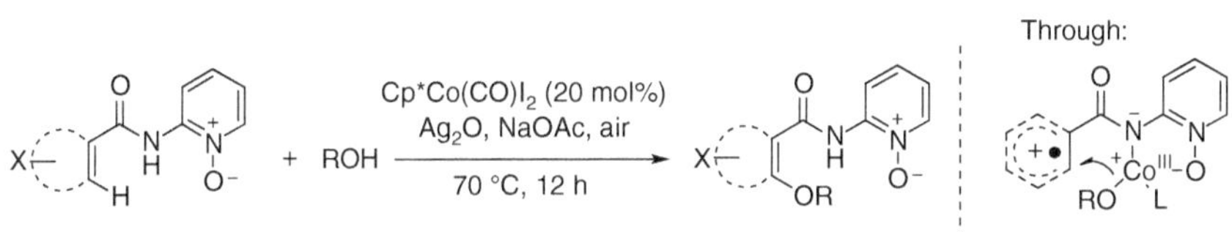

Scheme 4.2 Radical Co(II)/Co(III) cross-dehydrogenative C–O bond formation by Niu and Wei.

In 2016, You and coworkers reported an interesting example of cobalt-catalyzed dehydrogenative cross-coupling between two (hetero-)aromatic moieties (Scheme 4.3), utilizing Ag_2CO_3 as oxidant [7]. Importantly, the precious silver residue could be recycled by treatment with HNO_3 followed by Na_2CO_3, and thereafter reutilized in several runs.

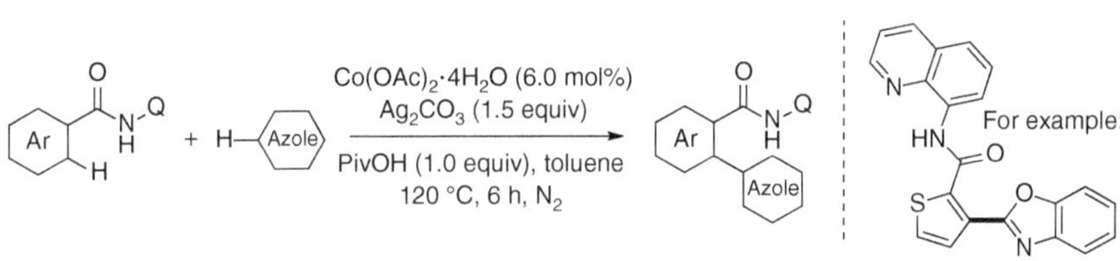

Scheme 4.3 You's dehydrogenative heterocyclic C–C bond formation.

The same year, Song, Niu, and coworkers developed a cobalt(II)-catalyzed C–H amination of benzamides with alkylamines, wherein $AgNO_3$ is the terminal oxidant of the reaction (Scheme 4.4) [8].

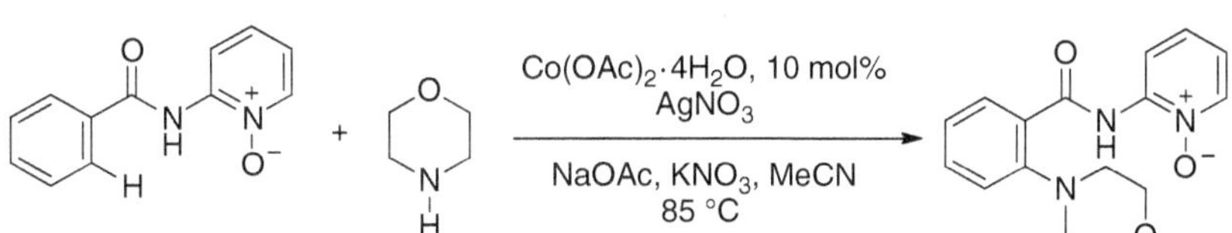

Scheme 4.4 Song's cocatalyzed dehydrogenative amination of benzamides. Source: Modified from Zhang et al., 2016.

In 2017, Lei and coauthors reported a cobalt-catalyzed reaction method to achieve the intramolecular oxidative C(sp^3)–H/N–H carbonylation of aliphatic amides with CO (Scheme 4.5) [9]. In this work, some alkyl amides were transformed into the corresponding succinimides. The authors also propose an active Co(III) species generated by oxidation through the silver salt.

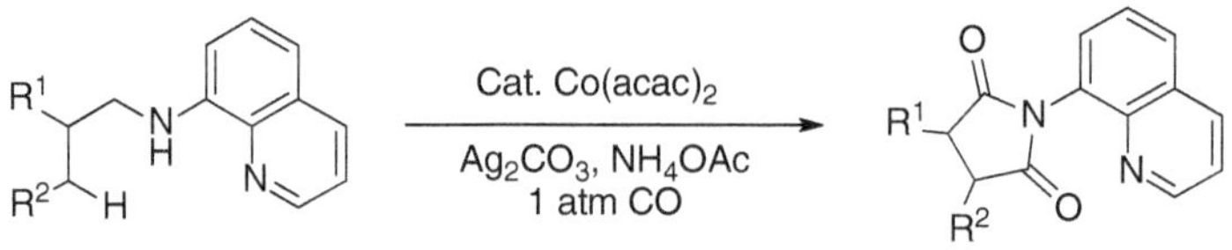

Scheme 4.5 Lei's oxidative C(sp^3)–H/N–H carbonylation reaction. Source: Modified from Zeng et al. [9].

Another great advantage of silver salts is their excellent propensity to capture halide ions from the coordination sphere of the active metal catalyst through transmetalation of more labile counter ions. This allows tuning the basicity and the coordinative (or noncoordinative) character of the catalyst's counter ions, with often dramatic effects on the overall activity. In 2012 for example, Cheng and coworkers illustrated this effect with their rhodium-catalyzed C—H bond activation of benzamides with boronic acids under very mild reaction conditions (Scheme 4.6) [10].

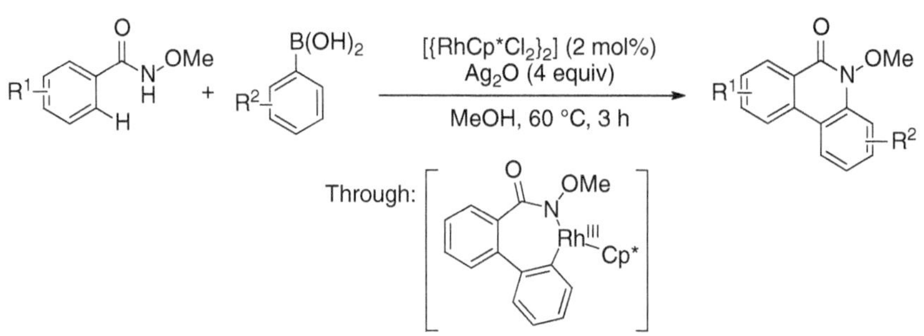

Scheme 4.6 Cheng's dehydrogenative reaction with benzamides and boronic acids. Source: Modified from Karthikeyan et al. [10].

In 2018, Glorius and coworkers utilized not one but two different silver salts: the highly cationic $AgSbF_6$ *and* the basic AgOAc in their RhCp*-catalyzed dehydrogenative C—O bond formation reaction, again clearly illustrating the highly multipurpose character of silver salts (Scheme 4.7) [11]. The same year, Blakey and coworker also proposed a RhCp*-catalyzed dehydrogenative C—O bond formation with a combination of silver salts (Scheme 4.8) [12].

Also in 2018, Lan, You and coworkers developed an impressive dehydrogenative *hetero* biaryl synthesis through double *hetero* chelate assistance under Rh-catalyzed C—H bond activation conditions. In this example, both silver and copper salts were utilized concomitantly. The fact that *homo* coupling biaryl products could be

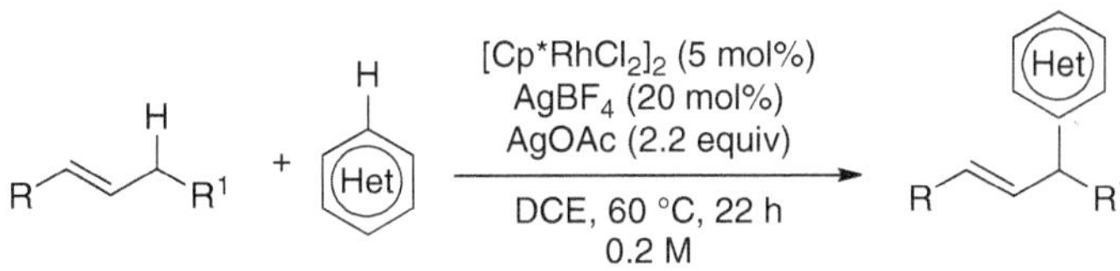

Scheme 4.7 Glorius's RhCp*-catalyzed dehydrogenative allylic arylation. Source: Lerchen et al. [11]. John Wiley & Sons.

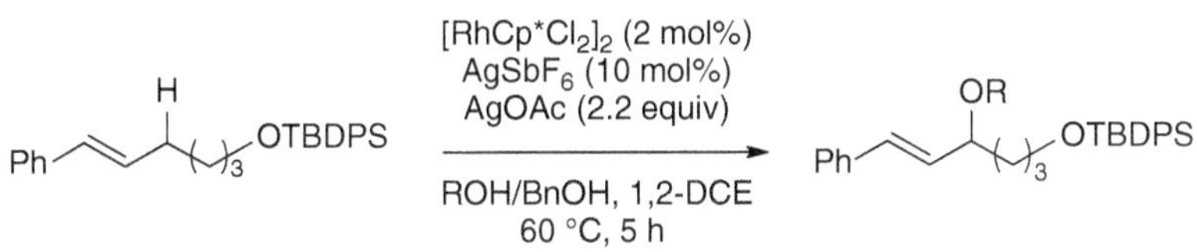

Scheme 4.8 Blakey's RhCp*-catalyzed dehydrogenative C—O bond formation. Source: Modified from Nelson and Blakey [12].

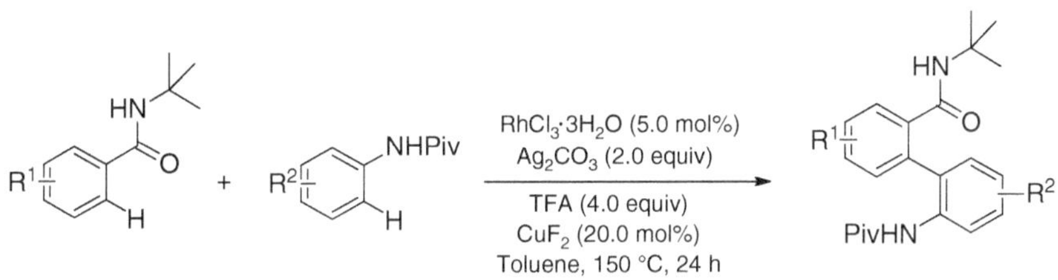

Scheme 4.9 You's dehydrogenative *hetero* biaryl synthesis. Source: Shi et al. [13]. John Wiley & Sons.

suppressed in favor of the *hetero* biaryl scaffold is particularly important for the organic synthesis toolbox (Scheme 4.9) [13].

Silver salts can also serve as oxidants in palladium-catalyzed C–H functionalizations. Here, we could give another example from You's group: the double-oxidative arylation of dihydroquinones (Scheme 4.10) [14], or a more recent C(sp^3)—H bond activation example of amino acids from Shi's group (Scheme 4.11) [15].

In 2018, the Yu group developed a series of long-range γ and δ C(sp^3)–H arylation methods of amino alkanes with transient DGs (Scheme 4.12) [16]. Interestingly, in spite of the use of preoxidized/prefunctionalized iodoarene coupling partners, the presence of several additional equivalents of silver trifluoroacetate (AgTFA) is

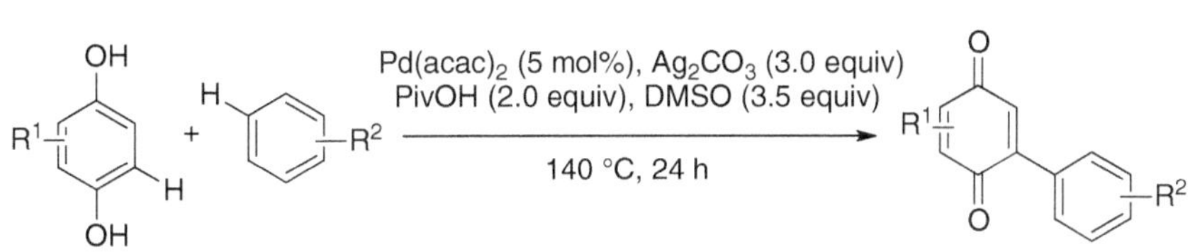

Scheme 4.10 You and Song's arylated quinone synthesis. Source: Zhang et al. [14]. Royal Society of Chemistry.

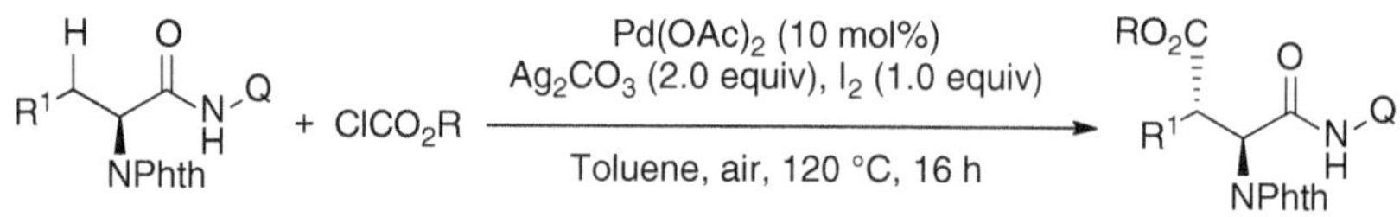

Scheme 4.11 Shi's amino-acid functionalization through Pd-catalyzed C(sp^3)–H activation. Source: Liao et al. [15]. Springer Nature Limited.

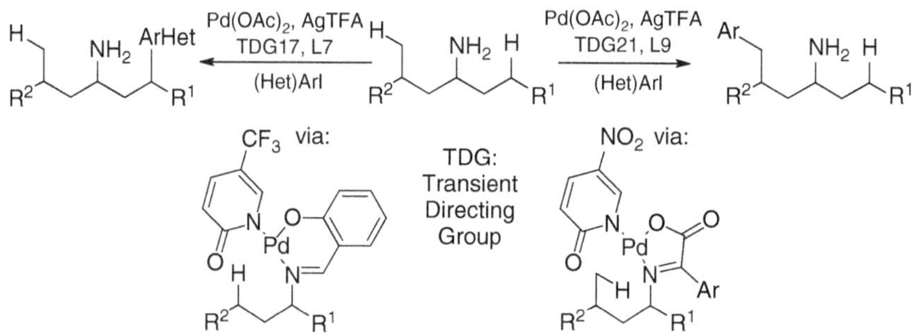

Scheme 4.12 Yu's γ and δ C–H arylation methods of amino alkanes. Source: Chen et al. [16]. American Chemical Society.

still necessary to achieve high yields. This constitutes another illustration of the importance of such silver salt additives in the development of new original C–H functionalizing methods, and gives an idea of the challenge should one wish to replace them with more sustainable/less onerous alternatives.

Silver salts are so powerful, versatile, and multifunctional that they can even promote cross-dehydrogenative couplings by themselves, without another transition metal catalyst. In 2013, Duan and coworker revealed a unique Ag(I)-mediated C–H/P–H oxidative annulation reaction between aromatic secondary phosphine oxides (SPOs) and internal alkynes to produce highly valuable functionalized phospole derivatives (Scheme 4.13) [17].

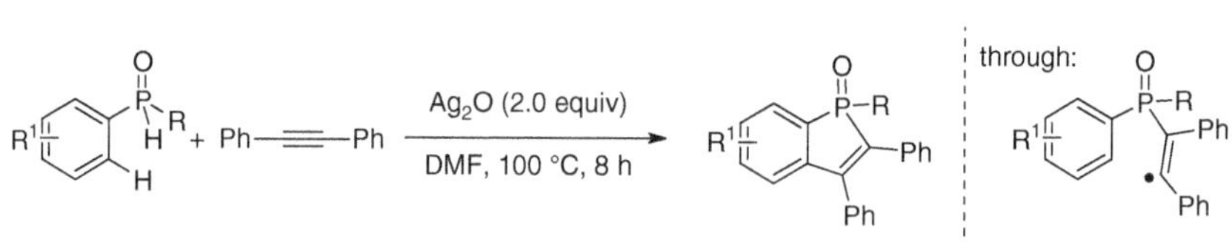

Scheme 4.13 Duan's reaction: Ag(I)-mediated dehydrogenative phosphole synthesis. Source: Modified from Chen and Duan [17].

4.2.2 Copper Salt Oxidants

Copper salts are historically among the most utilized metallic oxidants for cross-dehydrogenative couplings [5]. However, fewer copper-mediated oxidative C–H functionalizations have been developed in most recent years. Because of extensive past coverage of those cases [1], we shall only present a most recent example, from the Ritter group. In 2019, they reported an oxidative cyanation reaction of aromatic C—H bonds, in which the typical Cu(II) acetate can function as oxidant, in a mixture of dimethylformamide (DMF) and new exceptional 1,1,1,3,3,3-Hexafluor-2-propanol (HFIP) solvent (Scheme 4.14) [18].

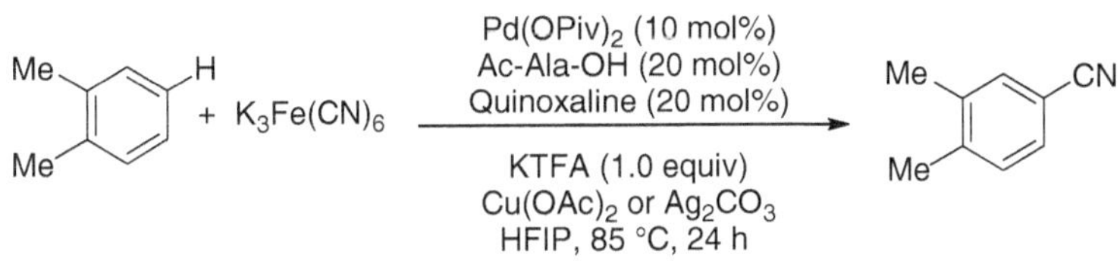

Scheme 4.14 Ritter's oxidative cyanation of aromatic C—H bonds. Source: Zhao et al. [18]. Elsevier.

4.2.3 Other Inorganic Oxidants

Of course, one can also conceive other metal salts in their higher oxidation states to assume a similar oxidizing role as Ag(I) and Cu(II) salts. For example, in 2018, Liu, Engle and coworkers developed a Pd-catalyzed "Fujiwara-like" oxidative olefination of olefins to produce diversely functionalized dienes, while using MnO_2 as terminal oxidant of the reaction (Scheme 4.15) [19]. It can be noted that they also performed the reaction under aerobic conditions.

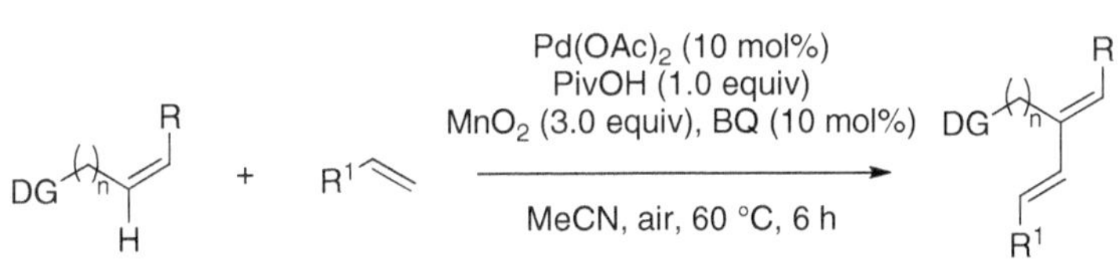

Scheme 4.15 Engle and Liu's dehydrogenative Fujiwara-like diene synthesis. Source: Modified from Liu et al. [19].

Persulfates and similar reagents have also been extensively utilized. For example, in 2017, Lei and coworkers developed a cross-dehydrogenative amination reaction of arenes with triazoles and related compounds, based on a Cu(I) catalyst and $(NH_4)_2S_2O_8$ as the oxidant (Scheme 4.16) [20]. The same year, the group of Xia also developed some elegant cross-dehydrogenative amination methods of phenols with a photocatalytic strategy and $(NH_4)_2S_2O_8$ as the terminal oxidant [21].

4.3 Organic Oxidants

The use of sacrificial carbon-containing oxidants is also a very popular and widespread oxidizing strategy in cross-dehydrogenative couplings. Indeed, they are cheap, usually quite soluble in the reaction medium, and often carry key functional

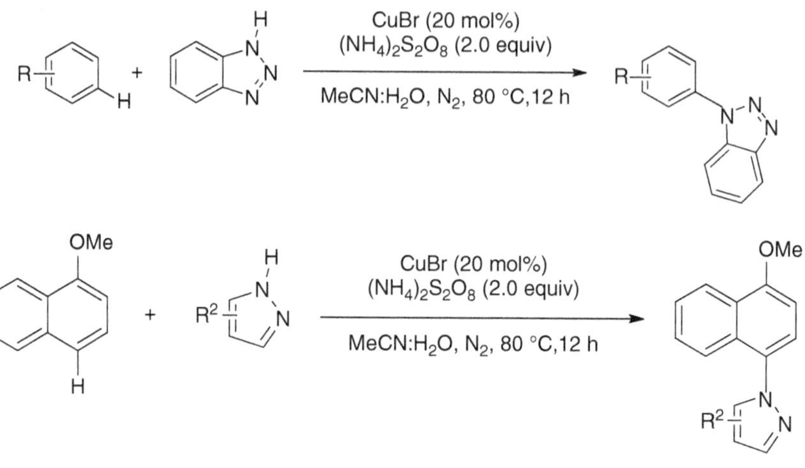

Scheme 4.16 Lei's cross-dehydrogenative amination with some azole derivatives. Source: Modified from H. Yi et al. [20].

groups able to directly coordinate to the active metal center. Of course, one of their drawbacks is that their scaffold is not incorporated into the product. Thus, organic oxidants do not usually score well in terms of atom efficiency. We have divided them into two main families: the organic peroxides and the quinone derivatives (Scheme 4.17).

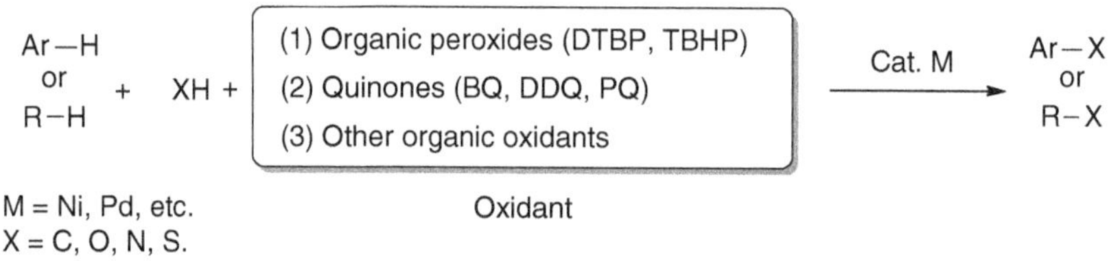

Scheme 4.17 Organic peroxides and quinones.

4.3.1 Organic Peroxides

Organic peroxide oxidants, such as di-*tert*-butyl peroxide (DTBP) or *tert*-butyl hydroperoxide (TBHP), are arguably particularly well suited as (re-)oxidant for first-row transition metal-catalyzed cross-dehydrogenative couplings, as these metals are generally well suited for SET events. Moreover, as an additional advantage, many of these metals have pronounced oxophilicity, thus facilitating those processes. Herein, we present a few selected recent examples. In 2015 for example, You and coworkers developed an α dehydrogenative $C(sp^3)$—$C(sp^3)$ bond formation coupling between protected amino acids and some benzylic positions, under Ni(II)-catalyzed conditions (Scheme 4.18) [22]. The authors proposed a Ni(III) active species, which would be generated from the peroxide oxidant.

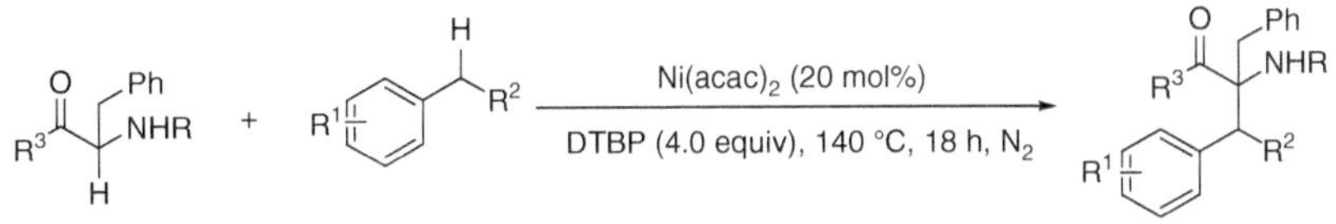

Scheme 4.18 Ni(II/III)-catalyzed $C(sp^3)-C(sp^3)$ dehydrogenative coupling from You. Source: Modified from Li et al., [22].

In a quite different example, Lei's group established in 2014 an iodine-catalyzed dehydrogenative $C(sp^3)$–H/S–H between benzylic positions and thiols, to furnish the corresponding thioethers in a single step (as opposed, for example, to a prior oxidation/nucleophilic substitution sequence, Scheme 4.19) [23]. The same year they also reported an interesting Ni(II)-catalyzed $C(sp^3)$–H arylation of tetrahydrofuran (THF) with boronic acids, utilizing DTBP (Scheme 4.20) [24]. The reaction has been proposed to go through an arene radical.

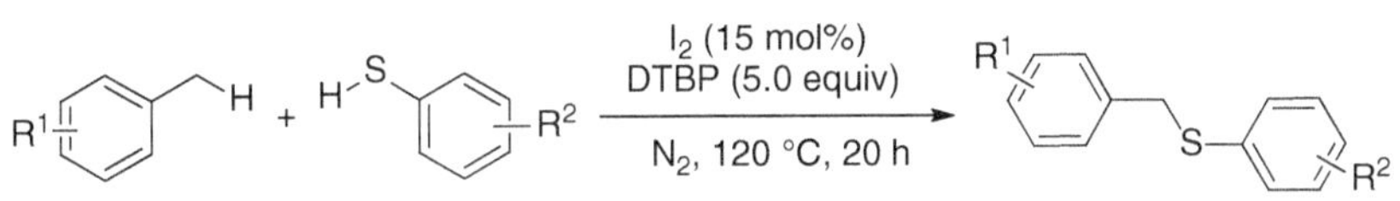

Scheme 4.19 Dehydrogenative thioether formation by Lei. Source: Modified from Yuan et al. [23].

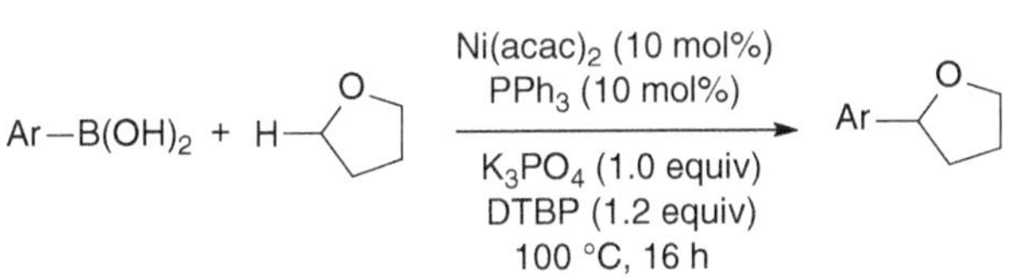

Scheme 4.20 Lei's $C(sp^3)$–H arylation of THF.

More recently, in 2017, the same group disclosed a palladium/copper cocatalyzed oxidative C–H/C–H carbonylation of diphenylamines utilizing the DTBP oxidant, toward acridones (Scheme 4.21) [25].

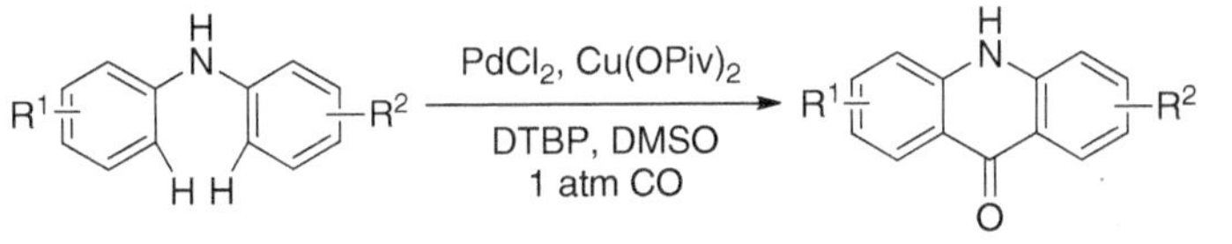

Scheme 4.21 Dehydrogenative acridone synthesis from Lei. Source: Wen et al. [25]. American Chemical Society.

Li, Song, and coworkers made use of TBHP for their Cu(II)-catalyzed C–H oxidative coupling and *ipso*-cyclization of *N*-arylpropiolamides with unactivated alkanes toward interesting *spiro*-compounds (Scheme 4.22) [26].

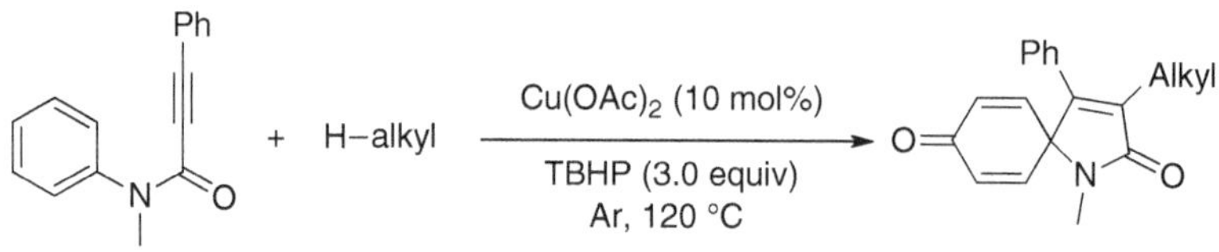

Scheme 4.22 Li and Song's *spiro*-[4,5]trienone synthesis. Source: Modified from Ouyang et al. [26].

In 2018, Lei and coworkers developed a versatile method for the direct dehydrogenative C(sp^3)–C(sp) cross-coupling of alkanes with terminal alkynes (Scheme 4.23) [27]. This reaction can be seen as a kind of oxidative Sonogashira reaction, in which a series of multiple catalysts act synergistically to allow the reaction to occur.

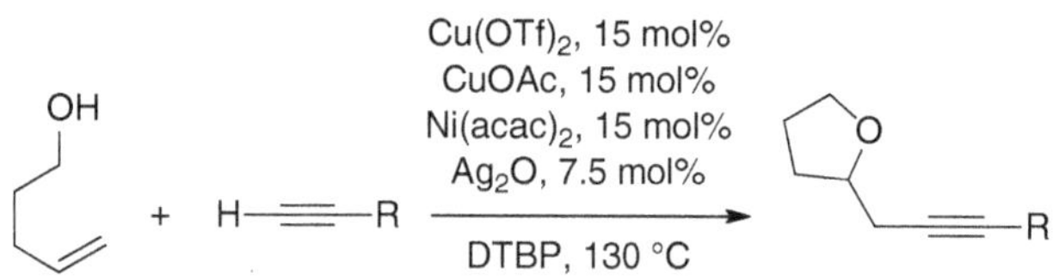

Scheme 4.23 Lei's Cu/Ni/Ag cocatalyzed oxidative-Sonogashira-like reaction. Source: Modified from Tang et al. [27].

Siegel and coworkers explored the reactivity of a somewhat more exotic organic peroxide: phthaloyl peroxide [28]. The philosophy of exploring unusual oxidants, or oxidants of unusual structure, makes complete sense if one searches for new reactivity, and this is one of the key messages that this chapter would convey (see Section 4.1). In this case, the direct oxidation of aromatic C—H bonds into the corresponding phenols is a difficult and long-sought reaction. The most efficient way to produce phenols is through the well-known Hock process, wherein an alkylated arene undergoes an oxidative rearrangement under strongly acidic conditions [29]. Beyond the harshness of those conditions, the Hock process also has the disadvantage of producing 1 equiv of acetone byproduct. The phthaloyl peroxide of Siegel allows to directly oxygenate the aromatic C—H bonds of arenes, producing the C–H benzoyloxilation product, which is only one small saponification step away from the desired phenol product (Scheme 4.24).

4.3.2 Quinones

The use of benzoquinone (BQ) is historically associated to palladium-catalyzed oxidative coupling reactions. One of the main advantages of quinones is their ability to bind to low-valent palladium intermediates, in particular Pd(0), which stabilizes them and prevents them from dropping out of the reaction mixture (in which case they might become less active). Recently, Luo, Lei and coworkers used BQ as the terminal oxidant in their elegant α C–H allylation of ketones through a transient enamine (Scheme 4.25) [30].

2,3-Dichloro-5,6-dicyano-1,4-benzoquinone (DDQ) is also a historic and widely utilized organic oxidant. As a recent example, we could cite Pumera's

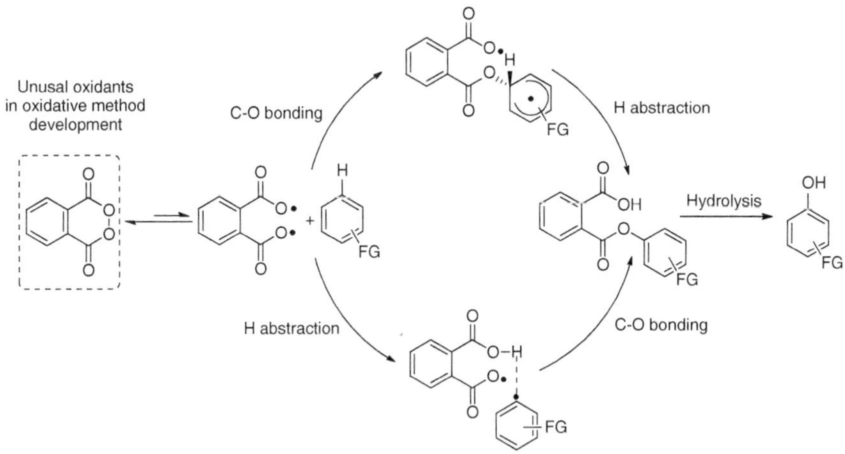

Scheme 4.24 Siegel's phenol synthesis with unusual phthaloyl peroxide.

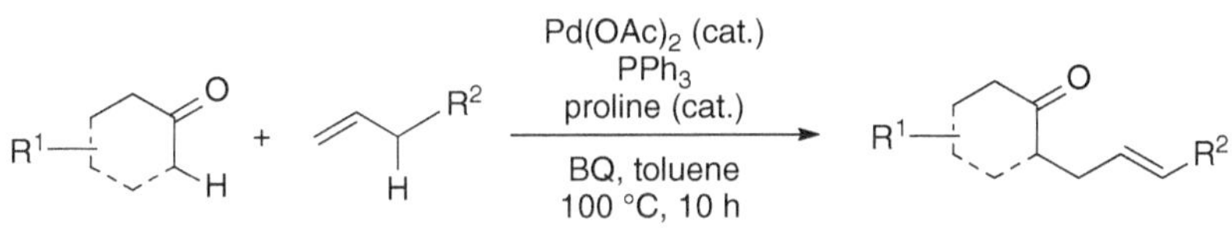

Scheme 4.25 Lei and Luo's C–H allylation of ketones. Source: Modified from Tang et al. [30].

Cu(II)-catalyzed cross-dehydrogenative coupling between some hydrogenated graphene derivative and tetrahydrothiophen-3-one, thereby providing new organic materials in a single step (Scheme 4.26) [31]. In general, the use of oxidative C—H bond functionalization techniques to synthesize new highly fused polycyclic and heterocyclic materials remains an underappreciated but very powerful approach to rapidly access original organic material scaffolds, in comparison to multistep classical cross-coupling methods. We expect such dehydrogenative methods to be increasingly applied in the synthesis of such materials, thereby considerably decreasing their price and sustainability footprint.

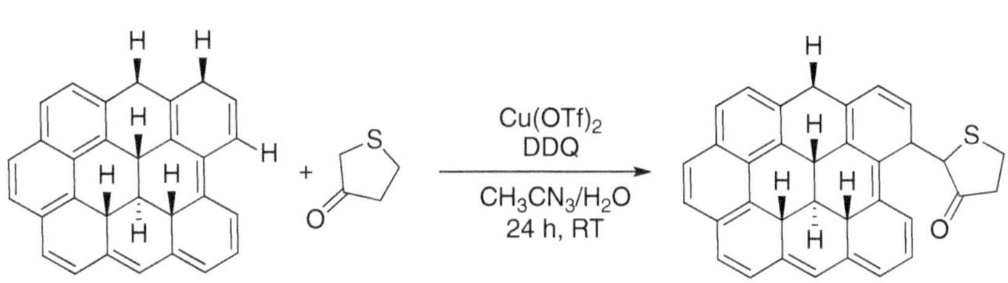

Scheme 4.26 Pumera's recent Cu(II)-catalyzed CDC reaction with DDQ. Source: Chua et al. [31]. John Wiley & Sons.

Interestingly, Liu, Floreancig and coworkers reported a detailed computational study on the mechanism and substituent effects of DDQ-mediated oxidative C–H functionalizations (Scheme 4.27) [32]. Density functional theory (DFT) calculations demonstrate the involvement of a hydride transfer within a charge transfer complex. We recommend this study for details.

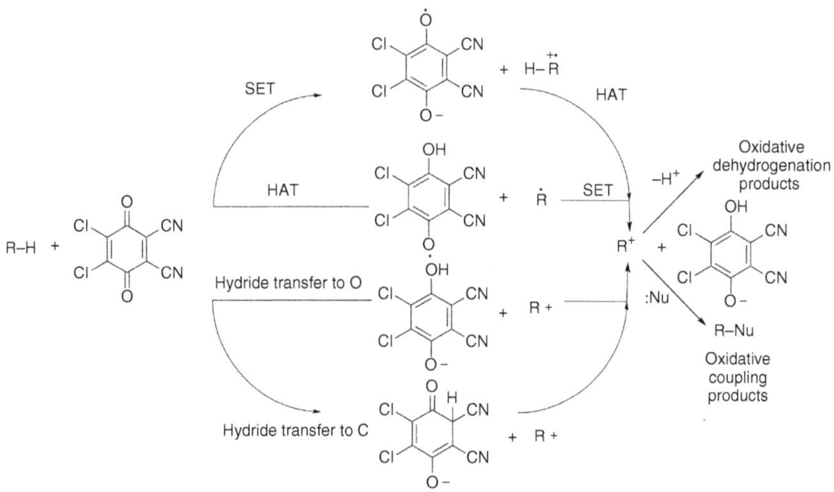

Scheme 4.27 Possible mechanisms of DDQ-mediated C–H cleavage according to Liu and Floreancig. Source: Rivera et al. [32]. American Chemical Society.

Quinone derivatives, and in particular DDQ, can not only be used as the terminal oxidant of the reaction, but also as cocatalysts, in which case another species bears the role of terminal oxidant. This strategy is often utilized when the desired terminal oxidant of a reaction, say, for example, highly sustainable O_2, is not reactive enough to directly engage substrates and/or catalysts. This again illustrates, as mentioned in the Section 4.1, the importance of structural and functional considerations when contemplating the design of the oxidant, in relation to the coupling's building blocks and catalyst. In the case of O_2 as the desired terminal oxidant, some quinones, and in particular copper salts such as $Cu(OAc)_2$, have been utilized as catalyst to propagate the oxidizing driving force of the coupling reaction. As a recent example, Chiang, Lei and coworkers reported a dehydrogenative C(sp^3)—N bond forming coupling reaction with DDQ *tert*-butyl nitrite (TBN, Scheme 4.28) [33].

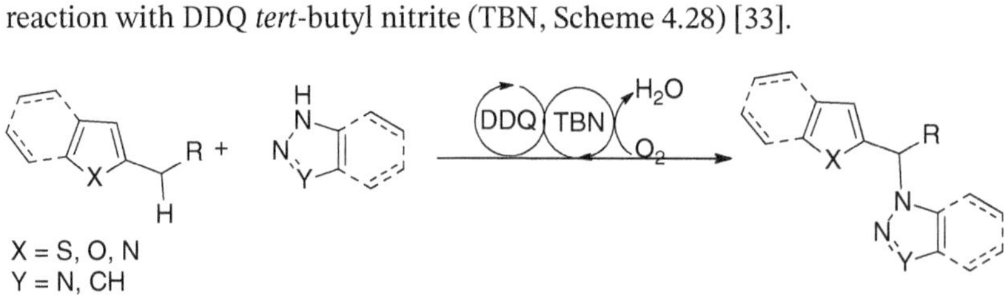

Scheme 4.28 Lei and Chiang's DDQ- and TBN-catalyzed aerobic C(sp^3)–N bond formation. Source: Song et al. [33]. American Chemical Society.

In 2018, Qing disclosed a new strategy for the direct introduction of a difluoromethyl group into heteroarenes. This mild and regioselective oxidative

C-H difluoromethylation method enables the convenient synthesis of a range of difluoromethylated heteroarenes in high yields. The use of 9,10-phenanthrenequinone (PQ) as an oxidant is critical to the success of this new difluoromethylation reaction (Scheme 4.29) [34]. In general, the direct difluoromethylation of C—H bonds is expected to become an important synthetic target in the near future because of the unique properties of this functional group.

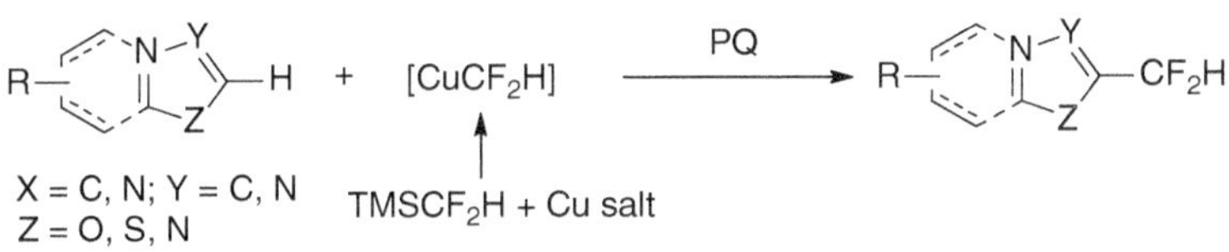

Scheme 4.29 Qing's oxidative C–H difluoromethylation of some heterocycles. Source: Zhu et al. [34]. American Chemical Society.

4.4 Internal Oxidants (DG^{ox})

About a decade ago, a new idea emerged in the field of C—H bond activation and functionalization. Indeed, instead of having the organic oxidant added externally, could any advantage be gained by constructing the oxidizing functionality within one of the building blocks? Of course, we do not mean by that the trivial case in which the building blocks are preoxidized and/or preactivated at the targeted positions, as in classical cross-coupling chemistry. The idea is rather to incorporate the oxidizing functionality *at a different position* than the targeted C—H bond, for example within the chelate-assisting DG (what is sometimes referred to as DG^{ox}). Very early, Glorius and Patureau recognized this novel idea as interesting, and discussed the very first examples in a short conceptual paper in early 2011 [35]. On first sight, this may appear as a counterproductive strategy to the sustainable chemistry expert. The gain is of course elsewhere than step and atom economy. As we explain in this chapter, the topographic structure of the oxidant has a large impact on the coupling reaction efficiency, because the oxidizing steps are often (increasingly) rate limiting. One could thus imagine a strategy in which the oxidizing functionality, say an electron-deficient N—O bond, would be covalently placed only Ångströms away from the metal catalyst's active site. When this concept appeared, the vast majority of C—H bond activation methods ran well above 80 °C, and often above 100 °C. This strategy allowed to bring C—H bond activation down to room temperature, and was one of the keys to a new generation of C—H bond functionalization methods, including enantioselective ones (not covered here). In this sense, the DG^{ox} strategy really highlights the central importance of the oxidation event. This chapter discusses some of the most recent examples (Scheme 4.30).

One of the pioneers of RhCp*-catalyzed C—H bond activation, Fagnou and coworkers reported in 2010 an interesting breakthrough: the external oxidant-free isoquinolone synthesis from *N*-methoxy benzamide and internal alkynes (Scheme 4.31) [36]. The term "external oxidant free" should not be

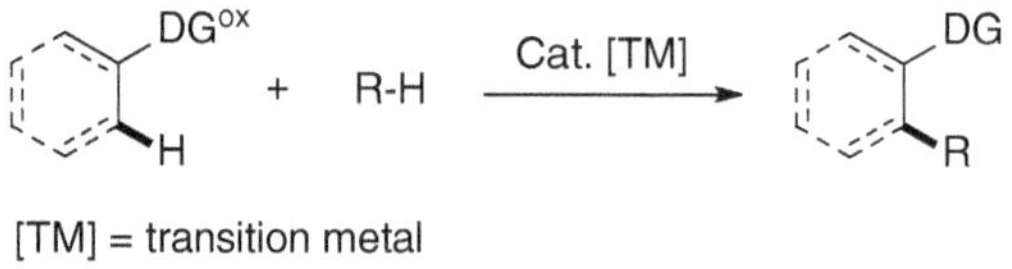

Scheme 4.30 DG^{ox} and internal oxidants.

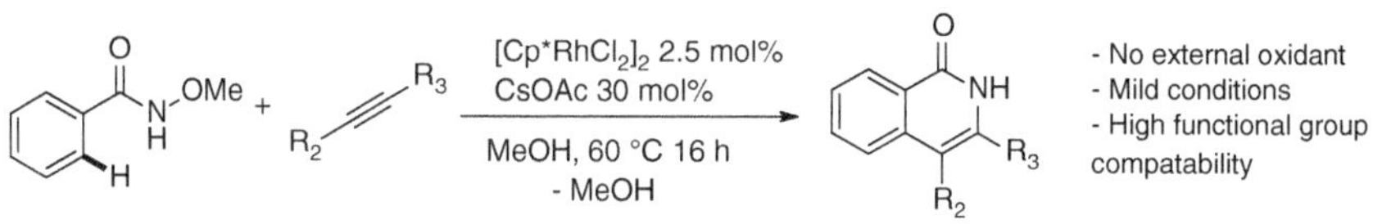

Scheme 4.31 Fagnou's DG^{ox}-mediated RhCp*-catalyzed isoquinolone synthesis. Source: Guimond et al. [36]. American Chemical Society.

understood beyond its factual meaning: in most cases, it is *not* an advantage to be "external oxidant free" if it means that the oxidizing functionality must be preinstalled in one of the substrates through additional synthetic steps. Many research groups nowadays utilize the term "external oxidant free" as if it was a per definition desirable feature. It would thus arguably be more appropriate to speak of very low reaction temperature (in the case of Fagnou's reaction, 60 °C is remarkably mild for 2010), and broad functional group tolerance, which is in part a consequence of the former argument.

Mechanistically, it is important to note that N—O bond cleavage generally occurs after the migratory insertion step, perhaps concomitant to the reductive elimination step, thus directly regenerating the active Rh(III) active catalyst (Scheme 4.32).

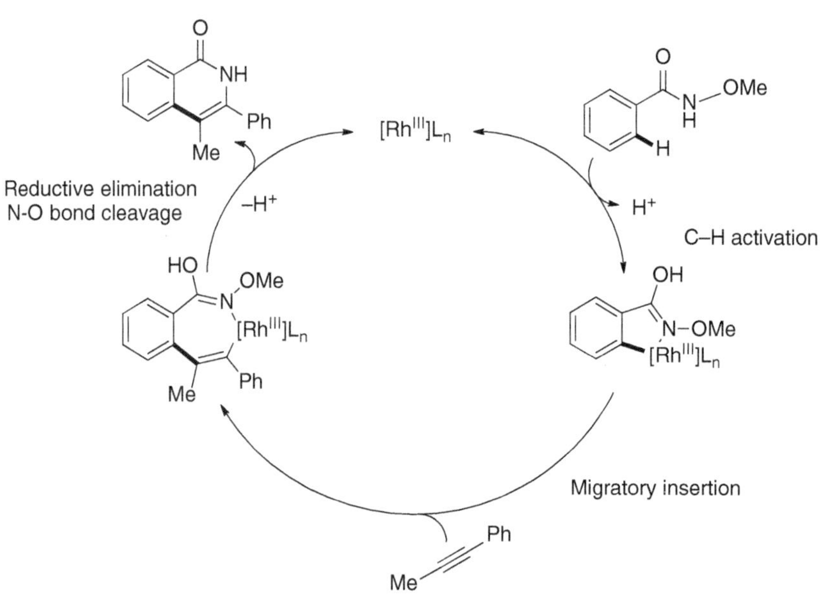

Scheme 4.32 General mechanism.

Finally, it should be stressed that one of the strengths of the DG^{ox} internal oxidant strategy is to limit reactivity to a single C–H oxidation event. Once the first C–H oxidative functionalization occurred, the DG no longer has the necessary oxidizing driving force to produce a second undesired C–H functionalization event.

In 2014, Glorius and colleagues reported a highly selective direct synthesis of unsymmetrical bis-isoquinolones from simple and accessible starting materials using RhCp* catalysis and a similar DG^{ox} strategy (Schemes 4.33 and 4.34) [37]. This reaction proceeded at 40 °C in air with good reactivity and excellent selectivity.

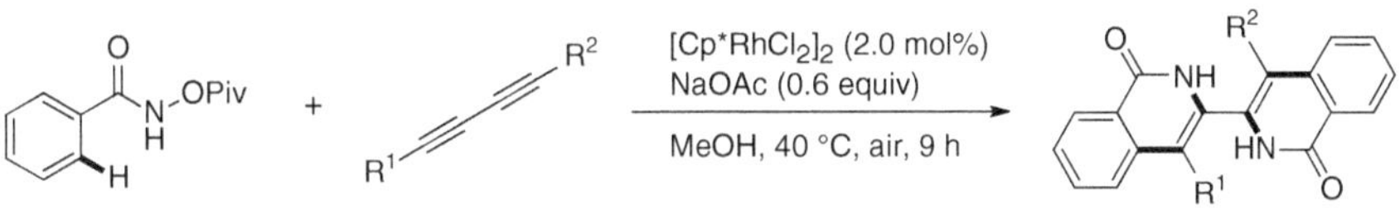

Scheme 4.33 Glorius's bis-isoquinolone synthesis. Source: Modified from Yu et al. [37].

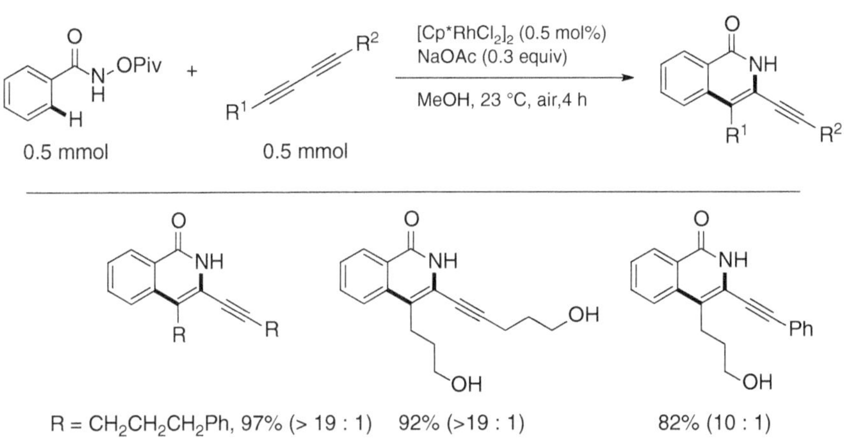

Scheme 4.34 Glorius's isoquinolone synthesis: very low reaction temperature toward high functional group tolerance and positional selectivity. Source: Yu et al. [37]. John Wiley & Sons.

In 2017, the same group developed a Cp*Co^{III}-catalyzed intramolecular C–H activation approach to access isoquinolones and pyridones [38]. Following a Cp*Co(III) catalytic cycle, this C–H activation approach employed alkyne-containing hydroxamic esters as a DG and internal oxidant to achieve these building blocks, wherein the alkyne coupling partner is also comprised (Scheme 4.35).

Having the alkyne coupling partner inside DG^{ox} allowed an absolutely regioselective intermediate by unsymmetrical terminal alkyne insertion. The proposed mechanism is depicted in Scheme 4.36, wherein reductive C—N bond formation and N—O bond cleavage would be closely connected.

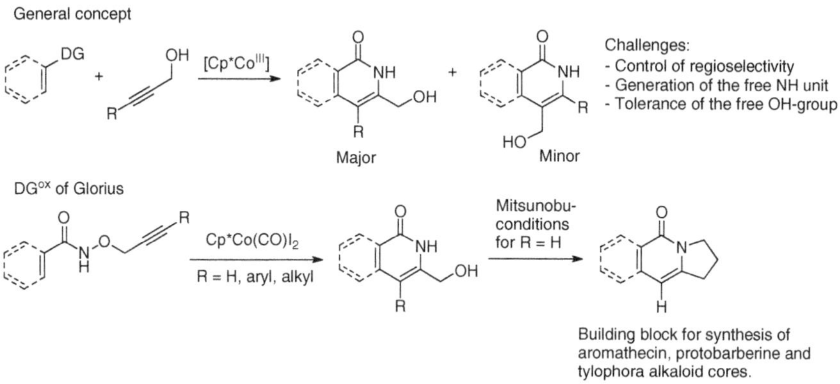

Scheme 4.35 The coupling partner inside the DG^{ox} by Glorius and coauthors.

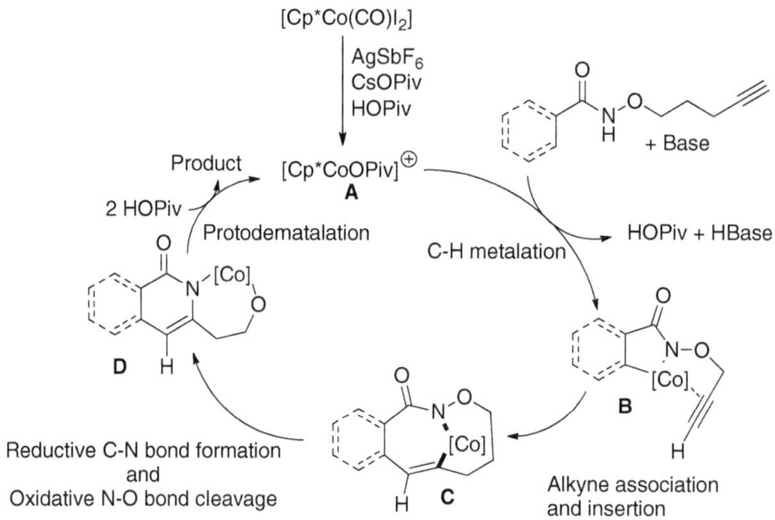

Scheme 4.36 Postulated mechanism.

In 2017, the Glorius group developed a C—H bond amination reaction by using a strained bicyclic olefin as cocatalyst proceeding by intramolecular amide transfer (Scheme 4.37) [39]. Using a Cp*Rh(III) metal catalyst, this C–H activation approach would observe a key Rh(V) intermediate that plays a crucial role in the catalytic cycle. The proposed mechanism for the reaction is shown in Scheme 4.38. Here again, it is interesting to note that N—O bond cleavage is proposed to occur only after C–H activation and olefin insertion steps. Should the reader be surprised by this, the reader should be reminded that N—O bonds are generally not as brittle as for example peroxides, and that they are quite easily handled, purified, and stored.

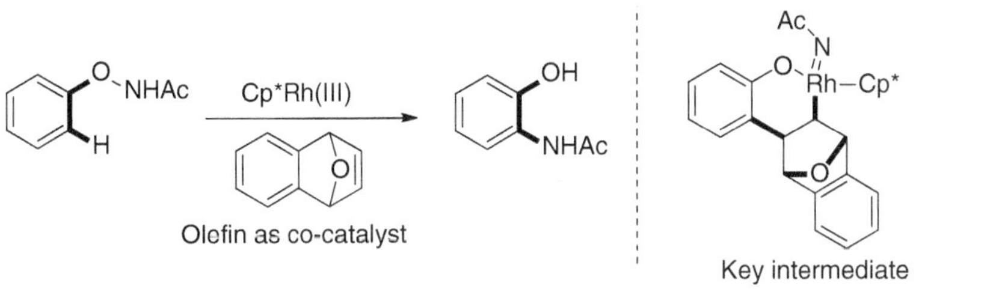

Scheme 4.37 Glorius's intramolecular RhCp*-catalyzed C–H amidation reaction. Source: Modified from Wang et al. [39].

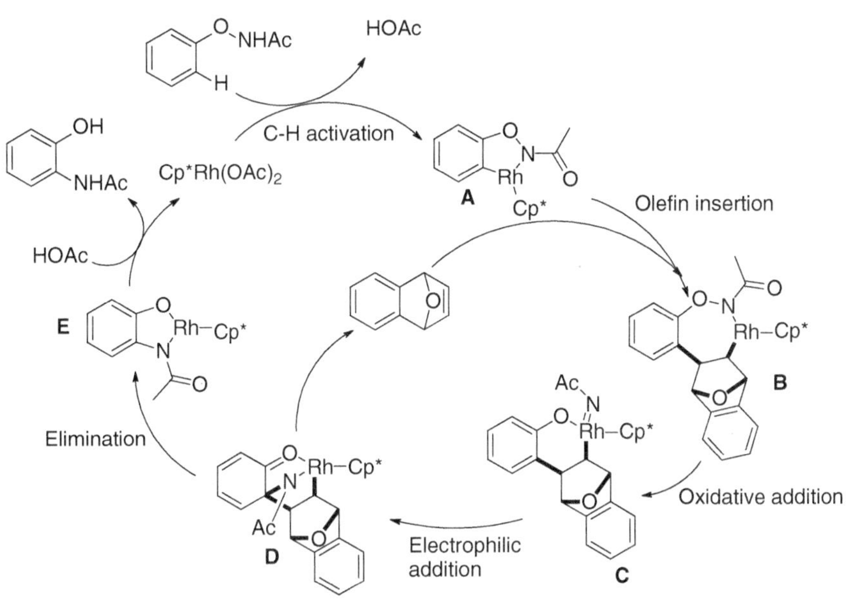

Scheme 4.38 Proposed mechanism for Glorius's intramolecular RhCp* catalyzed C–H amidation reaction.

4.5 Use of O_2 as an Oxidant

It is known since at least 1789 that this element in air called oxygen can oxidize substances, as opposed to a now-abandoned anterior theory that had the oxidizing element (the so-called phlogiston) contained within the substance itself, and released upon oxidation [40]. This understanding of aerobic oxidation is considered by many as the beginning of Chemistry as a rational science. Controlling this aerobic oxidation process is still at the heart of chemical research, in particular in the field of cross-dehydrogenative couplings. Some recent reviews have already covered progress in the field of O_2-mediated and aerobic cross-dehydrogenative coupling reactions [41, 42]. We shall therefore only go through a few most recent selected examples. In comparison to all the oxidizing strategies discussed above, O_2 is quite attractive because it is cheap, abundant, nontoxic, and atom economical. Moreover, its terminal reduction byproduct is H_2O, water. However, implementing O_2 as terminal oxidant entails a few drawbacks. Firstly, there is the risk of ignition of the solvent, especially when working on a preparative scale. This is not helped by the relative stability of O_2 and its triplet state, which reduce its reactivity toward organic (usually singlet) molecules, requiring in turn elevated temperatures. The lack of reactivity of O_2 toward many organic scaffolds implies the use of catalytic strategies capable of activating the oxidant. Copper salts have often been utilized for this purpose, as copper-oxo-species can form and help to activate the O—O bond. One can also rely on an organic solvent with pronounced radical ability, as we have done recently with cumene and similar systems. The advantage of using cumene, of which the tertiary cumyl radical intermediate is quite stabilized, is that it readily reacts with O_2 to form cumyl hydroperoxide, a more reactive species than O_2. The idea of utilizing a large excess of "facile" reductant (cumene solvent) to intercept some of the more reactive peroxide intermediates of O_2 is sometimes referred to, in other works [42], as the "Mukaiyama trick" [43]. We take this opportunity to recommend the cumene/O_2 system for any aerobic oxidations of organic compounds, with or without metal catalysis [44].

In 2015, we reported what is probably one of the simplest cross-dehydrogenative couplings there is, the O_2-mediated dehydrogenative phenothiazination of phenols (Scheme 4.39) [45]. This reaction is exceptional because it represents a rare case of cross-dehydrogenative amination method under metal free-conditions, and also because it operates under almost any oxidizing conditions [46]. The secret of the efficiency and specificity of this oxidative "click-like" reaction lies in the persistency of the N-centered radical intermediate.

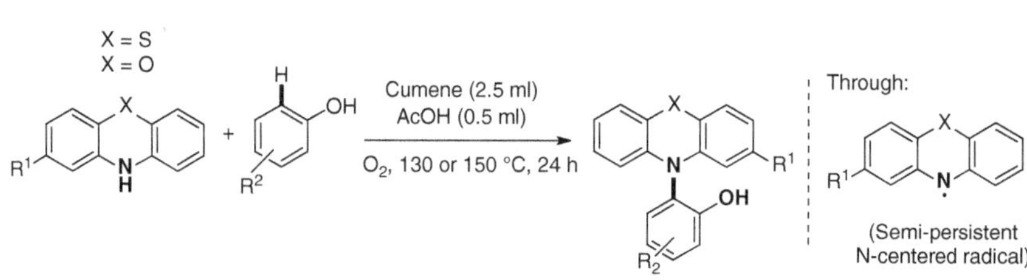

Scheme 4.39 O_2-mediated dehydrogenative phenothiazination of phenols. Source: Modified from Louillat-Habermeyer et al. [45].

Recently, in 2017, Lumb and coworkers have developed a bioinspired solution for coupling unprotected phenols with amines. They reported a simple one-pot cascade reaction using molecular oxygen in the presence of a copper catalyst and N,N′-Dibenzylethylenediamine (DBED) ligand to catalyze the oxidation of phenols to *ortho*-quinone intermediates (Scheme 4.40) [47]. The amine is then exposed to the quinone intermediate, which, in turn, triggers a condensation reaction followed by a radical isomerization to form the coupling product. The widespread applicability of this methodology was further demonstrated through the efficient synthesis of new C—N bonds in complex heterocycles such as benzoxazinone and immunosuppressive benzoxazoles.

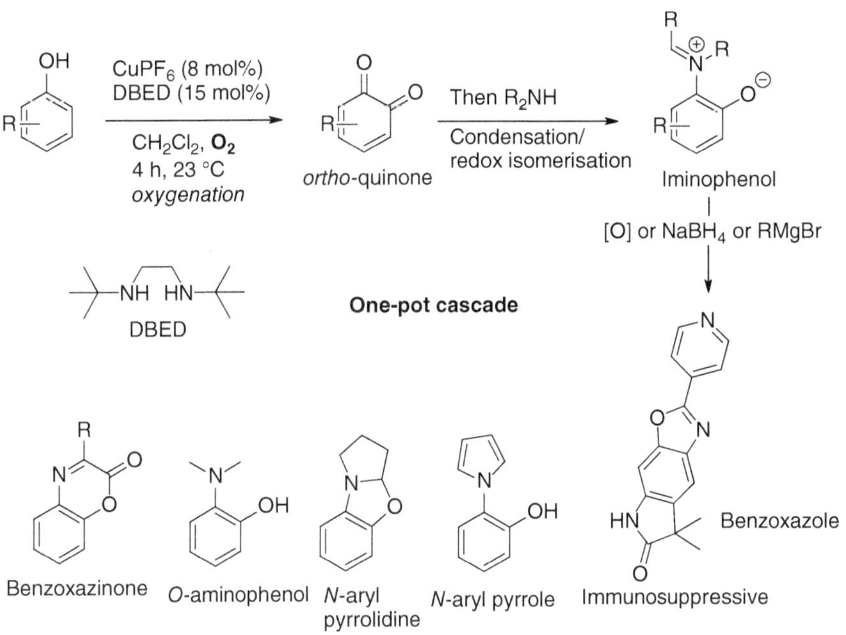

Scheme 4.40 O_2-mediated copper-catalyzed quinone intermediate generation from Lumb. Source: Modified from Esguerra and Lumb [47].

To further explore the potential of O_2 as a terminal oxidant, White and group members developed an aerobic linear allylic amination method to synthesize new C—N bonds, cocatalyzed by a Co(II) salt and BQ (Scheme 4.41) [48].

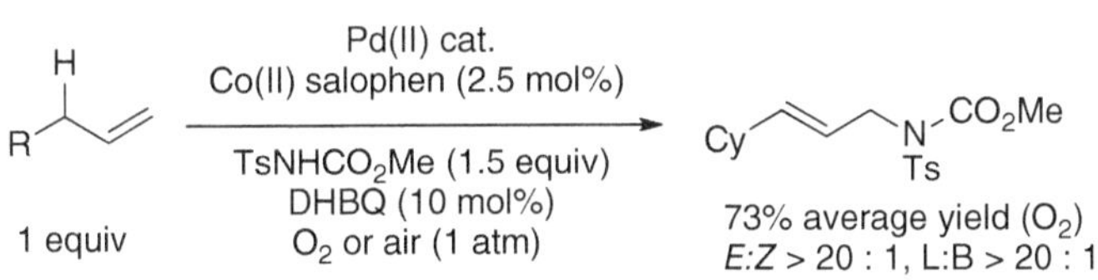

Scheme 4.41 White's O_2-mediated dehydrogenative allylic amination method. Source: Modified from Pattillo et al. [48].

As shown in Scheme 4.42, the reaction proceeds via redox-relay catalysis, in which O_2 triggers the oxidation of the cocatalyst, which then mediates the generation of quinone in the presence of Pd(0). This bulky quinone helps to increase the reaction yield by having a reduced ability to coordinate with the Pd(II) catalyst. The nitrogen nucleophile is activated by a Pd(II)-bis-sulfoxide complex, where bis-sulfoxide acts as a weakly coordinating, yet oxidatively stable ligand. The amine product is generated in higher yield (73%) as compared to the reaction where BQ is the terminal oxidant (65%), thereby suggesting an improved reaction efficiency and catalytic system with the help of a sustainable and environment-friendly oxidant.

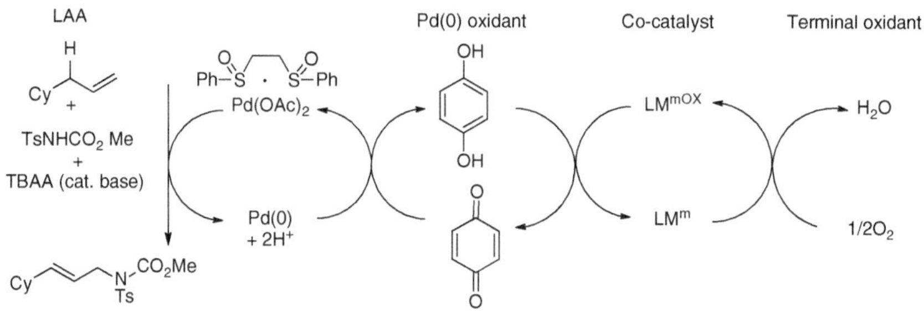

Scheme 4.42 Proposed mechanism for White's O_2 mediated dehydrogenative allylic amination method.

Recently, copper-catalyzed aerobic chemical oxidative functionalization has been developed to synthesize biologically valuable heterocycles like oxazoles [49]. The oxidative synthesis of 2,5,5-trisubstituted oxazoles was for example achieved by Chiba and coworkers, although utilizing prefunctionalized *N*-alkylamidine starting materials (Scheme 4.43a) [50]. Further modification to this synthesis was provided by Jiao and his group as they reported an oxidative dehydrogenative annulation reaction mediated by copper and oxygen [51]. Although the use of simple starting materials like amines and aldehydes is highly interesting, a large excess of copper salt was needed to afford the product (Scheme 4.43b).

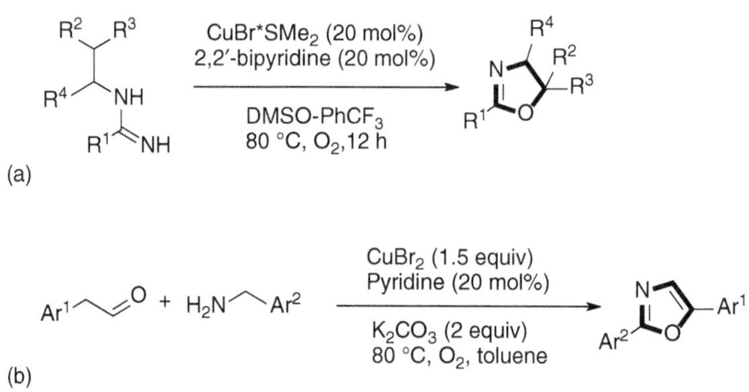

Scheme 4.43 O_2-mediated oxazole synthesis by Chiba and Jiao. (a) Copper-catalyzed intramolecular cyclization of acyclic percursors. Source: Modified from Wang et al. [50]. (b) Copper-mediated aerobic oxidative dehydrogenative annulation.

More recently, an interesting aerobic C—H bond functionalization approach was developed in 2018 (Scheme 4.44), in which, surprisingly, the oxazole's oxygen atom comes from the aerobic dioxygen [52]. 2,5,5-Substituted oxazoles were synthesized by using readily available reactants, like amines and alkynes, in the presence of a Cu catalyst and dioxygen activation. C—H bond functionalization by using O_2 as the sole oxidant achieved the transformation in a highly efficient and practical manner with large substrate scope.

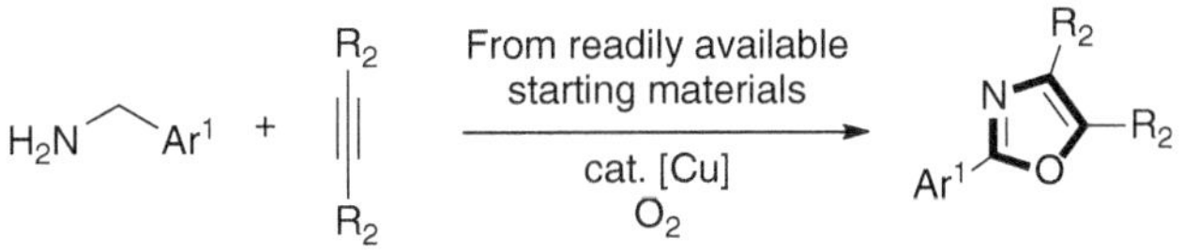

Scheme 4.44 O_2-mediated oxazole synthesis from amines and alkynes from Jiao.

The reaction is thought to proceed by the initial formation of an enamine intermediate through hydroamination of the alkyne-building block (Scheme 4.45). The enamine intermediate is then oxidized by Cu(II) and O_2 under basic conditions to form a metallodioxygen intermediate.

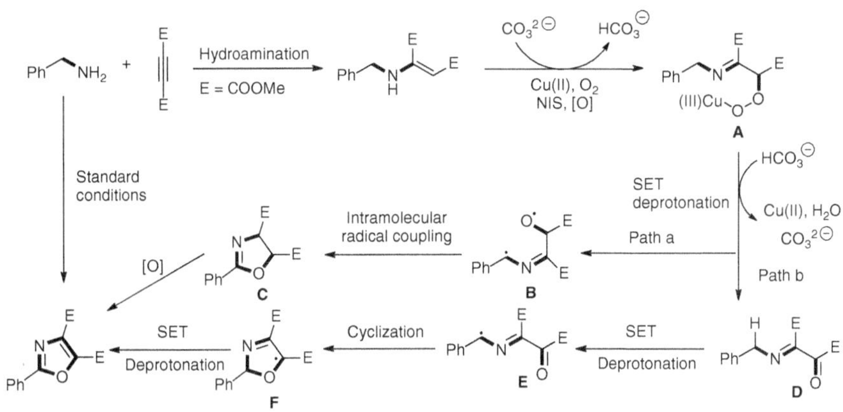

Scheme 4.45 Proposed mechanism for the O_2 mediated oxazole synthesis from amines and alkynes from Jiao.

The synthesis of α-keto amide motifs via oxidative functionalization has also received considerable interest [53–55] due to their importance in both natural products and functional materials [56]. A Cu-catalyzed aerobic coupling of electron-deficient anilines with aryl acetaldehydes was reported by Jiao and coworkers, wherein some of the oxygen atoms are also incorporated into the product (Scheme 4.46).

Employing a CuBr catalytic system triggered by O_2 as sole oxidant and following two Csp^3—H bond cleavages and one Csp^2—H bond cleavage, the corresponding carbonylated products were obtained in moderate-to-very-good yields. The protocol was utilized to synthesize an orexin receptor antagonist. The proposed mechanism is shown in Scheme 4.47. Initially, an imine intermediate **A** is formed by the condensation reaction of an aniline onto the aldehyde-building block. This is followed by a single-electron oxidation of **A** by O_2 to form a peroxy radical intermediate **B**, which then interacts with the oxidized copper Cu(II) species to form a Cu(III)

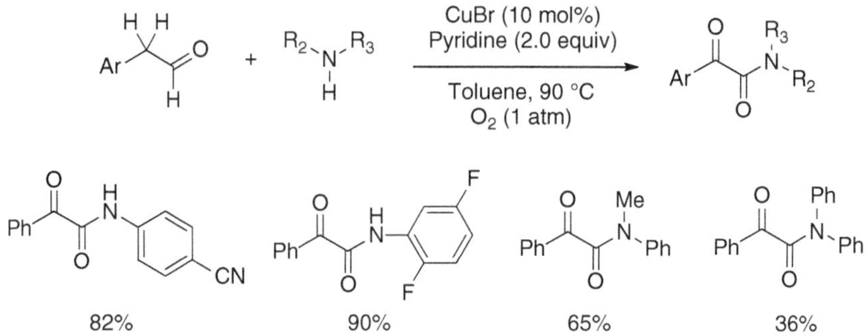

Scheme 4.46 O_2-mediated α-keto amide synthesis of Jiao.

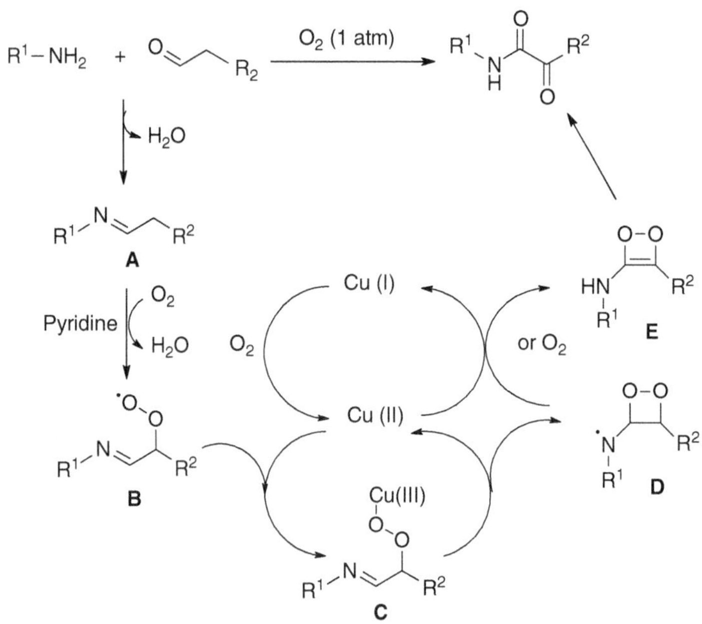

Scheme 4.47 Proposed mechanism for the O_2 mediated α-keto amide synthesis of Jiao.

metallointermediate **C**. The subsequent intramolecular addition onto the imine and homolysis of the N—Cu bond generates radical intermediate **D**. Further oxidation of **D** by Cu or O_2 forms intermediate **E**, which undergoes fragmentation to afford the final α-keto amide product.

4.6 Dehydrogenative Couplings with No Oxidant at All

In special cases, it is sometimes possible to perform cross-dehydrogenative couplings without a chemical oxidant. In those cases, H_2 gas generally evolves as a sole byproduct. This scenario can be viewed as having great sustainable character (smallest possible atomic waste), but is however rarely thermodynamically feasible. The few known cases (aside from electro-oxidative synthesis) usually operate at very high reaction temperature (100–200 °C) and are usually limited to intramolecular ring-closing processes. These usually employ precious metal catalysts, such as platinum salts, and the substrates have often hydridic character. Of course, this does not take into account recent developments in electro-oxidative dehydrogenative methods, which is not the topic of this chapter [57].

4.7 Conclusion

In conclusion, the field of C—H bond functionalization, and in particular of cross-dehydrogenative coupling reactions, has seen the use of very diverse classes of oxidants. In this context, apprehending reactivity through the view point of the oxidant holds the promise of many surprising discoveries to come. Indeed, because the nature of the oxidant has a decisive impact on the observed reactivity in C—H bond functionalization systems, it follows logically that actively designing the oxidant of the reaction should unlock completely unknown reactivity. Thus, we expect that many new reactivity concepts will arise in the near future through innovative oxidant design.

References

1 C—H bond functionalization: (a) Leitch, J.A. and Frost, C.G. (2017). *Chem. Soc. Rev.* 46: 7145. (b) Xue, X., Ji, P., Zhou, B., and Cheng, J. (2017). *Chem. Rev.* 117: 8622. (c) Shang, R., Ilies, L., and Nakamura, E. (2017). *Chem. Rev.* 117: 9086. (d) Ma, W., Gandeepan, P., Li, J., and Ackermann, L. (2017). *Org. Chem. Front.* 4: 1435. (e) Gensch, T., Hopkinson, M.N., Glorius, F., and Wencel-Delord, J. (2016). *Chem. Soc. Rev.* 45: 2900. (f) Li, S., Qin, L., and Dong, L. (2016). *Org. Biomol. Chem.* 14: 4554. (g) Sandtorv, A.H. (2015). *Adv. Synth. Catal.* 357: 2403. (h) Topczewski, J. and Sanford, M.S. (2015). *Chem. Sci.* 6: 70. (i) Thirunavukkarasu, V.S., Kozhushkov, S.I., and Ackermann, L. (2014). *Chem. Commun.* 50: 29. (j) Wencel-Delord, J. and Glorius, F. (2013). *Nat. Chem.* 5: 369. (k) Neufeldt, S.R. and Sanford, M.S. (2012). *Acc. Chem. Res.* 45: 936. (l) Arockiam, P.B., Bruneau, C., and Dixneuf, P.H. (2012). *Chem. Rev.* 112: 5879. (m) Song, G., Wang, F., and Li, X. (2012). *Chem. Soc. Rev.* 41: 3651. (n) Colby, D.A., Tsai, A.S., Bergman, R.G., and Ellman, J.A. (2012). *Acc. Chem. Res.* 45: 814. (o) Engle, K.M., Mei, T.-S., Wasa, M., and Yu, J.-Q. (2012). *Acc. Chem. Res.* 45: 788.

2 CDC reactions: (a) Guo, S.-R., Kumar, P.S., and Yang, M. (2017). *Adv. Synth. Catal.* 359: 2. (b) Varun, B.V., Dhineshkumar, J., Bettadapur, K.R. et al. (2017). *Tetrahedron Lett.* 58: 803. (c) Henry, M., Mostafa, M.A.B., and Sutherland, A. (2017). *Synthesis* 49: 4586. (d) Lakshman, M.K. and Vuram, P.K. (2017). *Chem. Sci.* 8: 5845. (e) Lv, L. and Li, Z. (2016). *Top. Curr. Chem.* 374: 38. (f) Chen, T., Zhang, J.-S., and Han, L.-B. (2016). *Dalton Trans.* 45: 1843. (g) Batra, A., Singh,

P., and Singh, K.N. (2016). *Eur. J. Org. Chem.* 2016: 4927. (h) Miao, J. and Ge, H. (2015). *Eur. J. Org. Chem.* 2015: 7859. (i) Krylov, L.B., VilQ, V.A., and TerentQev, A.O. (2015). *Beilstein J. Org. Chem.* 11: 92. (j) Girard, S.A., Knauber, T., and Li, C.-J. (2014). *Angew. Chem. Int. Ed.* 53: 74.

3 A selected reference on KIE determination and interpretation:Simmons, E.M. and Hartwig, J.H. (2012). *Angew. Chem. Int. Ed.* 51: 3066.

4 Some examples from our laboratory, in which high H/D scrambling occurs under catalytic conditions: (a) Louillat, M.-L. and Patureau, F.W. (2013). *Org. Lett.* 15: 164. (b) Louillat, M.-L., Biafora, A., Legros, F., and Patureau, F.W. (2014). *Angew. Chem. Int. Ed.* 53: 3505. c) Jones, A.W., Louillat-Habermeyer, M.-L., and Patureau, F.W. (2015). *Adv. Synth. Catal.* 357: 945. (d) Jones, A.W., Rank, C.K., Becker, Y. et al. (2018). *Chem. Eur. J.* 24: 15178.

5 Fujiwara, Y., Moritani, I., Danno, S. et al. (1969). *J. Am. Chem. Soc.* 91: 7166.

6 Guo, X.-K., Zhang, L.-B., Wei, D.-H., and Niu, J.-L. (2015). *Chem. Sci.* 6: 7059.

7 Tan, G.-Y., He, S., Huang, X.-L. et al. (2016). *Angew. Chem. Int. Ed.* 55: 10414.

8 Zhang, L.-B., Zhang, S.-K., Wei, D.-H. et al. (2016). *Org. Lett.* 18: 1318.

9 Zeng, L., Tang, S., Wang, D. et al. 2017 *Org. Lett.* 19: 2170.

10 Karthikeyan, J., Haridharan, R., Cheng, C.-H 2012 *Angew. Chem. Int. Ed.* 51: 12343.

11 Lerchen, A., Knecht, T., Koy, M. et al. (2018). *Angew. Chem. Int. Ed.* 57: 15248.

12 Nelson, T. and Blakey, S.B. (2018). *Angew. Chem. Int. Ed.* 57: 14911.

13 Shi, Y., Zhang, L.-Q., Lan, J.-B. et al. (2018). *Angew. Chem. Int. Ed.* 57: 9108.

14 Zhang, S., Song, F.-J., Zhao, D.-B., and You, J.-S. (2013). *Chem. Commun.* 49: 4558.

15 Liao, G., Yin, X.-S., Chen, K. et al. (2016). *Nat. Comm.* 7: 12901.

16 Chen, Y.-Q., Wang, Z., Wu, Y.-W. et al. (2018). *J. Am. Chem. Soc.* 140: 17884.

17 Chen, Y.-R. and Duan, W.-L. (2013). *J. Am. Chem. Soc.* 135: 16754.

18 Zhao, D., Xu, P., and Ritter, T. (2019). *Chem* 5: 97.

19 Liu, M.-Y., Yang, P.-S., Karunananda, M.K. et al. (2018). *J. Am. Chem. Soc.* 140: 5805.

20 Yi, H., Tang, Z.-L., Bian, C.-L. et al. (2017). *Chem. Commun.* 53: 8984.

21 See also:Zhao, Y., Huang, B., Yang, C. et al. (2017). *ACS Catal.* 7: 2446.

22 Li, K., Wu, Q., Lan, J., and You, J. (2015). *Nat. Comm.* 6: 8404.

23 Yuan, J.-W., Ma, X., Yi, H. et al. (2014). *Chem. Commun.* 50: 14386.

24 Liu, D., Liu, C., Li, H., and Lei, A.-W. (2013). *Angew. Chem. Int. Ed.* 52: 4453.

25 Wen, J.-W., Tang, S., Zhang, F. et al. (2017). *Org. Lett.* 19: 94.

26 Ouyang, X.-H., Song, R.-J., Liu, B., and Li, J.-H. (2016). *Chem. Commun.* 52: 2573.

27 Tang, S., Liu, Y.-H., Gao, X.-L. et al. (2018). *J. Am. Chem. Soc.* 140: 6006.

28 Yuan, C.-X., Liang, Y., Hernandez, T. et al. (2013). *Nature* 499: 192.

29 (a) Hock, H. and Lang, S. (1944). *Ber. Dtsch. Chem. Ges.* 77: 257. (b) "Phenol":Weber, M., Weber, M., and Kleine-Boymann, M. (2004). *Ullmann's Encyclopedia of Industrial Chemistry*. Weinheim: Wiley-VCH.

30 Tang, S., Wu, X.-D., Liao, W.-Q. et al. (2014). *Org. Lett.* 16: 3584.

31 Chua, C.-K., Sofer, Z., and Pumera, M. (2016). *Angew. Chem. Int. Ed.* 55: 10751.

32 Rivera, C.A.M., Floreancig, P.E., and Liu, P. (2017). *J. Am. Chem. Soc.* 139: 17935.

33 Song, C.-L., Dong, X., Yi, H. et al. (2018). *ACS Catal.* 8: 2195.

34 Zhu, S.-Q., Liu, Y.-L., Li, H. et al. (2018). *J. Am. Chem. Soc.* 140: 11613.

35 Patureau, F.W. and Glorius, F. (2011). *Angew. Chem. Int. Ed.* 50: 1977.

36 Guimond, N., Gouliaras, C., and Fagnou, K. (2010). *J. Am. Chem. Soc.* 132: 6908.

37 Yu, D.-G., de Azambuja, F., Gensch, T. et al. (2014). *Angew. Chem. Int. Ed.* 53: 9650.

38 Lerchen, A., Knecht, T., Koy, M. et al. (2017). *Chem. Eur. J.* 23: 12149.

39 Wang, X., Gensch, T., Lerchen, A. et al. (2017). *J. Am. Chem. Soc.* 139: 6506.

40 Lavoisier, A. (1789). *Traité élementaire de chimie.* Paris: Chez Cuchet.

41 Gulzar, N., Schweitzer-Chaput, B., and Klussmann, M. (2014). *Catal. Sci. Technol.* 4: 2778.

42 Sterckx, H., Morel, B., and Maes, B.U.W. (2019). *Angew. Chem. Int. Ed.* 58: 7946.

43 Markó, I.E., Giles, P.R., Tsukazaki, M. et al. (1996). *Science* 274: 2044.

44 Malekafzali, A., Malinovska, K., and Patureau, F.W. (2017). *New J. Chem.* 41: 6981.

45 Louillat-Habermeyer, M.-L., Jin, R., and Patureau, F.W. (2015). *Angew. Chem. Int. Ed.* 54: 4102.

46 Goswami, M., Konkel, A., Rahimi, M. et al. (2018). *Chem. Eur. J.* 24: 11936.

47 Esguerra, K.V.N. and Lumb, J.-P. (2017). *ACS Catal.* 7: 3477.

48 Pattillo, C.C., Strambeanu, I.I., Calleja, P. et al. (2016). *J. Am. Chem. Soc.* 138: 1265.

49 Tang, X., Wu, W., Zeng, W., and Jiang, H. (2018). *Acc. Chem. Res.* 51: 1092.

50 Wang, Y.-F., Chen, H., Zhu, X., and Chiba, S. (2012). *J. Am. Chem. Soc.* 134: 11980.

51 Xu, Z., Zhang, C., and Jiao, N. (2012). *Angew. Chem. Int. Ed.* 51: 11367.

52 Pan, J., Li, X., Qiu, X. et al. (2018). *Org. Lett.* 20: 2762.

53 Kumar, D., Vemula, S.R., and Cook, G.R. (2016). *ACS Catal.* 6: 4920.

54 Zhang, C. and Jiao, N. (2010). *J. Am. Chem. Soc.* 132: 28.

55 Zhang, C., Xu, Z., Zhang, L., and Jiao, N. (2011). *Angew. Chem. Int. Ed.* 50: 11088.

56 Liang, Y.-F. and Jiao, N. (2017). *Acc. Chem. Res.* 50: 1640.

57 Tang, S., Zeng, L., and Lei, A. (2018). *J. Am. Chem. Soc.* 140: 13128.

5

Electrochemical Reductive Transformations

Mahito Atobe[1] *and Toshio Fuchigami*[2]

[1] *Yokohama National University, Department of Chemistry and Life Science, 79-5 Tokiwadai, Hodogaya-ku, 240-8501, Yokohama, Japan*
[2] *Tokyo Institute of Technology, Department of Electronic Chemistry, 4259, Nagatsuta, Midori-ku, 226-8502, Yokohama, Japan*

5.1 General Characteristics of Electrochemical Reactions

Electrochemical reactions have following features, many of which cannot be achieved by other methods:

(1) Electrochemical reactions are typical heterogeneous reactions, and the reaction fields are specific since oxidation and reduction take place separately at different field.
(2) Umpolung (polarity inversion) is readily performed without the use of any reagents.
(3) The selectivity of electrochemical reactions is often different from that of ordinary organic reactions.
(4) Since electron is used as a reagent, the use of hazardous reagents can be avoided, i.e. electrochemical reactions are low-emission processes.
(5) Electrochemical reactions proceed under mild conditions such as room temperature and normal pressure.
(6) Electrochemical reactions can be started or stopped readily by on-off switch of the power supply, i.e. electrochemical reaction control is easy.
(7) The scale effect is generally small.

Electrochemical reactions are redox reactions through electron transfer between a substrate molecule and an electrode. The main reaction field is an electrode surface (solid–liquid interface) or near the electrode surface, and the surface has an extremely large electrical field, which is quite different from ordinary redox reactions on heterogeneous catalysts. Electrochemical reactions therefore take place in highly unique fields [1–7].

Using a chemical redox reaction as an example, the difference between an ordinary chemical reaction and an electrode reaction will be explained in detail. Figure 5.1a shows the reduction of substrate **B** by reducing reagent **A**. When an

Organic Redox Chemistry: Chemical, Photochemical and Electrochemical Syntheses, First Edition. Edited by Jun-Ichi Yoshida and Frédéric W. Patureau.

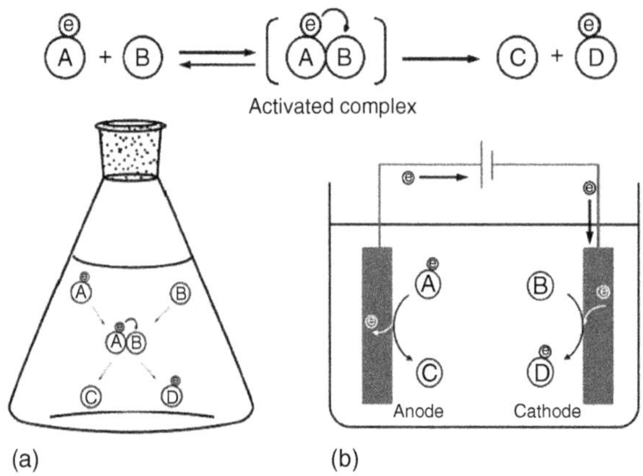

Figure 5.1 Difference between (a) chemical reaction and (b) electrochemical reaction.

activated complex is formed or **A** and **B** approach each other closely enough for electron transfer, electron transfer from **A** to **B** takes place. Next, reductant **A** is transformed to oxidized **C** while substrate **B** is transformed to reduced product **D**. The former electron transfer through the complex is called inner-sphere electron transfer or bonded electron transfer, while the latter one without any complex is called outer-sphere electron transfer or nonbonded electron transfer. Thus, both oxidation and reduction occur at the same place in the case of an ordinary chemical reaction, while oxidation and reduction occur at different places, such as an anode and cathode, respectively, in the case of an electrochemical reaction (Figure 5.1b). In other words, electrode electron transfer takes place separately due to the existence of the electrode interfaces, which is a significant characteristic feature that is different from ordinary chemical reactions.

5.2 Mechanism of Organic Electrochemical Reductions

In the case of organic electrochemical reactions, electron transfer generally does not take place cleanly, and pre and/or post reactions usually accompany the transfer. An organic electrochemical reaction consists of an electron transfer step as well as several chemical and physical steps.

In this section, we consider mechanism of an organic electrochemical reduction. Figure 5.2 illustrates each elementary reaction step of substrate S forming reduction product P via intermediate I. In step (a), mass transport of S from the bulk of an electrolytic solution to the cathode surface takes place by diffusion or migration. In step (b), pre-reactions such as desolvation, dissociation, and/or protonation of S take place to form intermediate I. However, such pre-reactions do not always take place. In step (c), the intermediate I adsorbs on the surface of the cathode to form intermediate I_{ad}. In step (d), electron transfer from the cathode to I_{ad} generates intermediate I'_{ad}. In step (e), desorption of I'_{ad} followed by subsequent chemical reaction (f) proceeds to provide a reduction product P that diffuses to the bulk of the electrolytic solution, and then the sequential reaction is completed. Intermediate I may

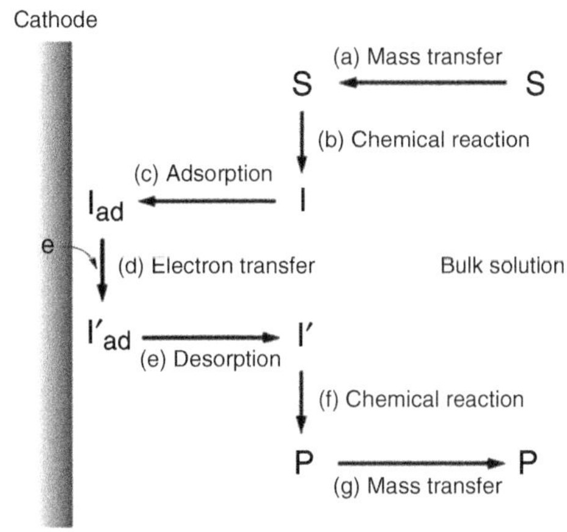

Figure 5.2 Elementary processes of electrode reaction.

undergo an electron transfer reaction without an adsorption step (c) and also the order of sequential steps (e) and (f) may be reversed, i.e. I'_{ad} undergoes subsequent reaction and then desorption of the resulting product occurs. Thus, the electrochemical reduction is typical in a heterogeneous system, and mass transfer steps (a) and (g) as well as adsorption and desorption steps (c) and (e) are involved, which is quite different from homogeneous reactions.

If the electron transfer process is abbreviated as E and the chemical process is abbreviated to C, and the organic electrochemical reduction can be shown using these abbreviations. For example, the electrochemical reduction illustrated in Figure 5.2 can be shown by the sequence CEC (pre-chemical reaction → electron transfer process → follow-up chemical reaction), and adsorption and desorption steps are usually disregarded unless they are important.

In order to clarify the electrochemical reduction mechanism, electrochemical analyses such as coulometry and voltammetry in addition to ordinary organic mechanistic studies are necessary to obtain information such as the number of electrons transferred, redox potentials, and detection of the reaction intermediates.

5.3 Characteristics of Organic Electrochemical Reductions

5.3.1 Umpolung

The polarity inversion of chemical bonding can be readily carried out in electrochemical reductions [6, 7]. In other words, electrophiles can be converted electroreductively to nucleophiles without use of any reagents. Such polarity inversion can be widely used for organic synthesis. For example, alkyl halides are inherently electrophilic reagents. In order to convert them to nucleophilic reagents, they have to be transformed to Grignard reagents or lithium compounds using Mg or Li metal. However, such polarity inversion (umpolung) can be readily achieved in one step by cathodic reduction of the alkyl halides as shown in Eq. (5.1). When the basic products are isolated after Grignard reaction, the work up is troublesome, and it may be rather difficult to separate the product from aqueous alkaline solution due to insoluble $Mg(OH)_2$. In sharp contrast, in the case of electrochemical reductions, the product isolation is rather easy and the severe waste problem does not occur because there is no use of metals. In addition, electrochemical reductions do not require easily flammable ethereal solvents, and alternative solvents like acetonitrile can be used as electrolytic solvents.

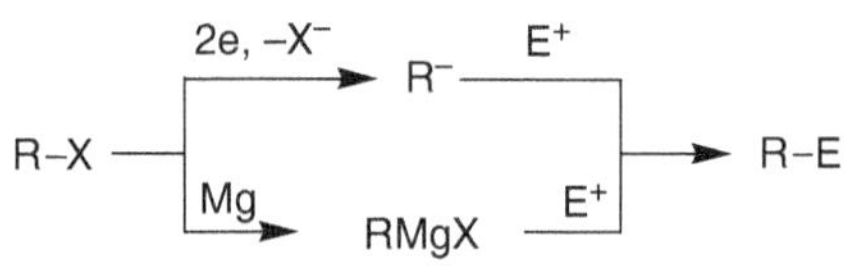

(5.1)

When alkyl substituents are introduced to phosphine, silicon, and stannum compounds, their corresponding chloro compounds (R_3ZCl, R_2PCl) and alkyl Grignard reagents are usually used. However, electrochemical reduction of these chloro compounds generates the anionic intermediates, which can readily react with alkyl halides (R′X) to provide the corresponding alkyl-substituted products (Eq. (5.2)).

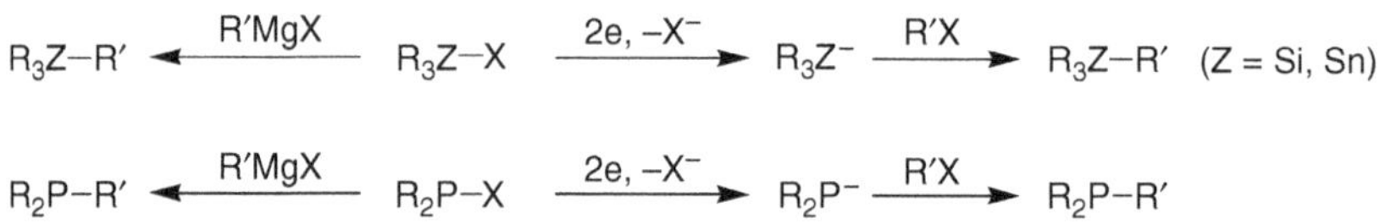

(5.2)

Furthermore, industrialized electroreductive hydrodimerization of acryronitrile shown in Eq. (5.3) is also a typical example of umpolung using cathodic reduction.

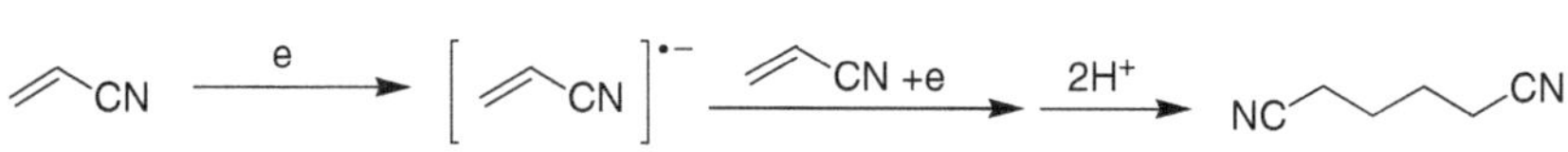

(5.3)

5.3.2 Selectivity

The selectivity of electrochemical reductions is rather complicated since it is controlled by many factors, such as electrode materials, applied potential, current density, electrolytic solvents, supporting electrolytes, electric field, adsorption orientation of substrate or intermediate species at the electrode surface and so on [4–8]. Heterogeneous electrochemical reductions therefore often exhibit different selectivity from ordinary homogeneous chemical reductions. In particular, when cathodically generated reactive species or intermediate reacts with reagents before it diffuses from the cathode into the solution, the stereo- and regioselectivities of the product are often quite different from those of ordinary chemical reductions. Typical examples of the selectivity of electrochemical reductions are described in the following sections.

5.3.2.1 Chemoselectivity

Chemoselectivity in ordinary chemical reductions is controlled by the choice of reagents, but it is quite difficult to achieve high chemoselectivity when multiple similar functional groups exist in a single molecule. On the other hand, chemoselectivity in electrochemical reductions can be achieved by the control of applied potential based on the difference between the redox potentials of functional groups. Since the applied potential is adjusted precisely using a potentiostat, high selectivity can be readily achieved. For instance, even though phenylimino and alkylimino groups exist in the same molecule, as shown in Eq. (5.4), the phenylimino group can be predominantly reduced by constant potential electrolysis. This is because the phenylimino group is more easily reduced than the alkylimino group as a result of the electron-withdrawing phenyl group. However, such selective reduction cannot be achieved by ordinary reducing reagents.

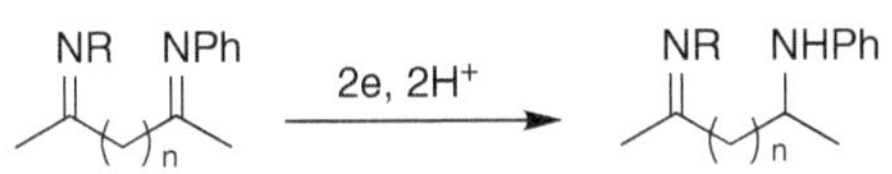

(5.4)

Similarly, in the case of the molecule with three halogen atoms as shown in Eq. (5.5), the halogen atom at the α-position to the carbonyl group is most easily reducible, thereby this halogen can be predominantly reduced at constant potential electrolysis.

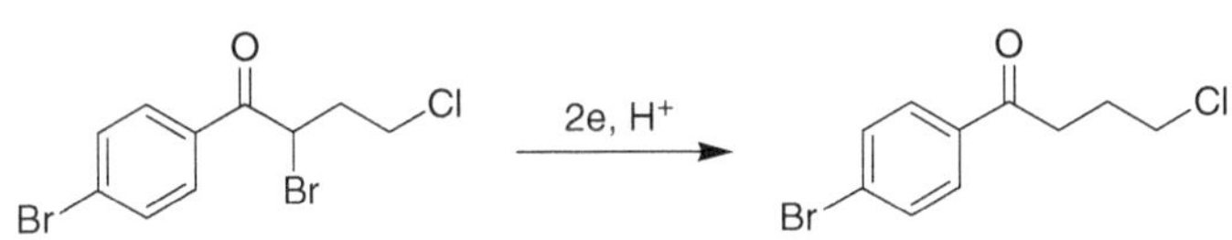

(5.5)

5.3.2.2 Reaction Pathway Selectivity

It is known that the reaction pathway is greatly changed depending on applied potentials. As shown in Eq. (5.6), one-electron and two-electron reduction products are obtained, respectively depending on the applied cathode potentials.

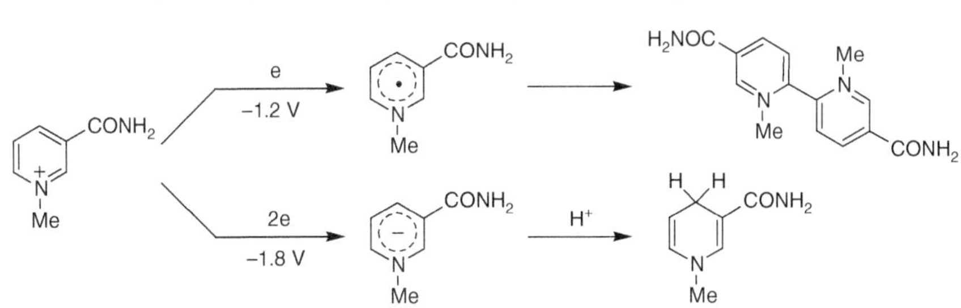

(5.6)

5.3.2.3 Regioselectivity

As shown in Figure 5.3, tetrachloropicolinic acid is regioselectively dechlorinated by cathodic reduction. The high regioselectivity is attributed to the controlled

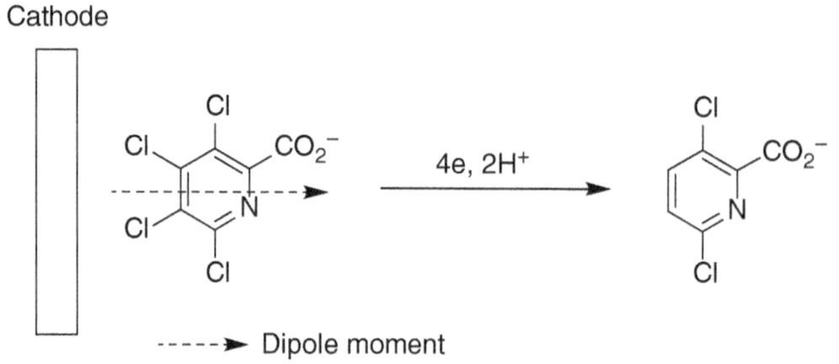

Figure 5.3 Regioselective cathodic dechlorination.

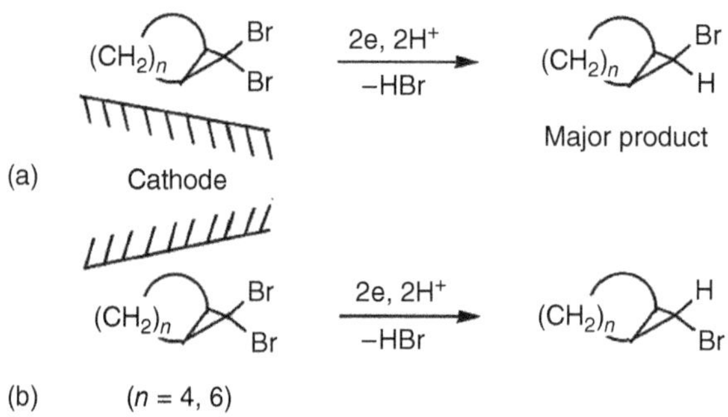

Figure 5.4 Stereoselective reduction based on orientation of substrate on the cathode.

orientation of the substrate at the cathode surface owing to the dipole moment of the molecule [9].

5.3.2.4 Stereoselectivity

Simple stereocontrol is achieved by steric hindrance between substrate and cathode. As we can easily see in Figure 5.4, the orientation of bicyclic *gem*-dibromocyclopropane at the cathode surface as shown in (a) is favored over that of the substrate shown in (b) owing to steric repulsion between the cathode and the 6- or 8-membered ring fused to cyclopropane. Therefore, the *exo*-bromide near the cathode is more easily reducible than the *endo*-bromide, and consequently the *endo*-bromide product is mainly formed [10].

5.3.2.5 Selectivity Depending on Electrode Materials

There are many examples of cathode materials that greatly affect product selectivity and stereoselectivity.

(a) *Product selectivity*: Acetone is reduced at a lead cathode in an acidic aqueous solution to give the corresponding alcohol, isopropyl alcohol, while the reduction with zinc and copper cathodes provides the corresponding alkane, propane [11].
(b) *Stereoselectivity*: The reduction of 2-methylcyclohexanone at zinc and copper cathodes in aqueous NaOH solution provides mainly the corresponding

trans-alcohol, while that with tin cathode provides *cis*-alcohol as a major product [12].

5.4 Electroauxiliaries

Electroauxiliaries are functional groups that facilitate electron transfer and control the reaction pathways of electrogenerated reactive species to provide desired products selectively. Although synthetic applications of electroauxiliaries in anodic oxidation are widespread, those for cathodic reduction are still rare [13].

5.4.1 Electroauxiliaries Based on Readily Electron-Transferable Functional Groups

It is known that carbonyl groups work as electroauxiliaries in cathodic processes. As shown in Eq. (5.7), the first electron transfer takes place at the easily electron accepting carbonyl group, and then adjacent carbon-heteroatom bond cleavage takes place selectively [14].

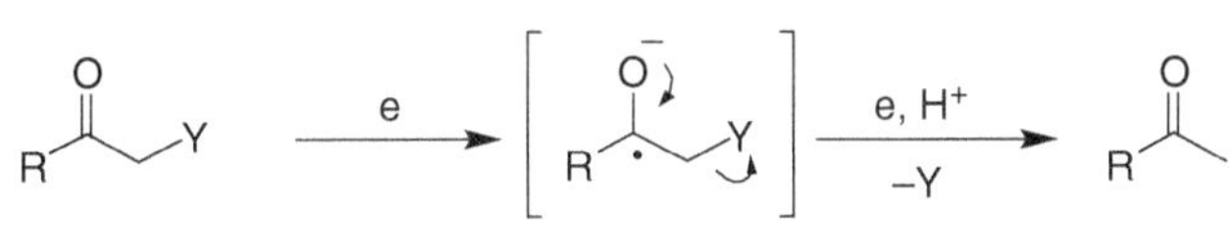

(Y = OR′, SR′, NR′R″) (5.7)

Based on this principle, selective C—O bond cleavage at benzylic and allylic positions can be easily achieved, as shown in Eq. (5.8).

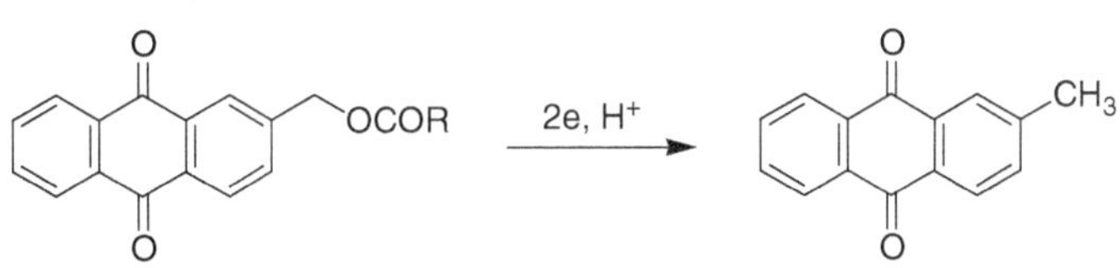

(5.8)

Although the C—F bond is not so easily cleaved by reduction owing to its larger bond energy compared to that of the C—H bond, fluorine attached to a benzylic position or α to carbonyl and imino groups is readily removed by cathodic reduction. This is also quite similar to the cases shown in Eqs. (5.7) and (5.8). As shown in Eq. (5.9), electron transfer first takes place at the aromatic ring or carbonyl and imino groups followed by β-elimination of the fluoride ion [15]. When the cathodic reaction shown in Eq. (5.9) is carried out in the absence of Me_3SiCl, further cathodic reduction takes place to result in successive elimination of fluorine atoms, providing complicated products.

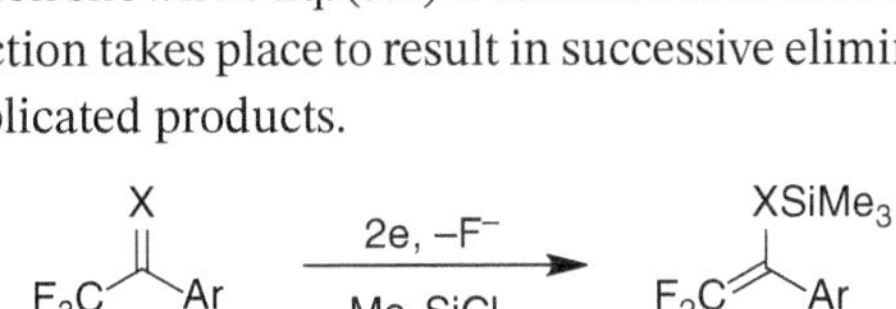

(X = O, NR) (5.9)

5.4.2 Electroauxiliaries Based on Coordination Effects

This effect is quite important from an electron transfer aspect. The electron transfer reaction in solution is generally facilitated by the stabilization of the resulting radical ion and ionic intermediates by the coordination of solvent molecules or counter ions in the solution. For instance, the reduction potential of the metal cation usually becomes more negative in solvents with large donor numbers because the positive nature of the metal cation is decreased by solvation in such solvents.

It is well known that the reduction potentials of ketones, imines, cyano, and nitro compounds shift in the positive direction by proton addition. The polarography of the reduction of nitro compounds has been intensively studied. When the pH value of the solution is made more acidic by 1 pH unit, the reduction potential becomes more positive by about 58 mV. The positive shift of the reduction potential is due to the protonation of unsaturated functional groups resulting in a positively charged form, as shown in Figure 5.5.

Furthermore, reduction potentials of ketones often shift to positive in the presence of Lewis acids. This is due to the coordination of the Lewis acid to the oxygen atom of the carbonyl group resulting in a decrease in the electron density of the carbonyl group or coordination of the Lewis acid to the anion radical intermediate generated by one-electron reduction of the carbonyl group resulting in stabilization of the intermediate, as shown in Figure 5.6.

It is also known that the second reduction potentials of diketones like anthraquinone and dinitrobenzenes shift greatly in the positive direction in the presence of alcohols. This is due to the formation of hydrogen bonds between dianion intermediates generated by two-electron reduction and the alcohols.

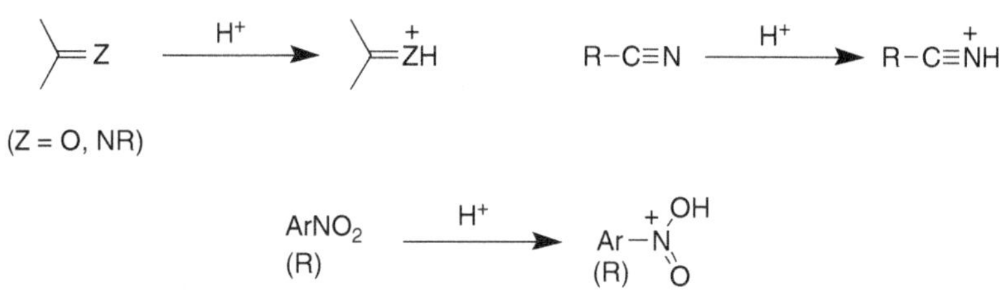

Figure 5.5 Anodic shift of reduction potential owing to protonation.

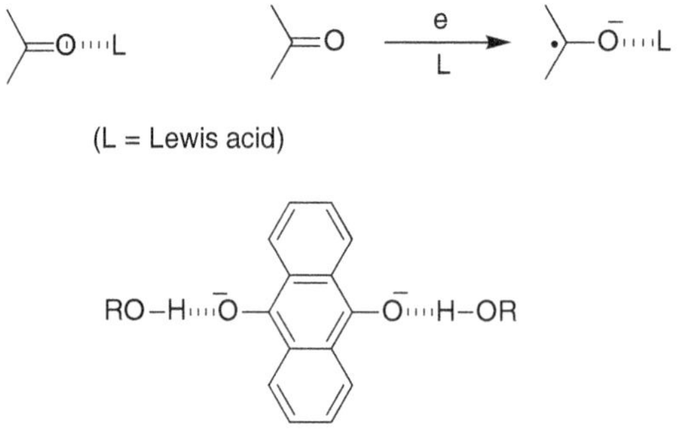

Figure 5.6 Anodic shift of reduction potential owing to coordination with Lewis acid and hydrogen bonding.

5.5 Reaction Pattern of Organic Electrochemical Reductions

Although cathodic reactions are limited to reduction, various follow-up chemical reactions often take place after organic electron transfer.

From a synthetic viewpoint, organic electrochemical reductions are classified into various reaction types, and examples of cathodic reduction are given here.

5.5.1 Transformation Type of Functional Group

$$\mathrm{R{-}NO_2} \xrightarrow[-\mathrm{H_2O}]{2e,\ 2\mathrm{H^+}} \mathrm{R{-}NO} \xrightarrow{2e,\ 2\mathrm{H^+}} \mathrm{R{-}NHOH} \xrightarrow[-\mathrm{H_2O}]{2e,\ 2\mathrm{H^+}} \mathrm{R{-}NH_2} \tag{5.10}$$

$$\mathrm{RR'C{=}O} \begin{cases} \xrightarrow{2e,\ 2\mathrm{H^+}} \mathrm{RR'CH{-}OH} \\ \xrightarrow[-\mathrm{H_2O}]{4e,\ 4\mathrm{H^+}} \mathrm{RR'CH_2} \end{cases} \tag{5.11}$$

$$\underset{(\mathrm{H})}{\mathrm{Ar{-}COOR}} \xrightarrow{4e,\ 4\mathrm{H^+}} \mathrm{Ar{-}CH_2OH} + \mathrm{ROH} \tag{5.12}$$

$$\mathrm{R{-}CONR'_2} \xrightarrow{4e,\ 4\mathrm{H^+}} \mathrm{R{-}CH_2NR'_2} + \mathrm{H_2O} \tag{5.13}$$

$$(\mathrm{R{-}CO})_2\mathrm{NR'} \xrightarrow{8e,\ 8\mathrm{H^+}} (\mathrm{R{-}CH_2})_2\mathrm{NR'} + 2\mathrm{H_2O} \tag{5.14}$$

$$\mathrm{Ar{-}N_2^+Cl^-} \xrightarrow{4e,\ 4\mathrm{H^+}} \mathrm{Ar{-}NH{-}NH_2} + \mathrm{HCl} \tag{5.15}$$

$$\mathrm{R_2C{=}N{-}R'} \xrightarrow{2e,\ 2\mathrm{H^+}} \mathrm{R_2CH{-}NH{-}R'} \tag{5.16}$$

$$\mathrm{R_2C{=}N{-}OH} \xrightarrow{4e,\ 4\mathrm{H^+}} \mathrm{R_2CH{-}NH_2} + \mathrm{H_2O} \tag{5.17}$$

$$\mathrm{R{-}C{\equiv}N} \xrightarrow{2e,\ 2\mathrm{H^+}} [\mathrm{R{-}CH{=}NH}] \xrightarrow{2e,\ 2\mathrm{H^+}} \mathrm{R{-}CH_2NH_2} \tag{5.18}$$

$$\text{Ar–SO}_2\text{H} \xrightarrow{4e,\ 4H^+} \text{Ar–SH} + 2\text{H}_2\text{O} \tag{5.19}$$

$$\text{–C≡C–} \xrightarrow{2e,\ 2H^+} \text{–CH=CH–} \xrightarrow{2e,\ 2H^+} \text{–CH}_2\text{–CH}_2\text{–} \tag{5.20}$$

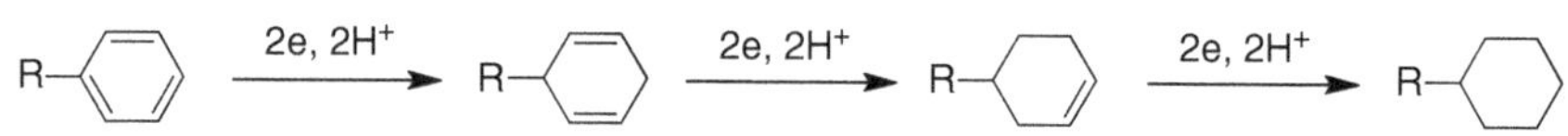

(5.21)

Equations (5.10)–(5.21) are shown using stoichiometric formula including elementary reaction steps, but only final products are shown and elementary reaction steps are omitted in the following examples.

5.5.2 Addition Type

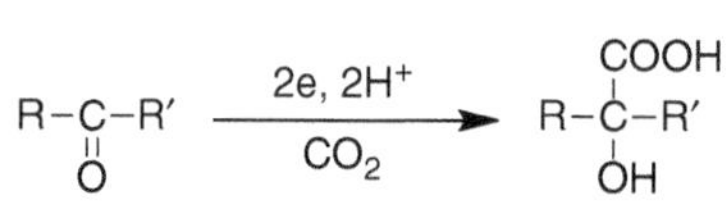

(5.22)

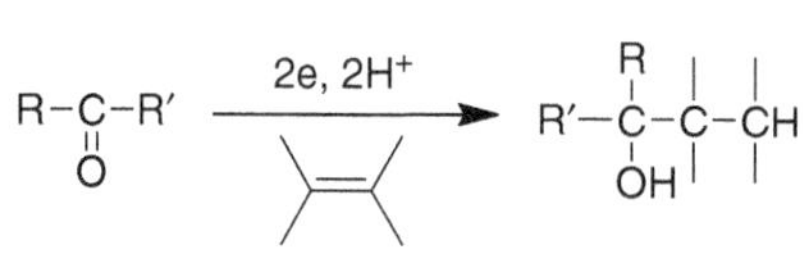

(5.23)

$$\text{H}_2\text{C = CH–CH = CH}_2 \xrightarrow[2CO_2]{2e,\ 2H^+} \text{HOOC–CH}_2\text{–CH = CH–CH}_2\text{–COOH} \tag{5.24}$$

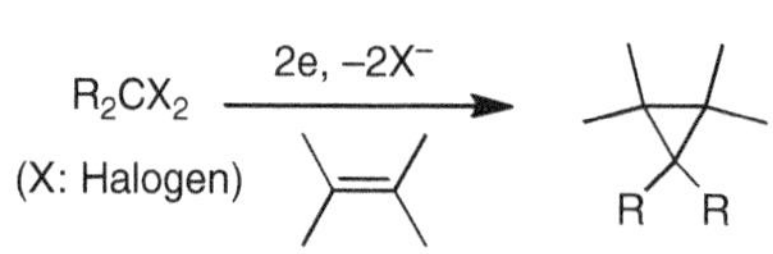

(5.25)

5.5.3 Insertion Type

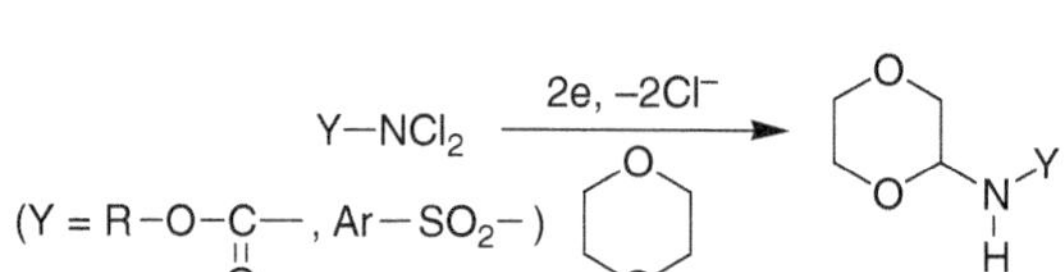

(5.26)

5.5.4 Substitution Type

$$\text{R–X} \xrightarrow[\text{H}^+]{2e,\ -\text{X}^-} \text{R–H}$$

(X = I, Br, Cl) (5.27)

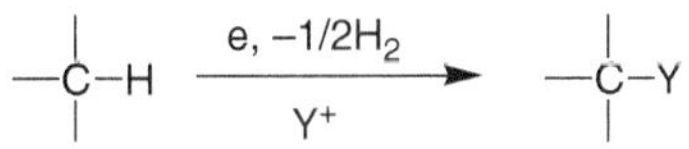

(5.28)

5.5.5 Substitutive Exchange Type

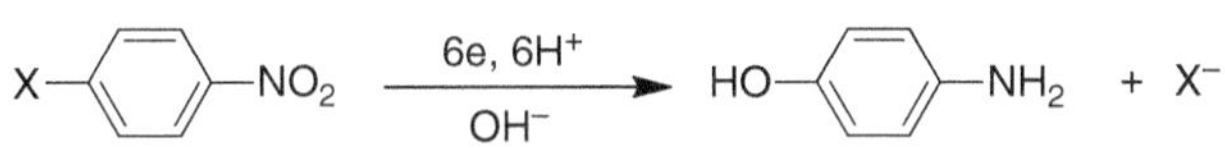

(X = halogen, OR) (5.29)

In Eq. (5.29), the transformation type of functional group also occurs.

5.5.6 Elimination Type

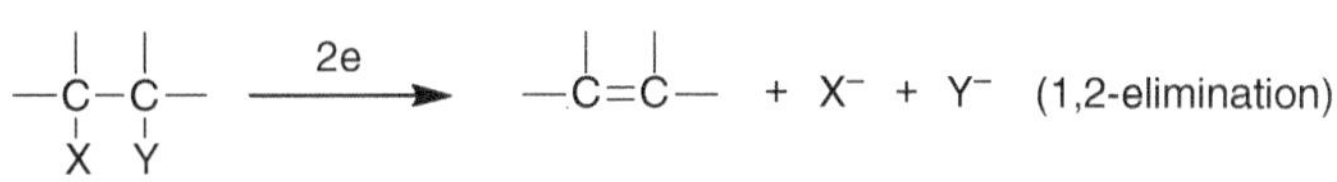

(X, Y = halogen, RCOO, RSO_3, RS, HS, HO; X = Y = —O–C(=O)–O—) (5.30)

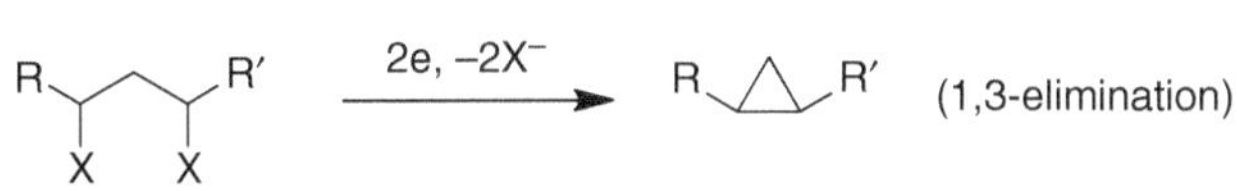

(5.31)

5.5.7 Dimerization Type

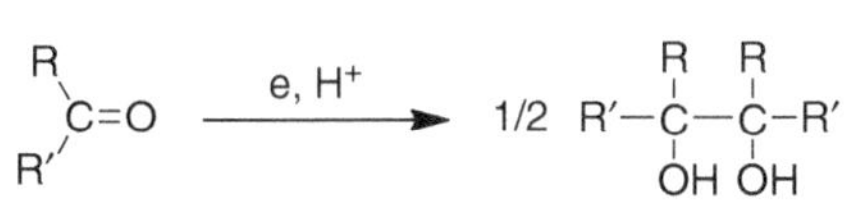

(5.32)

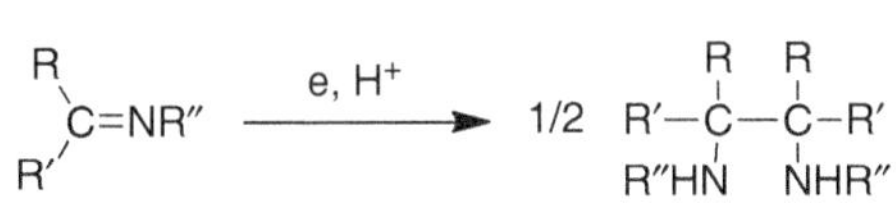

(5.33)

(Y = CN, COOR, $CONR_2$, RC(=O)—) (5.34)

$$\mathrm{R{-}X} \xrightarrow{\mathrm{e}} \tfrac{1}{2}\ \mathrm{R{-}R} + \mathrm{X^-} \qquad (\mathrm{X} = \text{halogen}) \tag{5.35}$$

$$\begin{aligned}\mathrm{Ar{-}NO_2} &\xrightarrow[-3/2\mathrm{H_2O}]{3\mathrm{e},\ 3\mathrm{H^+}} \tfrac{1}{2}\ \mathrm{Ar{-}N{=}N(\rightarrow O){-}Ar}\\ &\xrightarrow[-1/2\mathrm{H_2O}]{\mathrm{e},\ \mathrm{H^+}} \tfrac{1}{2}\ \mathrm{Ar{-}N{=}N{-}Ar}\\ &\xrightarrow[-1/2\mathrm{H_2O}]{\mathrm{e},\ \mathrm{H^+}} \tfrac{1}{2}\ \mathrm{Ar{-}NH{-}NH{-}Ar}\end{aligned} \tag{5.36}$$

5.5.8 Crossed Dimerization

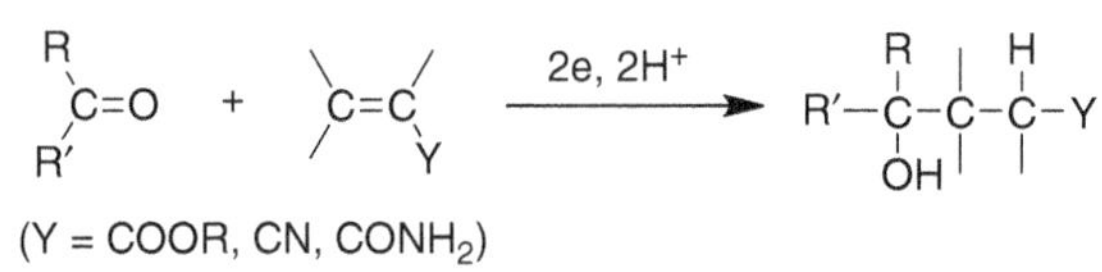

(5.37)

In Eq. (5.37), the ratio of crossed dimerization to non-crossed dimerization can be increased, but this is not so easy in the case of Eq. (5.38).

$$\mathrm{R{-}COOH} + \mathrm{R'{-}COOH} \xrightarrow[-2\mathrm{CO_2}]{-2\mathrm{e},\ -2\mathrm{H^+}} \mathrm{R{-}R'} \tag{5.38}$$

5.5.9 Cyclization Type

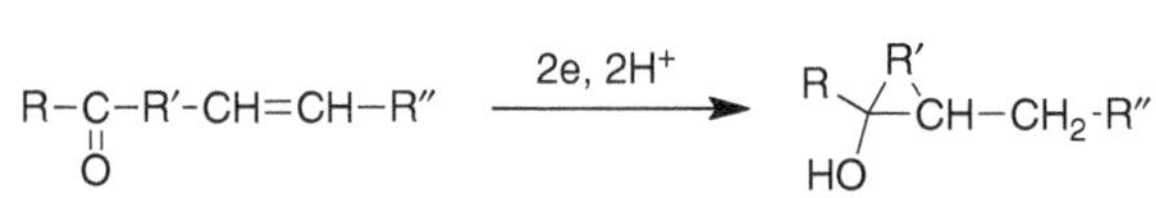

(5.39)

5.5.10 Polymorphism Formation Type

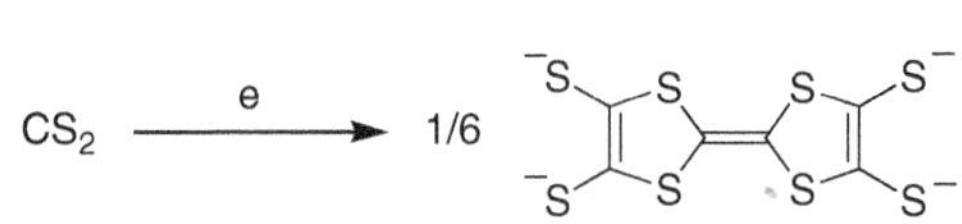

(5.40)

5.5.11 Polymerization Type

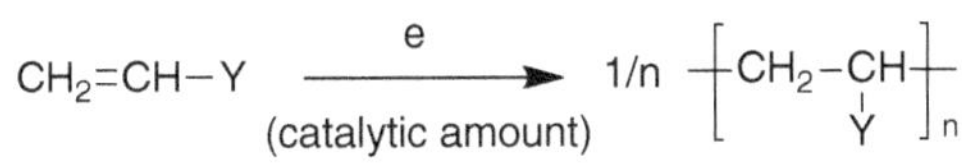

(Y = COOR, CN, $CONH_2$, Ph, AcO, alkyl) (5.41)

The reaction shown in Eq. (5.41) is initiated by cathodic reduction of activated olefin and a catalytic amount of electricity is enough to facilitate the polymerization in a chain reaction mechanism.

5.5.12 Cleavage Type

$$\text{R–S–S–R} \xrightarrow{2e,\ 2H^+} 2\ \text{R–SH} \tag{5.42}$$

5.5.13 Metalation Type

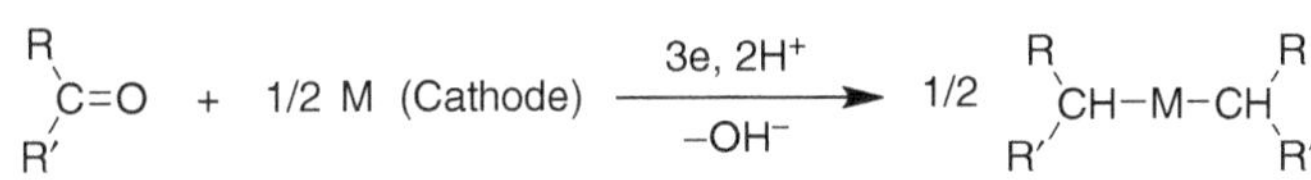

(M = Hg, Pb) (5.43)

In Eq. (5.43), metal used as an electrode is incorporated in the product.

5.5.14 Asymmetric Synthesis Type

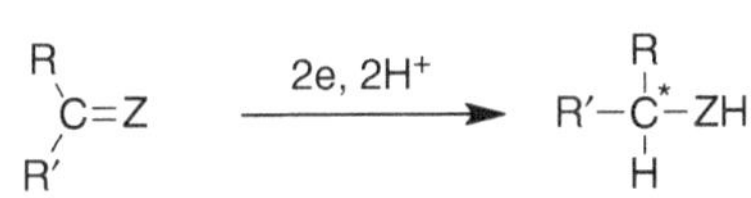

(X = O, NR″) (5.44)

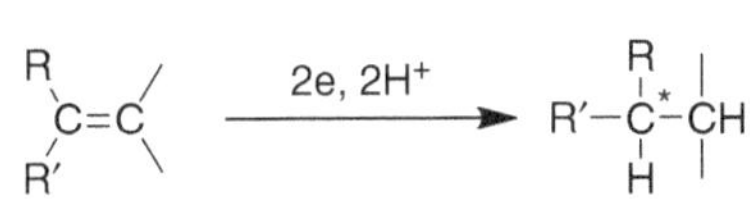

(5.45)

As a chiral source, chiral supporting salts, chiral solvents, chiral adsorbents and chiral modified electrodes are used.

5.6 Electrochemically Generated Reactive Species

Cathodically generated reactive species often undergo follow-up reactions in an adsorbed state or near the cathode surface to provide products. Since these reactive species are affected by cathode itself and strong electric field (c. $10^7 \sim 10^8\ \text{V cm}^{-1}$), their reactivity and behavior are often quite different from the same reactive species generated by other methods. Useful synthetic reactions can be developed utilizing such unique cathodically generated reactive species. A variety of ionic and radical species are generated by cathodic reduction. Cathodically generated reactive species are classified into carbon species and heteroatom ones, and their generation methods are explained below.

5.6.1 Cathodically Generated Carbon Species

5.6.1.1 Reduction of Alkyl Halides

The ease of reduction of alkyl halides is related to the bond energy of C—X, and hence iodide compounds are the most easily reduced, while chloride compounds are the most difficult to reduce. One-electron and two-electron reduction of alkyl halides generate alkyl radical and alkyl anion, respectively.

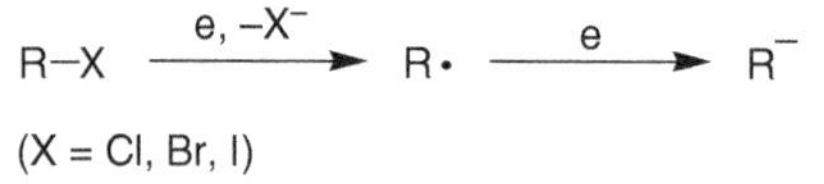

(5.46)

5.6.1.2 Reduction of Ketone and Imine

In the case of cathodic reduction of ketone and imine in aqueous solution, the generated reactive species as well as reaction mechanism are changed by the pH of the solution. In an acidic solution, the oxygen atom of the ketone and the nitrogen atom of the imine are protonated, therefore their reduction potentials shifts to the positive side, and their one-electron reduction generates neutral radicals. In contrast, in an alkaline solution, the protonation of ketone and imine does not occur due to low proton concentration. In this case, the radical anion is generated first, and then the dianion is formed.

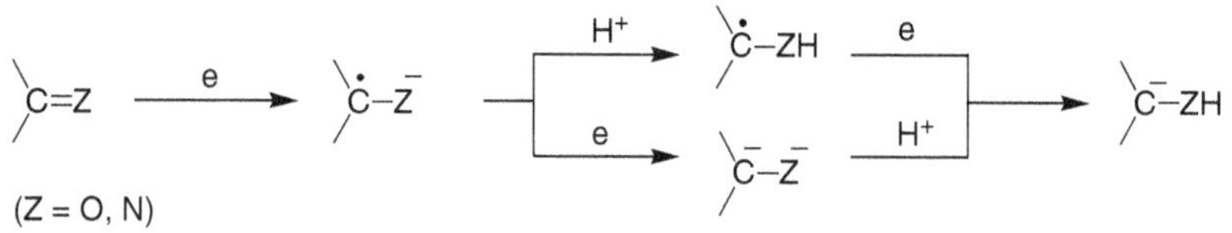

(5.47)

5.6.1.3 Reduction of Activated Olefin and Conjugated Olefin

Activated olefins are readily reduced because the electron-withdrawing group attached to the double bond decreases the electron density of the double bond. Isolated olefins are not so easy to reduce, but conjugated olefins are reducible. One-electron reduction of the olefin generates a radical anion.

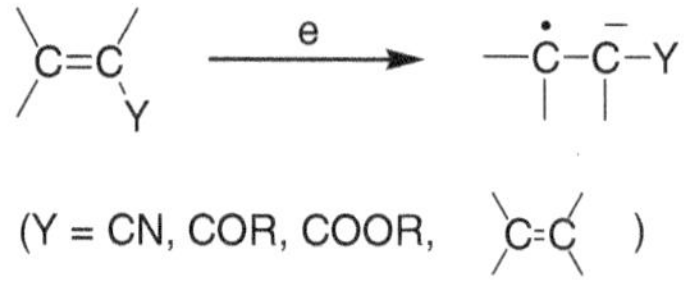

(5.48)

5.6.1.4 Reduction of Active Hydrogen Compounds

One-electron reduction of active hydrogen compound generates the corresponding anion, eliminating the active hydrogen atom as hydrogen gas.

$$X{-}CH_2{-}Y \xrightarrow{e,\ -1/2H_2} X{-}\overset{-}{C}H{-}Y$$

(X, Y = CN, COR, COOR) (5.49)

5.6.1.5 Reduction of *gem*- and *vic*-Dihalogeno Compounds

Two-electron reduction of these compounds generates carbene and benzyne.

$$>CX_2 \xrightarrow{2e,\ -2X^-} >C:$$

(X = Cl, Br, I) (Carbene) (5.50)

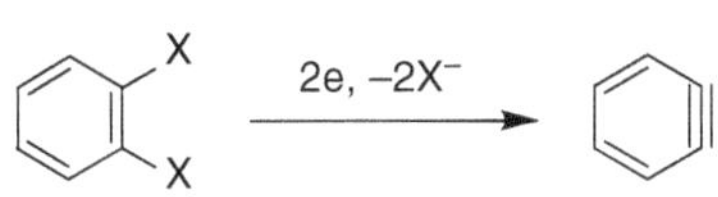

(X = Cl, Br, I) (Benzyne) (5.51)

5.6.2 Cathodically Generated Heteroatom Species

5.6.2.1 Cathodically Generated Nitrogen Species

Two-electron reduction of *N*,*N*-dihalo compounds generates the corresponding nitrenes which are used as electrophiles.

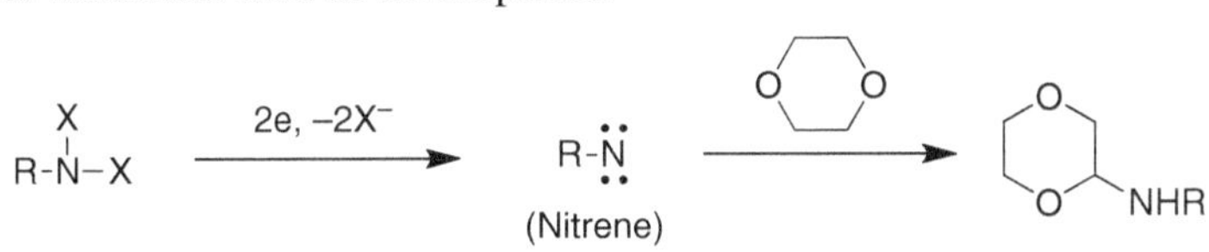

(5.52)

Two-electron reduction of amines with an electron-withdrawing group generates the corresponding aminyl anions. These species react with alky halides to afford alkylated amines.

$$R^1R^2N{-}H \xrightarrow{2e,\ -1/2H_2} R^1R^2N^- \xrightarrow{R^3X} R^1R^2NR^3$$

(Aminyl anion) (R^1 or R^2: Electron-withdrawing group)

(5.53)

5.6.2.2 Reduction of Alcohol and Carboxylic Acid

One-electron reduction of these compounds generates the corresponding anions, eliminating a hydrogen molecule.

$$YOH \xrightarrow{-e,\ -1/2H_2} YO^-$$

(Y = Ar, R, RCO) (5.54)

5.6.2.3 14-Family and 15-Family Element Species

Reactive species are generated predominantly at heteroatoms by cathodic reduction, as shown in Eqs. (5.55) and (5.56).

$$R_3Y{-}X \xrightarrow{2e,\ -X^-} R_3Y^- \quad (Y = Si, Sn) \tag{5.55}$$

$$R_2Z{-}X \xrightarrow{2e,\ -X^-} R_2Z^- \quad (Z = P, As, Sb) \tag{5.56}$$

5.7 Advanced Methodology for Electrochemical Reductive Transformations

Development of new methodology to achieve highly selective reactions is one of the most important area in organic synthetic chemistry. As already described, electrochemical reactions have their own specific factors for controlling selectivity, therefore both electrochemical and ordinary chemical factors make the control of electrochemical reactions more complicated. Electrodes are of great importance for both electron transfer interfaces and reaction fields. As described earlier, an electrode has a function to control chemical reaction pathway through adsorption and orientation of the substrate molecule to the electrode surface. Although hydrogen and oxygen overpotentials could be criteria for the selection of suitable electrode materials to achieve the desired electrochemical reaction in an aqueous solution, those overpotentials are not proper criteria in aprotic solutions. Hence, it is not so easy to predict suitable electrode materials for desired electrochemical reaction in aprotic solvents. However, many novel electrolytic methodologies have been developed in order to achieve high selectivity for the desired reactions. In this section, relatively new electrolytic methodologies for electrochemical reductive transformations, which have been already established and widely used, will be described in detail. Although there are so many applications of such electrochemical reductive transformations in organic syntheses, limited examples are given in this section [1–7, 16].

5.7.1 Electrocatalysis for Reductive Transformations

5.7.1.1 Direct and Indirect Electrochemical Reductions

Electrochemical reduction is classified as direct or indirect electroreduction. The former is based on direct electron transfer between substrate molecule and cathode, which is a simple and common electroreduction method. The latter is based on electron transfer using reductive mediators dissolved in an electrolytic solvent, as

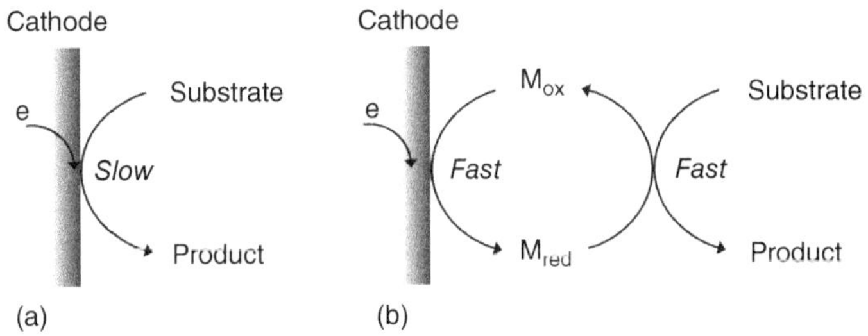

Figure 5.7 Principle of indirect electroreduction using mediator. (a) Direct electroreduction. (b) Indirect electroreduction.

Figure 5.8 Current-potential curve at direct and indirect electroreduction.

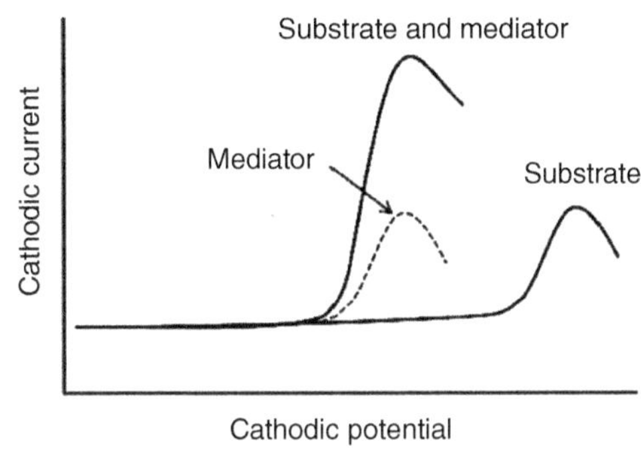

shown in Figure 5.7. Indirect electroreduction is also classified into in-cell, where electroreduction is carried out in the presence of both substrate and mediator, and ex-cell, where electroreduction is used for only regeneration of mediator. In the former case, the reduction potential of the mediator should be lower than that of the substrate. When the reduction potential of mediator is higher than that of substrate, the latter type is employed. In the former case, a catalytic amount of reductive mediator should be enough, while in the latter case, a quantitative or excess amount of mediator is necessary. When the heterogeneous electron transfer between reductive mediator and cathode, as well as the reductive reaction with the substrate is fast enough (Figure 5.7b), a significantly enhanced catalytic current can be obtained due to decreased reduction potential (decreased activation energy), as shown in Figure 5.8. In other word, a large electrocatalytic current can be obtained.

5.7.1.2 Kinds of Mediators for Reductive Transformations

Characteristics and proper choice of reductive mediators are of importance for the selective reduction. Many types of reductive mediators are available for these reductive transformations and the reduction potentials of mediators can also be tuned by their molecular design.

Examples of reductive mediators are shown as follows:

> $Ni^{2+}/Ni(0)$, Co^{2+}/Co^{+}, Cr^{3+}/Cr^{2+}, $Mo^{2+}/Mo(0)$, Ti^{3+}/Ti^{2+}, Sn^{4+}/Sn^{2+}, $Sn^{2+}/Sn(0)$, $Mg^{2+}/Mg(0)$, $Pd^{2+}/Pd(0)$, naphthalene, anthracene, phenanthrene, pyrene, fullerene C60, benzonitrile, phthalonitrile, anthraquinone, nitrobenzene, viologen, superoxide ion.

Table 5.1 Reduction potentials of aromatic mediators for indirect cathodic reduction.

Compound	$E_{1/2}$ (V vs. SCE)
Phthalonitrile	−1.7
4-Methoxybenzoquinone	−1.8
Anthracene	−2.0
Methyl benzoate	−2.2
Benzonitrile	−2.2
Chrysene	−2.5
Naphthalene	−2.5

The reduction potentials of aromatic mediators for indirect cathodic reduction are shown in Table 5.1.

5.7.1.3 Electrorechemical Reductive Transformations Using Mediators

The aromatic compounds such as anthracene and naphthalene are chosen to have a reduction potential less negative than that of alkyl halides and the aromatic radical anions are formed at the cathode surfaces. Electron transfer, in solution, from the radical anion to the alkyl halides causes C—X bond dissociation and also regenerates the aromatic compound, as shown in Eq. (5.57). Quite recently, novel mediators such as carboranes with characteristic of boron atom have also been developed for highly efficient cathodic reductive dehalogenation [17].

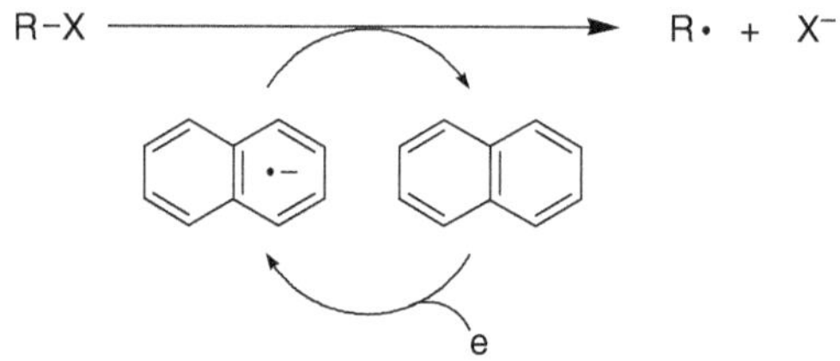

(5.57)

Transition metal complex mediators have various reactivities and many advantages. Their redox potentials and the selectivity of the desired reaction can be controlled by changing ligand. Their synthetic application is mainly based on the reactivity of the low valent state generated by cathodic reduction of the mediators [3, 18]. Typically, Co(III) complex (vitamin B_{12}) is readily reducible, optically active, non-toxic, and inexpensive. It is reduced at −0.9 V vs. SCE to form Co(I) complex, which undergoes oxidative addition to alkyl halide to form alkyl Co(III) complex as an intermediate. The resulting intermediate is reduced at more negative potential, −1.5 V vs. SCE, to generate alkyl radical or anion, and Co(I) complex is regenerated simultaneously (reductive elimination). Since the resulting alkyl radical or anion undergoes conjugate addition, efficient Michael addition can be achieved under neutral conditions, as shown in Figure 5.9 [19]. This mediatory reaction is widely applicable to various halogeno compounds like allyl halides, vinyl

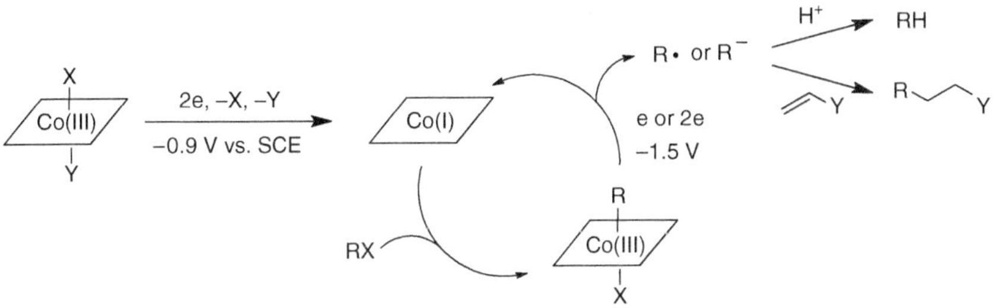

Figure 5.9 Electrosynthesis using Co(III) complex, Vitamin B12 as mediator. Source: Modified from Scheffold et al. [19].

halides, α-halo ethers, and so on. As shown in Eq. (5.58), even halogeno compounds with non-protected hydroxyl and carbonyl groups can be used, which is one of the advantages of this mediatory system [20].

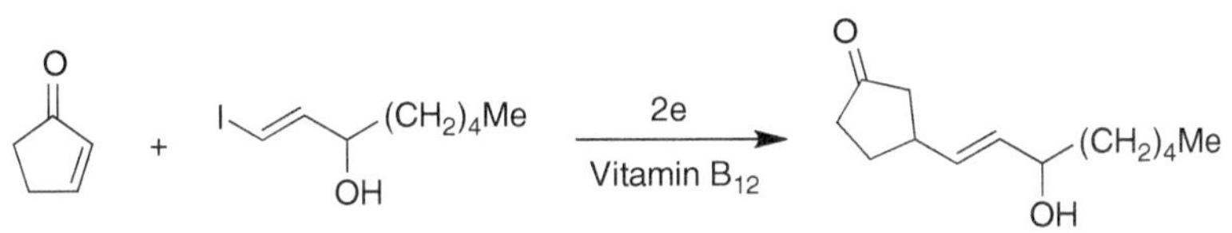

(5.58)

On the other hand, β-bromodiester and ω-bromoalkyl acrylate undergo 1,2-rearangement and intramolecular cyclization, respectively, to form large cyclic lactones by their cathodic reduction using hydrophobic Vitamin B12 mediator under UV irradiation [21].

Homo-coupling products are obtained from aromatic halides using Pd(0) complex as well as Ni(0) complex as a mediator [22]. The yields and turnover of the Pd(0) complex are generally superior to those using Ni(0) complex. The mechanism proposed is shown in Figure 5.10. Aromatic halide reacts with Pd(0) complex to generate an aryl Pd intermediate, which is reduced cathodically followed by reaction with one more aryl halide molecule to form diaryl Pd complex, resulting in reductive elimination to give a homo-coupling product. When this reaction is performed in the presence of CO_2, aromatic carboxylic acids are obtained in high yields [23].

Figure 5.10 Homo-coupling of aryl halide using Pd(0) complex.

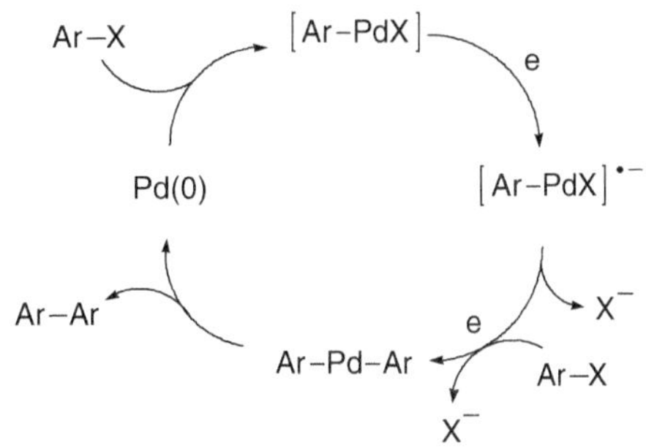

5.7.2 Electrogenerated Bases

Acid and base play very important roles in organic synthesis. It is well known that when aqueous solution containing a neutral supporting electrolyte is electrolyzed in a divided cell, the anolyte becomes acidic while the catholyte becomes alkaline. This is because the hydroxide ions are consumed at anode while protons are consumed at cathode. In a similar manner, some acid is generated in anolyte while some base is generated in catholyte during electrolysis in an organic solvent. Even in an undivided cell, the vicinity of an anode becomes acidic while that of a cathode becomes basic during electrolysis.

Anionic species generated cathodically act not only as nucleophiles but also as bases, and have interesting reactivities in organic synthesis. The inventor of cathodic hydrodimerization process of acrylonitrile, Baizer, demonstrated that cathodically generated anion radical of hindered azobenzene (Figure 5.11) could be utilized as a useful base for various organic synthesis, and he named such bases electrogenerated bases (EGBs) [24].

There are two main methods for the generation of EGBs, as shown in Eq. (5.59). One method is cathodic reduction of compounds with an unsaturated moiety in aprotic solvents containing quaternary ammonium salt as a supporting electrolyte to generate the radical anions or dianions, and the other one is cathodic reductive deprotonation of active hydrogen compounds to generate the corresponding anions.

$$\mathrm{Y=Z} \xrightarrow[\mathrm{R_4N^+X^-}]{\mathrm{e}} \mathrm{[Y=Z]^{\bullet -}\ R_4N^+}$$

$$\mathrm{Y{-}H} \xrightarrow[\mathrm{R_4N^+X^-}]{\mathrm{e,\ -1/2\ H_2}} \mathrm{Y^-R_4N^+}$$

$$\mathrm{(Y, Z = C, O, S, N, P)} \tag{5.59}$$

Cathodic reductive deprotonation of 2-pyrrolidone, a hindered phenol like 2,6-*t*-butyl-4-methylphenol, and triphenylmethane generates the corresponding anions. Since EGBs have quaternary ammonium cation (Q^+), the anions formed by the treatment with EGB have high reactivity (Figure 5.12), and hence this methodology is highly useful for organic synthesis.

In particular, EGB derived from 2-pyrrolidone is a versatile base and applicable to organic synthesis such as Stevens rearrangement, selective α-monoalkylation

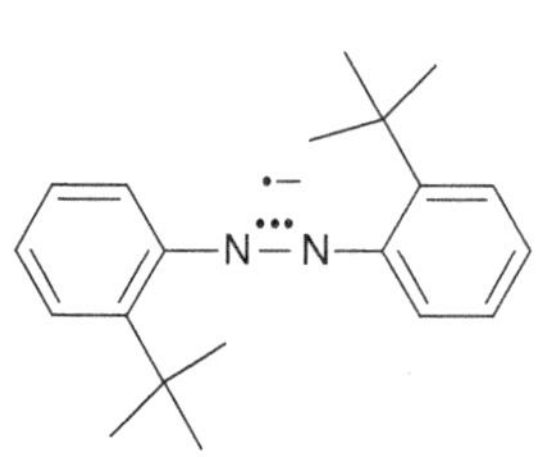

Figure 5.11 Electrogenerated base of hindered azobenzene.

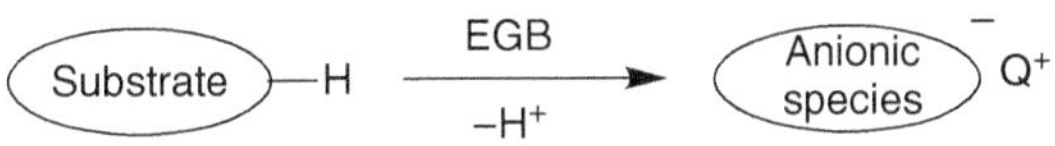

Figure 5.12 Reactive anion derived from electrogenerated base (EGB) ($Q^+ = Et_4N^+$).

of α-(aryl)acetate esters, and C-monoalkylation of 1,3-diketones [25, 26]. This EGB has also been demonstrated to be a highly efficient base for the synthesis of organofluorine compounds. For instance, it is known that trifluoromethyl anion is so unstable that it undergoes α-elimination of fluoride anion to generate fluorocarbene, but stable trifluoromethyl anion can be generated by the treatment of fluoroform with this EGB, and consequently trifluoromethylation of aromatic aldehydes and ketones is realized to give the trifluoromethylated alcohols, as shown in Eq. (5.60) [27].

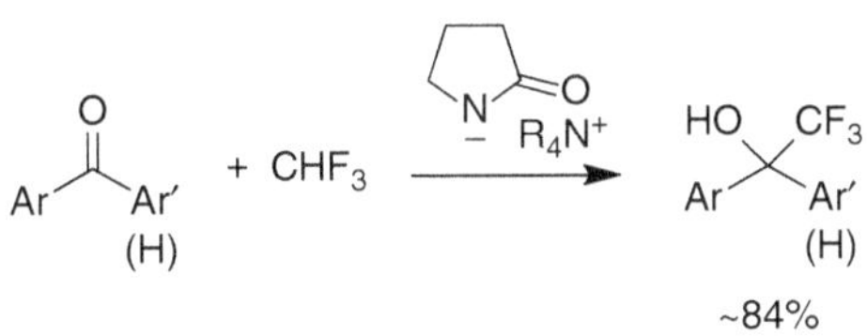

(5.60)

Generally, it is quite difficult to generate the enolate anion with an α-CF_3 group since the enolate anion readily undergoes decomposition such as β-elimination of fluoride ion. However, pyrrolidone-derived EGB enables the generation of stable trifluoromethylated enolate anion, and alkylation can be performed in good yield without elimination of fluorine atom, as shown in Eq. (5.61) [28].

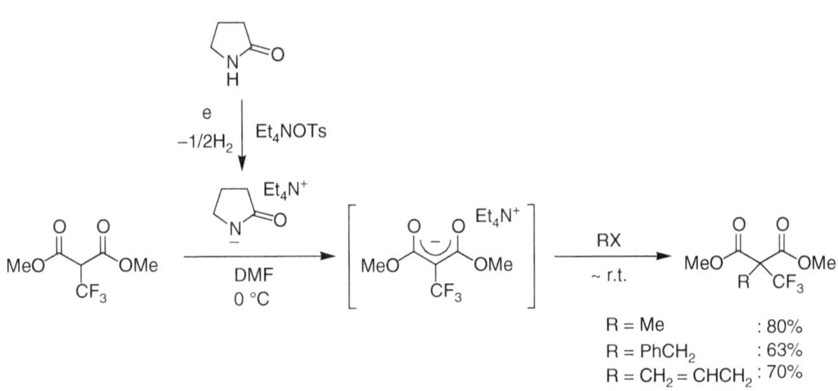

(5.61)

Furthermore, this EGB catalyzes ring-opening polymerization of *N*-carboxyanhydrides of α-amino acid (NCA) to provide poly(amino acids) in a short time in excellent yield, as shown in Eq. (5.62). The yield is much higher and reaction time is relatively short compared to reactions using conventional base with a metal cation like Na^+ [29].

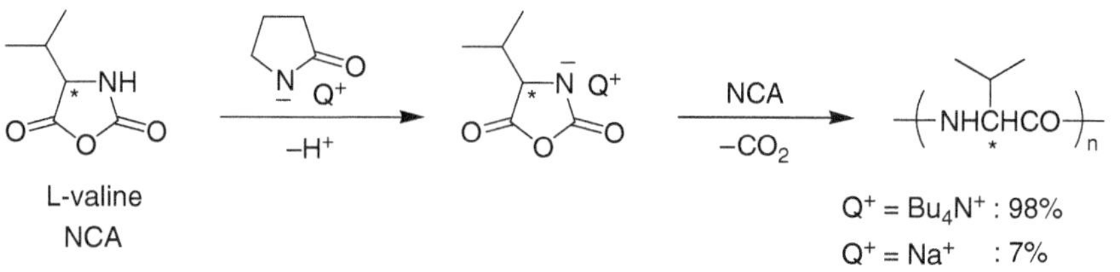

(5.62)

EGB derived from hindered phenol catalyzes selective double Aldol condensation reaction, as shown in Eq. (5.63) [30].

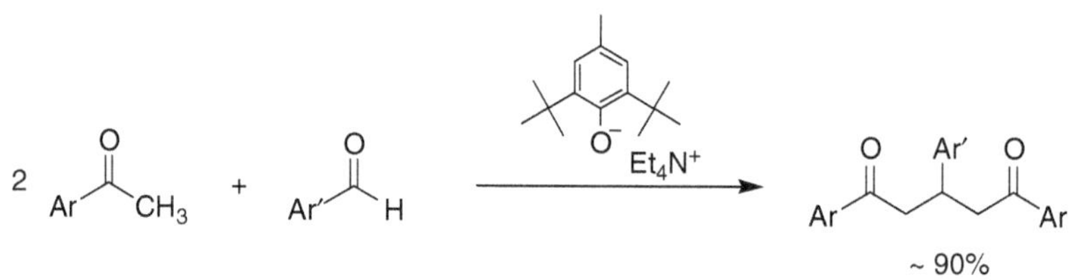

(5.63)

5.8 Conclusions

Organic electrosynthesis is expected to be a typical green chemistry process since it does not require any hazardous reagents and produces less waste than conventional chemical synthesis. This viewpoint has prompted organic electrochemists as well as organic chemists to make great efforts to develop new systems not only of electroreductive synthesis but also of electrooxidative synthesis in order to achieve green and sustainable chemistry. In fact, a number of successful new green organic electrolytic systems have been developed to date, as illustrated in this chapter. We believe that cutting-edge developments in organic electrochemistry will be achieved through hybridization with various other scientific fields.

References

1 Torii, S. (2006). *Electroorganic Reduction Synthesis*, Vols 1 and 2. Weinheim: Kodansha and Wiley-VCH Verlag GmbH.

2 Bard, A.J. and Stratmann, M. (2004). *Organic Electrochemistry*, Encyclopedia of Electrochemistry, vol. 8 (ed. H.J. Schaefer). Weinheim: Wiley-VCH Verlag GmbH.

3 Lund, H. and Hammerich, O. (eds.) (2001). *Organic Electrochemistry*, 4e. New York: Marcel Dekker, Inc.

4 Grimshaw, J. (2000). *Electrochemical Reactions and Mechanisms in Organic Chemistry*. Amsterdam: Elsevier.

5 Fry, A.J. (1989). *Synthetic Organic Electrochemistry*. New York: Wiley Interscience.

6 Shono, T. (1984). *Electroorganic Chemistry as a Tool in Organic Synthesis*. Berlin: Springer-Verlag.

7 Fuchigami, T., Atobe, M., and Inagi, S. (2015). *Fundamentals and Applications of Organic Electrochemistry*. West Sussex: Wiley.

8 Fuchigami, T., Nonaka, T., and Schafer, H.J. (2003). Encyclopedia of electrochemistry. In: *Organic Electrochemistry*, vol. 8 (eds. A.J. Bard and M. Stratmann). Weinheim: Wiley-VCH, Verlag GmbH.

9 Edamura, F., Kyriyacou, D., and Love, J. (1980). US Patent 4, 217, 185; *Chem. Abstr*. (1981). 94: 22193.

10 (a) Fry, A.J. and Moor, R.H. (1968). *J. Org. Chem.* 33: 1283–1284. (b) Erickson, R.E., Annino, R., Sainlor, M.D., and Zon, G. (1969). *J. Am. Chem. Soc.* 91: 1767–1770.

11 Sekine, T., Yamura, A., and Sugino, K. (1965). *J. Electrochem. Soc.* 112: 439–443.

12 Nonaka, T., Wachi, S., and Fuchigami, T. (1977). *Chem. Lett.* 6: 47–50.

13 Yoshida, J., Kataoka, K., Horcajada, R., and Nagaki, A. (2008). *Chem. Rev.* 108: 2265–2299.

14 Kandeel, Z., Nonaka, T., and Fuchigami, T. (1986). *Bull. Chem. Soc. Jpn.* 59: 338–340.

15 (a) Uneyama, K. and Kato, T. (1998). *Tetrahedron Lett.* 39: 587–590. (b) Uneyama, K., Naeda, K., Kato, T., and Katagiri, T. (1998). *Tetrahedron Lett.* 39: 3741–3744.

16 (a) Hammerich, O. and Speiser, B. (eds.) (2004). *Organic Electrochemistry*, 5e. CRC/Tayler & Francis. (b) Rifi, M.R. and Covitz, F.H. (1974). *Introduction to Organic Electrochemistry Techniques and Applications in Organic Synthesis*. New York: Marcel Dekker. (c) Mann, C.K. and Barnes, K.K. (1970). *Electrochemical Reactions in Nonaqueous Systems*. New York: Marcel Dekker.

17 Hosoi, K., Inagi, S., Kubo, T., and Fuchigami, T. (2011). *Chem. Commun.* 47: 8632–8634.

18 (a) Saveant, J.M. (1980). *Acc. Chem. Res.* 13: 323–329. (b) Steckhan, E. (1986). *Angew. Chem.* 98: 681–699. (c) Torii, S. (1986). *Synthesis*: 873–886. (d) Steckhan, E. (1987). *Electrochemistry I*, Topics in Current Chemistry, vol. 142, 1–69. Berlin: Springer-Verlag.

19 Scheffold, R., Dike, M., Dike, S. et al. (1980). *J. Am. Chem. Soc.* 102: 3642–3644.

20 Scheffold, R., Rytz, G., Walder, L., and Orlinski, R. (1983). *Pure Appl. Chem.* 55: 1791–1797.

21 Shimakoshi, H., Nakazato, A., Hayashi, T. et al. (2001). *J. Electroanal. Chem.* 507: 170–176.

22 Torii, S., Tanaka, H., and Morisaki, K. (1985). *Tetrahedron Lett.* 26: 1655–1658.

23 Torii, S., Tanaka, H., Hamatani, H. et al. (1986). *Chem. Lett.*: 169–170.

24 Baizer, M.M. (1984). *Tetrahedron* 45: 935–969.

25 Shono, T., Kashimura, S., Ishizaki, K., and Ishige, O. (1983). *Chem. Lett.*: 1311–1312.

26 Shono, T., Ishifune, M., Ishige, O. et al. (1990). *Tetrahedron Lett.* 31: 7181–7184.

27 Shono, T., Ishifune, M., Okada, T., and Kashimura, S. (1991). *J. Org. Chem.* 56: 2–4.

28 Fuchigami, T. and Nakagawa, Y. (1987). *J. Org. Chem.* 52: 5276–5277.

29 Komori, T., Nonaka, T., and Fuchigami, T. (1986). *Chem. Lett.* 15: 11–12.

30 Fuchigami, T., Awata, T., Nonaka, T., and Baizer, M.M. (1986). *Bull. Chem. Soc. Jpn.* 59: 2873–2879.

6

Electrochemical Redox-Mediated Polymer Synthesis

Naoki Shida[1] *and Shinsuke Inagi*[1,2]

[1] *Tokyo Institute of Technology, School of Materials and Chemical Technology, Department of Chemical Science and Engineering,*
[2] *PRESTO, Japan Science and Technology Agency (JST), 4-1-8 Honcho, Kawaguchi, Saitama 332-0012, Japan*

6.1 Introduction

Electrochemistry has been recognized as a powerful methodology for the synthesis of polymeric materials. In this chapter, we summarize significant accomplishments in the field of electrochemical redox-mediated polymer synthesis.

A representative and conventional example of the electrosynthesis of polymers is the electropolymerization of aromatic monomers to yield conjugated polymers. Due to the conducting nature of conjugated polymers, the successive growth of polymer films is achieved on an electrode in the form of a solid-state reaction. The mechanism and unique examples of the electropolymerization of conjugated polymers are summarized in Section 2. In oxidative electropolymerization, aromatic monomers/oligomers are oxidized and undergo continuous coupling reactions to give an insoluble film on the electrode surface. This method has been widely accepted as one of the most efficient ways to prepare conjugated polymers. The reductive polymerization of conjugated polymers is much less pronounced than its oxidative counterpart; however, several impressively efficient systems have been reported.

Polymer reaction, which is a chemical transformation using a polymer as a precursor material, is another way to obtain polymers with various architectures. The idea of incorporating functionalities after polymerization is especially referred to as post-functionalization. In Section 3, we explain the art of electrochemical post-modification of polymeric materials. Electrochemical post-functionalization has not been very well explored due to the low reaction efficiency in homogeneous solutions, which results from slower diffusion compared to small organic molecules. On the other hand, recent research has revealed that electrochemical polymer reactions proceed well when the polymers are fixed on the electrode surface.

Electrochemical synthesis of nonconductive polymers is more challenging than that of conducting polymers due to the passivating nature of these polymers, which

Organic Redox Chemistry: Chemical, Photochemical and Electrochemical Syntheses, First Edition.
Edited by Jun-Ichi Yoshida and Frédéric W. Patureau.

hampers electron transfer on the electrode. However, several groups have reported significant achievements in the electrosynthesis of nonconductive polymers. The electrochemistry contributes to generate or quench the active species by changing their oxidation state, by which the electrochemical switching of reactions is realized. The synthesis of nonconjugated polymers is summarized in Section 4.

6.2 Synthesis of Conducting Polymers by Electrochemical Redox

6.2.1 Electrochemical Redox Behavior of Conducting Polymers

Conducting polymers are composed of aromatic monomers polymerized in a conjugated manner [1]. Conjugated polymers are poorly conductive in the neutral state, while they exhibit electrical conductivity by the removal or addition of electrons, known as doping (Scheme 6.1). Such a doping process has been achieved by both chemical and electrochemical methods. Electrochemical doping is typically performed using conducting polymers fixed on an electrode. Electrochemical doping injects holes or electrons directly into a polymer film, followed by concomitant incorporation of counter ions from an electrolyte solution to compensate for the excess charge. The doped state of conducting polymer is relatively stable compared to the oxidized/reduced state of common small organic molecules due to the delocalization of charges over the π-system. Doping of the conducting polymer film changes the electronic structure of the polymer chains, resulting in a dramatic color change [2]. Application of an opposite bias to the doped conducting polymers reverses the reaction to give the neutral polymer (de-doping). These features make conducting polymers promising materials for electrochromic devices, organic capacitors, and actuators [3, 4].

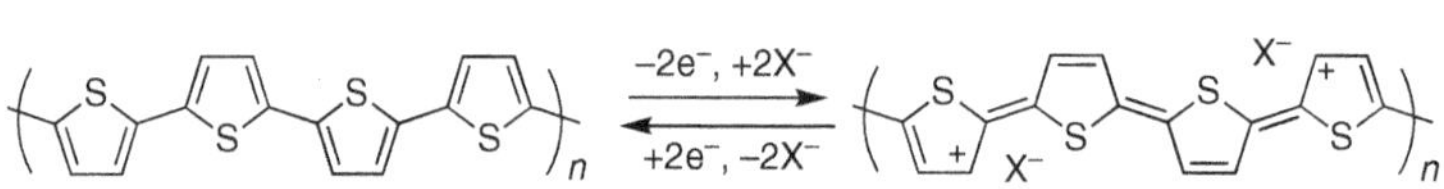

Scheme 6.1 Electrochemical doping and dedoping behavior of polythiophene.

6.2.2 Oxidative Electropolymerization of Aromatic Monomers

Electron-rich aromatic and heteroaromatic monomers undergo one-electron oxidation on an electrode surface to give radical cationic species. Generated radical cations couple with other radical cations, followed by further oxidation and deprotonation to aromatize and give dimeric species (Scheme 6.2). A dimer is more easily oxidized due to their expanded π-systems compared to that of the original monomer; therefore, generated dimers are immediately oxidized to undergo further oligomerization, by which conjugated polymers are formed. In this methodology, poor solvents for the conjugated polymers are typically used to induce precipitation of the polymer film on the electrode surface. Electrochemical doping of the polymer film occurs along with oxidation of the monomer/oligomer, so that the reaction surface maintains electrical conductivity. The polymer film continues to grow as long as the electrolysis continues.

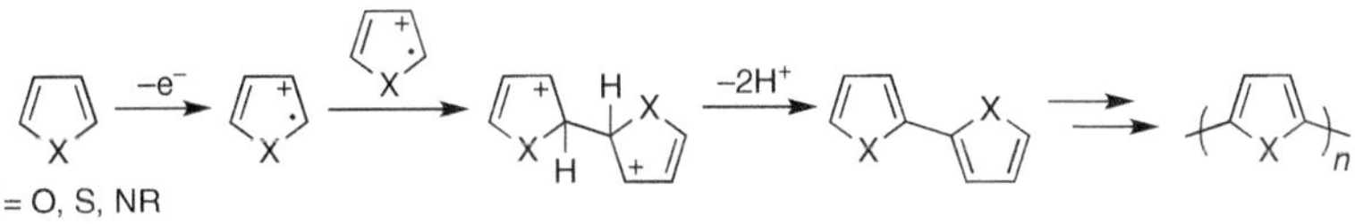

Scheme 6.2 Plausible mechanism for the anodic electropolymerization of heteroaromatic monomers.

Electropolymerization is commonly performed by the potential sweep method, constant potential electrolysis, and constant current electrolysis. Potential sweep polymerization is beneficial because it allows the polymerization to be monitored over the sweep cycles. In cyclic voltammetry (CV) measurement during the polymerization, the oxidation-onset potential typically shifts to lower potential upon the cycle due to the extension of the π-system. The redox current also increases by cycles, due to a greater amount of polymers deposited on the electrode after cycles, so that more doping/dedoping currents are observed.

The electrolyte is usually composed of an oxidation-tolerant combination of solvent and supporting salt. Acetonitrile (MeCN) is commonly used as a solvent due to its wide potential window in the positive direction, high polarity, and the lower solubility of conjugated polymers in MeCN. Alkylammonium salts with less coordinating anions such as ClO_4^-, BF_4^-, and PF_6^- are widely used as supporting electrolyte salts due to their solubility in common organic solvents.

Instead of common electrolytes, room-temperature ionic liquids are also employed for electropolymerization [5]. The use of ionic liquids has some advantages for electropolymerization, such as (i) fast film growth, (ii) higher compatibility between the film and the electrode, and (iii) higher density and smoother surface of the product film. These features result in higher electric conductivity and electrochemical capacitance, and good reversibility of the redox cycle of the film [5].

Another interesting medium for electropolymerization is the boron trifluoride–ether complex (BF_3OEt_2) [6]. BF_3OEt_2 forms complexes with aromatic monomers to reduce their aromaticity, which results in lowering of the oxidation potential. Thus, this medium is useful for the anodic polymerization of aromatic monomers with high oxidation potentials. The reduced oxidation potential helps to avoid overoxidation of the product polymer.

6.2.3 Electrochemical Copolymer Synthesis

Polymers composed of multiple distinct types of monomer units are called copolymers, and they are of significant interest in polymer science due to their various functionalities derived from each monomer component. For instance, a π-conjugated copolymer composed of electron-rich and -poor segments in an alternative manner is called a donor–acceptor alternating copolymer, which is

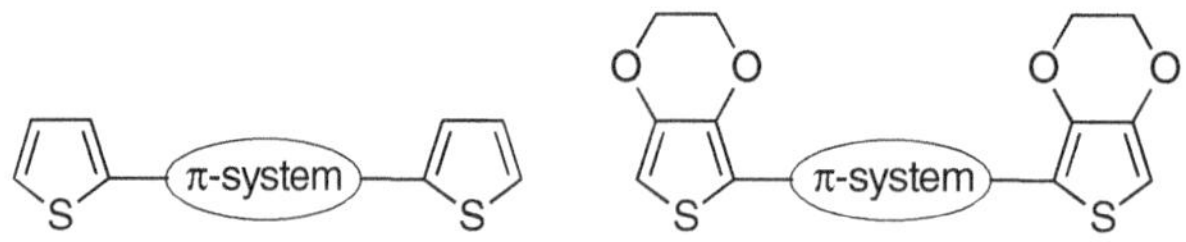

Figure 6.1 General design of electropolymerizable aromatic monomers with dithienyl or bis(3,4-ethylenedioxythienyl) motifs.

one of the most general designs to access narrow band-gap organic semiconductors [7].

Several researchers have reported the preparation of conducting copolymer films by anodic oxidation. The simplest way is to mix two different monomers. A copolymer obtained by this method showed intermediate optical properties of each homopolymer, which could be tuned by changing the feed ratio of each monomer [8–10]. However, the monomers have oxidation potentials close to each other; otherwise, polymerization of one monomer becomes dominant.

Another strategy to access copolymers is to synthesize monomers with alternating monomer units. One major method is to substitute aromatic monomers with two thienyl- or 3,4-ethylenedioxythienyl-groups as terminal units (Figure 6.1). These groups function as an electropolymerizable moiety, as well as the donor unit in the polymer backbone [11–13]. The polymer that results from this method is supposed to be an alternating copolymer of dithienyl- or bis(3,4-ethylenedioxythienyl)-units and the spacer aromatic unit, and there are in principle no defects, unlike copolymerization from a mixture solution of two different monomers.

Ruhlmann and coworkers reported the stepwise electrochemical synthesis of an alternating copolymer of porphyrin and viologen (Figure 6.2) [14]. This method is based on the anodic oxidation of porphyrin and nucleophilic substitution by 4,4′-bipyridine. A zinc-porphyrin and bipyridine were used for the polymerization. First, the zinc-porphyrin was oxidized and substituted by 4,4′-bipyridine under constant potential electrolysis conditions. The product was not overoxidized due to the

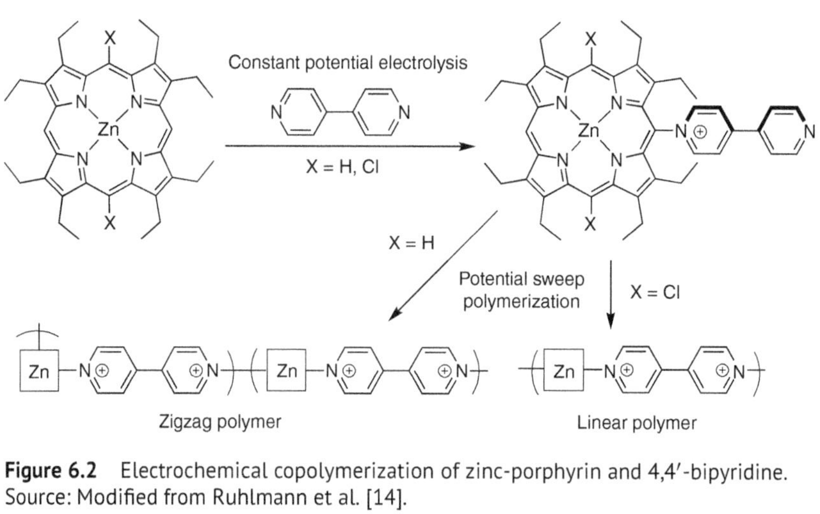

Figure 6.2 Electrochemical copolymerization of zinc-porphyrin and 4,4′-bipyridine. Source: Modified from Ruhlmann et al. [14].

electron-withdrawing nature of the pyridinium substituent. The obtained product (AB-monomer) was then subjected to potential sweep polymerization. The reduction wave derived from the viologen moiety successively increased with the scanning cycles, which suggests the formation of the alternating copolymer. The morphology of the obtained polymer was strongly affected by the substituent at the *meso*-position of the zinc-porphyrin. A hydrogen-substituted (X = H) polymer resulted in an aggregated macrostructure, as observed using scanning electron microscopy, while a chlorine-substituted (X = Cl) polymer gave a fiber-like structure. The former was derived from the statistical substitution at the porphyrin, which is expected to give a zigzag polymer structure. In the latter case, the diameter of the fiber was ca. 20 Å, which was close to the diameter of the octaethylporphyrin (19 Å) and suggested the formation of a polymer with a linearly connected porphyrin backbone [14].

6.2.4 Reductive Electropolymerization of Aromatic Monomers

Aromatic dihalides are commonly used monomers for reductive electropolymerization. Cathodic reduction of those monomers generates radical anion species, which polymerize with the concomitant release of the halides. The electrogenerated Ni(0)-catalyzed dehalogenative polymerization of dihalogenated quinoline derivatives was reported [15].

A unique synthesis of poly(*p*-phenylenevinylene) (**PPV**) by electroreduction was developed by Utley and Gruber (Scheme 6.3) [16]. They employed a *p*-xylene derivative (**PX**) as a monomer. The reaction mechanism is shown in Scheme 6.3. Two-electron reduction of **PX** generates a quinodimethane-intermediate, which reacts to give poly(*p*-xylene) (**PPX**). The polymerization proceeds in a radical mechanism, evidenced by suppression of the reaction through the addition of a radical-trapping reagent. Further two-electron reduction of **PPX** gave a conjugated **PPV** backbone.

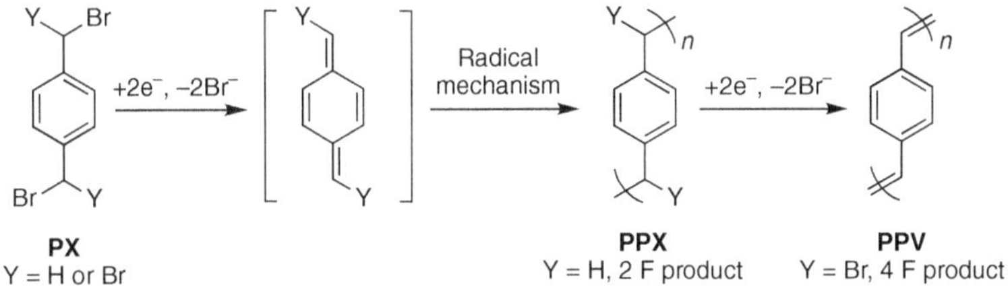

Scheme 6.3 Electroreductive synthesis of **PPV** from *p*-xylene derivative.

6.2.5 Polysilane Synthesis by Cathodic Reduction

Polysilanes, which are composed of linearly connected silicones, are known as conducting polymers and are conventionally synthesized by the Kipping method, where Na metal is used to reduce dichlorosilanes [17]. An equivalent electrochemical reaction can avoid the use of such a metal reductant. Although the direct electrochemical reduction of dichlorosilanes was difficult, the use of a sacrificial

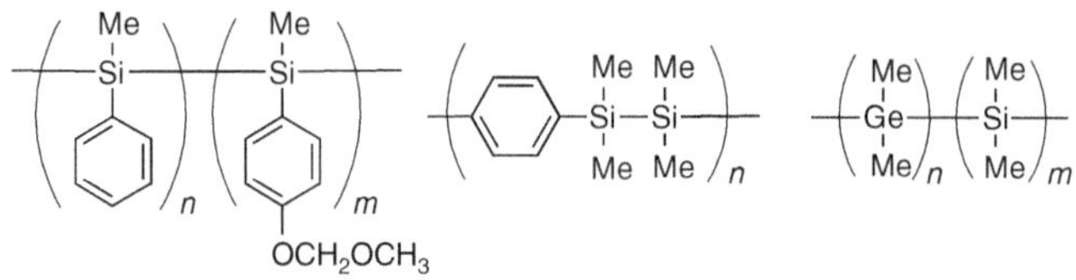

Figure 6.3 Electrochemically synthesized polysilanes and derivatives. Source: Modified from Kashimura et al. [19].

Mg anode promoted the electropolymerization of dichlorosilane monomers [18]. In this system, Mg^{2+} ions produced by dissolution of the sacrificial anode are reduced to form reactive Mg involving Si—Si bond formation. Cu and Al are also feasible as sacrificial anode materials. This strategy has also enabled the synthesis of a variety of poly(carbosilane)s and polygermanes and their copolymers (Figure 6.3) [19].

6.2.6 Electropolymerization Under Nonconventional Conditions

The electropolymerization processes discussed here are all conducted in a conventional batch-type electrolysis cell with two- (anode + cathode) or three-electrode (anode + cathode + reference electrode) configurations. Several reports have revealed that electropolymerization under special reaction conditions produces characteristic features with regard to morphology, electronic properties, and efficiency.

Atobe and coworkers reported the use of supercritical fluoroform as a solvent for the electropolymerization of thiophene and pyrrole, which resulted in the formation of denser and smoother films than those prepared using MeCN as the solvent [20]. They also reported the use of microflow chemistry for the electropolymerization of 3-hexylthiophene. The polymerization in a flow-type electrochemical cell proceeded without film formation, and the molecular weight of the polymers could be controlled by the choice of electrodes, solvent, electrolyte, and flow rate [21].

The electropolymerization of 3,4-ethylenedioxythiophene (EDOT) under special conditions has occasionally resulted in interesting outcomes. Wang and coworkers reported the electropolymerization of EDOT at the liquid–liquid interface of dichloromethane and water to prepare a free-standing composite film of poly(3,4-ethylenedioxythiophene) (PEDOT)-carbon nanotubes [22]. Akagi and coworkers reported the electropolymerization of EDOT in chiral nematic liquid crystal. Under this condition, the obtained PEDOT film had helical morphology, and calcination of the PEDOT gave a helical carbon and graphite film [23]. Lee and coworkers examined the electropolymerization of EDOT under the application of magnetic field, which yielded PEDOT with a more extended chain [24]. Inagi and coworkers reported the electropolymerization of EDOT under bipolar electrochemical conditions with the application of an alternating current, which yielded PEDOT fibers from the terminus of Au wire used as bipolar electrodes [25]. They also reported the in-plane growth of a PEDOT film on a glass substrate, unlike the out-of-plane film growth experienced under conventional electropolymerization conditions [26].

6.3 Post-Functionalization of Conducting Polymers by Electrochemical Redox

6.3.1 Functionalization of Conducting Polymers by Anodic Substitution

The electrosynthesis of various polymers via post-functionalization is described in this section. Post-functionalization is a concept to incorporate functional groups into a polymer backbone [27]. Post-functionalization can diversify one precursor polymer into a variety of polymers, depending on the functional groups and degree of functionalization, so that a polymer library can generally be constructed in an efficient manner. Post-functionalization can also introduce some substituents that may hinder the polymerization of the corresponding monomer units, i.e. incorporation of such functional groups is only achieved by post-functionalization.

Electrolysis is a promising methodology for the post-functionalization of polymeric materials because it allows control of the degree of reaction by monitoring the charge passed during the reaction. Due to the difficulty of electrolyzing polymers in homogeneous solution, the electrochemical post-functionalization of polymers has been performed using films fixed on an electrode. Therefore, nonconductive polymer films are not compatible with this idea. In this context, conducting polymer films have been the main interest for electrochemical post-functionalization [28].

Pioneering work on the post-functionalization of conducting polymer was reported by Pickup and coworkers [29]. They reported the substitution of poly(3-methylthiophene) (P3MT) by nucleophilic attack with chloride ions (Scheme 6.4). When Et_4NCl was used as a supporting electrolyte and a chlorine source, nucleophilic substitution of the oxidized P3MT-backbone by chloride ions occurred to give 4-chloro-substituted P3MT. This reaction is induced by the umpolung of P3MT upon electrochemical oxidation. As explained in the previous section, conducting polymers such as P3MT exhibit stable redox in the presence of less coordinating anions such as ClO_4^-, BF_4^-, and PF_6^-. On the other hand, CV measurement of P3MT shows an irreversible oxidation wave with the oxidation cycle, which suggests the chemical reaction occurs after electron transfer. The introduction of Cl-groups was determined by X-ray photoelectron spectroscopy (XPS) analysis, where C—Cl bond formation was confirmed [30]. The incorporation of electron-withdrawing chlorine atoms into the main chain of the conjugated

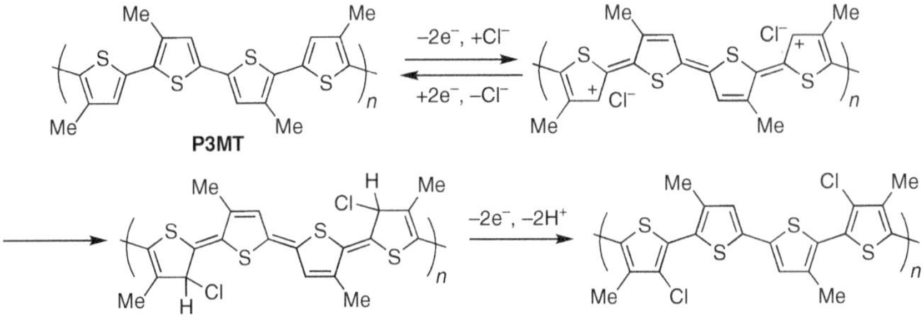

Scheme 6.4 Anodic chlorination of P3MT.

polymer changed the electronic properties of P3MT. Cl-modified P3MT had a slightly higher oxidation potential than the original polymer. Bromination and alkoxylation were also achieved in a similar manner [30, 31]. The introduction of methoxy groups was especially meaningful because the corresponding monomer, 3-methoxythiophene, is inactive for electropolymerization. On the other hand, iodination was not successful due to the poor nucleophilicity of iodide. Apart from thiophene derivatives, polypyrrole was also shown to be applicable for this methodology.

Bélanger and coworkers reported the chlorination of poly(3-(4-fluorophenyl) thiophene) in a similar manner to P3MT [32]. They also demonstrated the cleavage of C—Cl bonds by application of a cathodic potential, which resulted in the complete formation of the initial Cl-free polymer by H-transfer from the electrolyte. Iyoda and coworkers reported the electrochemical pyridination of poly(3-hexylthiophene) [33]. The introduction of pyridine groups was confirmed by Fourier-transform infrared (FT-IR) spectroscopy and CV measurements. Fabre and Simonet reported the anodic cyanation of PEDOT [34], where incorporation was up to 1.4 CN groups per 1,4-dimethoxybenzene unit, as evidenced by elemental analysis and XPS [35].

These reports determined the formation of desired structures via solid-state analysis. Inagi and Fuchigami have actively reported the electrochemical transformation of conjugated polymers and their structural characterization by solution nuclear magnetic resonance (NMR) analysis [28]. To solubilize the conjugated polymers, the alkylated monomers were designed and polymerized by common transition-metal-catalyzed crosscoupling polymerization. For example, the obtained polymer, poly(9,9′-dioctylfluorene-*alt*-thiophene), was subjected to the electrochemical oxidation reaction (Figure 6.4a) [36]. Chloride or bromide ions were used to achieve successful electrochemical halogenation of the fluorene-thiophene-alternating copolymers. ^{1}H NMR showed the disappearance of the corresponding peak, which supported the formation of the desired structure. The soluble nature of the polymer allowed analysis of the resulting polymer by gel-permeation chromatography (GPC). GPC analysis suggested there was no significant change in the molecular weight between the precursor polymer and resultant polymer. This result suggested that no side reactions such as coupling of polymers proceeded [36].

They also reported the fluorination of a fluorene-containing conjugated polymer (Figure 6.4b). For anodic fluorination, amine- or ammonium-based HF-containing ionic liquids are used as a fluorine source [37]. This fluorination method was established by Fuchigami and coworkers, and is known as one of the most successful processes in organic electrosynthesis because it can enable fluorination of numerous organic compounds under mild, safe, and environmentally benign conditions. Introduction of the thioether group as an electroauxiliary group in the precursor polymer was a key to apply this fluorination process in polymer reactions [38, 39].

The electrochemical polymer reaction has also been utilized for random and block copolymers [40]. A series of poly(3-alkylthiophene) (P3AT) and poly(3-alkylselenophene) (P3AS) and their random and block copolymers were

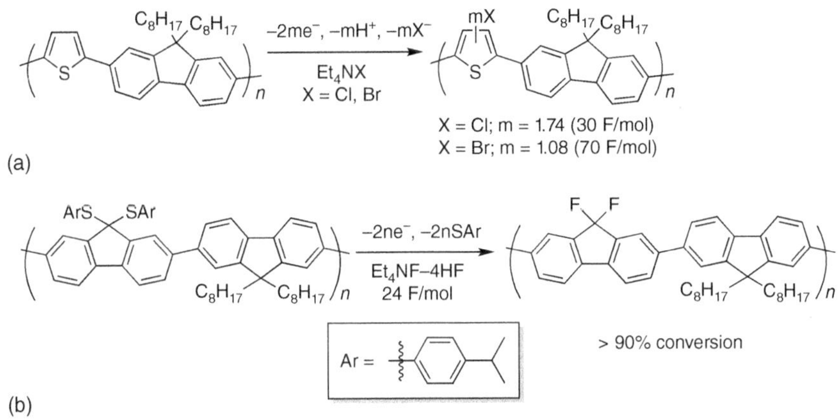

Figure 6.4 Electrochemical halogenation of fluorene-containing conjugated polymers. (a) Chlorination and bromination of a thiophene-fluorene alternating copolymer and (b) fluorination of a fluorene-based polymer. Source: Modified from Inagi et al. [36].

systematically synthesized by Kumada catalyst-transfer polycondensation. The Faraday efficiency for anodic chlorination was higher in the case of P3AS than P3AT. The random copolymer showed similar results as the polymer blend of each segment. A block copolymer of P3HT-*b*-P3HS showed 40% chlorination after electrolysis at 8 F/mol, and the selenophene-containing segment was found to contain more chlorine atoms. When the effect of the side chain was compared, a branched side chain completely hindered the reaction. Based on this result, they prepared a block copolymer of poly(3-(2-ethylhexyl)thiophene) and poly(3-hexylselenophene), P3HS-*b*-P3EHS, and segment-selective polymer reaction was achieved for the first time (Scheme 6.5) [40].

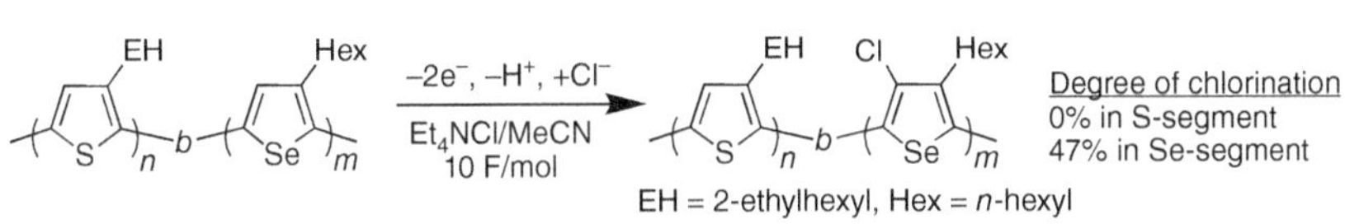

Scheme 6.5 Segment-selective anodic chlorination of rod–rod block copolymer. Source: Modified from Shida et al. [40].

In anodic oxidation, the commonly faced problem is that better nucleophiles have lower oxidation potential due to their electron-rich nature, which results in the undesired oxidation of nucleophiles. To avoid this problem, several strategies have been proposed, including the cation pool method [41], flow electrosynthesis [42], and the development of redox mediators [43]. However, these methods have not been utilized for the electrochemical polymer reaction. Instead, a unique reaction system using BF_3OEt_2 was reported by Inagi and coworkers [44]. BF_3OEt_2 is known to destabilize aromatic monomers and lower the oxidation potential. The same was evident for the conjugated polymers. When BF_3OEt_2 was used as a solvent for the oxidation of P3HT, the oxidation potential was significantly lowered.

The oxidation potential of chloride ions also increased due to complexation with BF_3OEt_2. As a consequence, competitive oxidation of chloride ions was suppressed and the Faraday efficiency was increased significantly. This system also worked with BF_3OEt_2 added to the MeCN solvent.

6.3.2 Cathodic Reduction and Paired Reactions

Electrochemical polymer reactions have been achieved by the cathodic reduction mechanism. Poly(9,9-dioctylfluorene-*alt*-9-fluorenone) ($M_n = 4100$) was chemically synthesized and used for cathodic reduction (Scheme 6.6) [45]. The polymer was fixed on a zinc cathode, which has a high overpotential toward proton reduction. Pr^iOH was used as a protic solvent. The polymer film underwent reductive doping, then quantitative hydrogenation and dehydroxylation of the fluorenone moiety occurred ($4e^-/4H^+$ process) to give the fluorene structure after the passage of 16 F/mol of charge under constant current conditions. The degree of hydrogenation was tunable by changing the passage of charge.

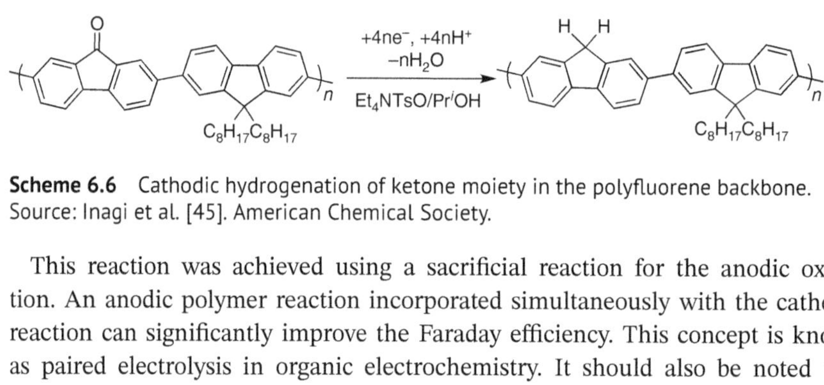

Scheme 6.6 Cathodic hydrogenation of ketone moiety in the polyfluorene backbone. Source: Inagi et al. [45]. American Chemical Society.

This reaction was achieved using a sacrificial reaction for the anodic oxidation. An anodic polymer reaction incorporated simultaneously with the cathodic reaction can significantly improve the Faraday efficiency. This concept is known as paired electrolysis in organic electrochemistry. It should also be noted that the reaction of two polymers at once and isolation of the products is unfeasible in homogeneous reaction systems, which indicates the uniqueness of this idea. As a proof of concept, a fluorenol-containing conjugated polymer, poly(9-fluorenol-*alt*-9,9′-dioctylfluorene) **P1**, was prepared and fixed on both the anode (Pt) and cathode (Zn). Electrolysis was conducted in Et_4NOTs/Pr^iOH electrolyte solution. Anodic oxidation gave the fluorenone-containing polymer **P2**, while cathodic reduction gave fluorene-containing **P3**. Quantitative conversion was observed after passing 16 F/mol of charge (Scheme 6.7) [46].

6.3.3 Functionalization of Polyaniline by the CRS Method

Polyaniline is a widely studied conducting polymer due to its multiple electrochromic nature derived from oxidative doping and the protonation of amino groups. One oxidized state of polyaniline, known as the emeraldine base (EB) state, is reactive to nucleophiles. Han and coworkers developed the concurrent reduction and substitution (CRS) method, where an EB state of polyaniline is reduced to the leucoemeraldine base (LB) state with concomitant incorporation of nucleophiles

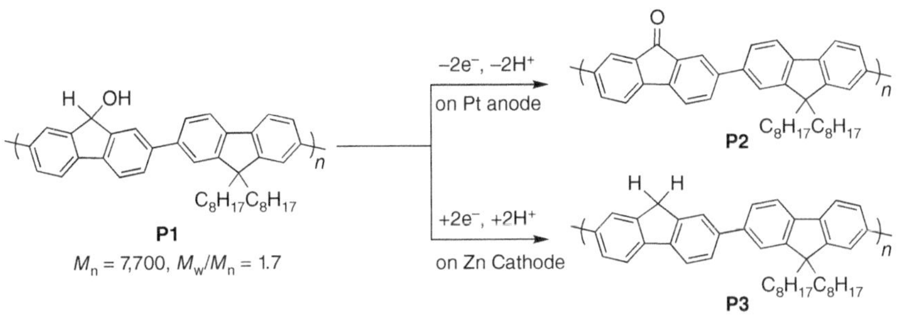

Scheme 6.7 Paired electrolysis of fluorene-containing conjugated polymer. Source: Inagi et al. [46]. John Wiley & Sons.

such as dialkylamines [47, 48]. The EB state of polyaniline prepared on the electrode surface by electropolymerization was reacted with fluoride ion derived from Bu_4NF in methanol to give fluorine-modified polyaniline in the main chain (Scheme 6.8). Repetitive treatment of the oxidation/CRS cycle enabled the quantitative fluorination of polyaniline (one fluorine atom per unit) [49]. The fluorinated-polyaniline obtained exhibited better conductivity than its precursor. Polymerization of the fluorinated aniline-monomer does not give a high-molecular-weight equivalent, which suggests the advantage of post-functionalization.

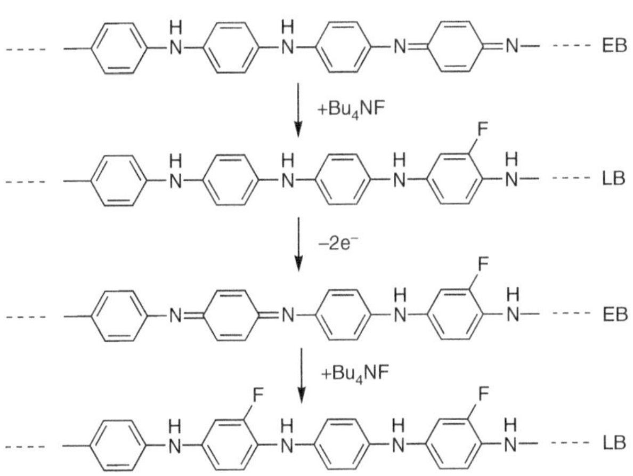

Scheme 6.8 Electrochemical fluorination of polyaniline via the CRS method.

6.3.4 Oxidation-Induced Intramolecular Cyclization of Conducting Polymer

Higher planarity of the polymer backbone can improve the conductivity, while such monomers tend to be less soluble in organic solvents, which hampers the polymerization process. Thus, post-polymerization intramolecular cyclization is an attractive concept for the design of highly functional conjugated polymers.

Song and Swager reported intramolecular cyclization triggered by anodic oxidation in the main chain of a conjugated polymer (Scheme 6.9) [50]. They designed a monomer composed of two bithiophene terminal groups for electropolymerization and binaphthol at the center. Anodic electropolymerization successfully afforded a polymer film on the electrode under potential sweep polymerization conditions. The obtained polymer film exhibited a reversible redox nature in the relatively low potential region, while oxidation became irreversible in the higher potential region due to oxidation-induced intermolecular cyclization at the binaphthol group to give a *peri*-xanthenoxanthene motif in the polymer backbone. In situ conductivity measurement of the pre- and post-cyclization polymers showed a clear difference in their electronic properties. The post-cyclization polymer had a lower conductivity onset potential and the maximum conductivity was almost doubled. UV–Vis measurement supported the formation of the desired structure.

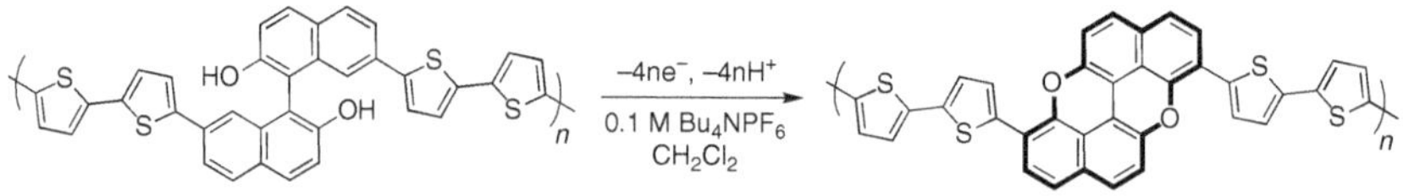

Scheme 6.9 Intramolecular cyclization of binaphthol moiety to give *peri*-xanthenoxanthene motif. Source: Modified from Song and Swager [50].

6.3.5 Electrogenerated Transition-Metal Catalysts for Post-Functionalization

Electrogenerated active catalysts have been used for the post-modification of conducting polymer films. Larsen and coworkers reported that Cu(I) species generated by the electroreduction of Cu(II) can catalyze the azide-alkyne cycloaddition reaction, i.e. the click reaction. This method is especially called the electroclick reaction [51]. A PEDOT derivative bearing azido group (PEDOT-N_3) at the side chain was prepared by electropolymerization of the corresponding monomer. The substrate was immersed in a solution containing Cu(II) salt, alkynes, and electrolyte, and then cathodic current was passed through the PEDOT film. Spectroscopic analysis revealed that the on-surface click reaction proceeded to give the functional surface, due to introduction of the alkynes. They also reported the electroclick reaction using a copper electrode with a slit, which generated a concentration gradient of the Cu(I) catalyst by diffusion to give a functionalized surface in a gradient manner [52]. Inagi and coworkers reported the electroclick reaction using a bipolar electrode to give a molecular gradient with various functionalities [53].

6.4 Synthesis of Nonconjugated Polymers by Electrochemical Redox

6.4.1 Electropolymerization of Electroactive Polymers

Electrochemical reaction occurs at the solid–liquid interface; therefore, the continuous growth of a polymer film by electrochemical means requires a polymer that can pass electricity. Therefore, the electrochemical synthesis of nonconjugated polymers is more challenging. However, nonconjugated polymers that bear redox-active groups exhibit conductivity by the hopping of charge through the neighboring redox-active groups.

Vinylpyridine–metal complexes are known to be polymerized by cathodic reduction. Murray and coworkers reported a pioneering result for this process in the 1980s [54]. Zhong and coworkers have extended this idea and applied it to various polypyridine complexes [55]. The propagation of this polymerization is considered to be a radical mechanism. The monomers are composed of transition metals (Ru, Os, Fe, Co, Cr, and Ir) and ligands (pyridine, bipyridine, terpyridine) with vinyl groups. Scheme 6.10 shows the structure and polymerization of a representative ruthenium complex, $[Ru(dvbpy)(bpy)]^{2+}$ (bpy = 2,2′-bipyridine, dvbpy = 5,5′-divinyl-2,2′-bipyridine). This complex exhibited one oxidation and three distinct reduction waves during CV measurements. Reductive polymerization proceeded under potential sweep conditions, even under scan up to the first reduction wave, which indicated a one-electron reduction was sufficient to trigger the polymerization. The resultant film showed an identical redox couple of $Ru^{II/III}$, which indicated the redox center was intact after the polymerization.

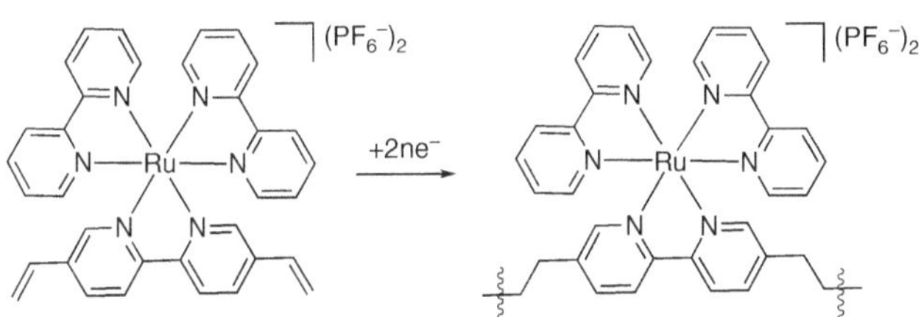

Scheme 6.10 Polymerization of $[Ru(dvbpz)(bpz)_2]$ by cathodic reduction.

6.4.2 Electrochemical Redox-Controlled Polymerization

Small electrogenerated active molecules can act as initiators for the ionic and radical polymerization of vinyl monomers. For instance, methyl radicals generated by Kolbe electrolysis of acetate can initiate radical polymerization. Although the mechanism is interesting, the heterogeneous nature of initiator generation in this method makes it difficult to access high-molecular-weight polymers.

In modern polymer science, the development of living radical polymerization such as atom transfer radical polymerization (ATRP) has been significant [56]. ATRP has expanded the polymer library and sparked a wide range of polymers with complex architectures [56]. In ATRP, the catalytic active species is Cu(I), which reduces the polymerization terminus (typically haloalkanes) by a one-electron process to generate a propagating group (carbon radical) and Cu(II)X (X represents a halogen).

An equilibrium exists between dormant and active species, and [Cu(I)]/[Cu(II)] determines the polymerization rate. The ratio of [Cu(I)] and [Cu(II)] can be tuned electrochemically, known as electrochemical atom transfer radical polymerization (*e*ATRP), was proposed by Matyjaszewski and coworkers in 2011

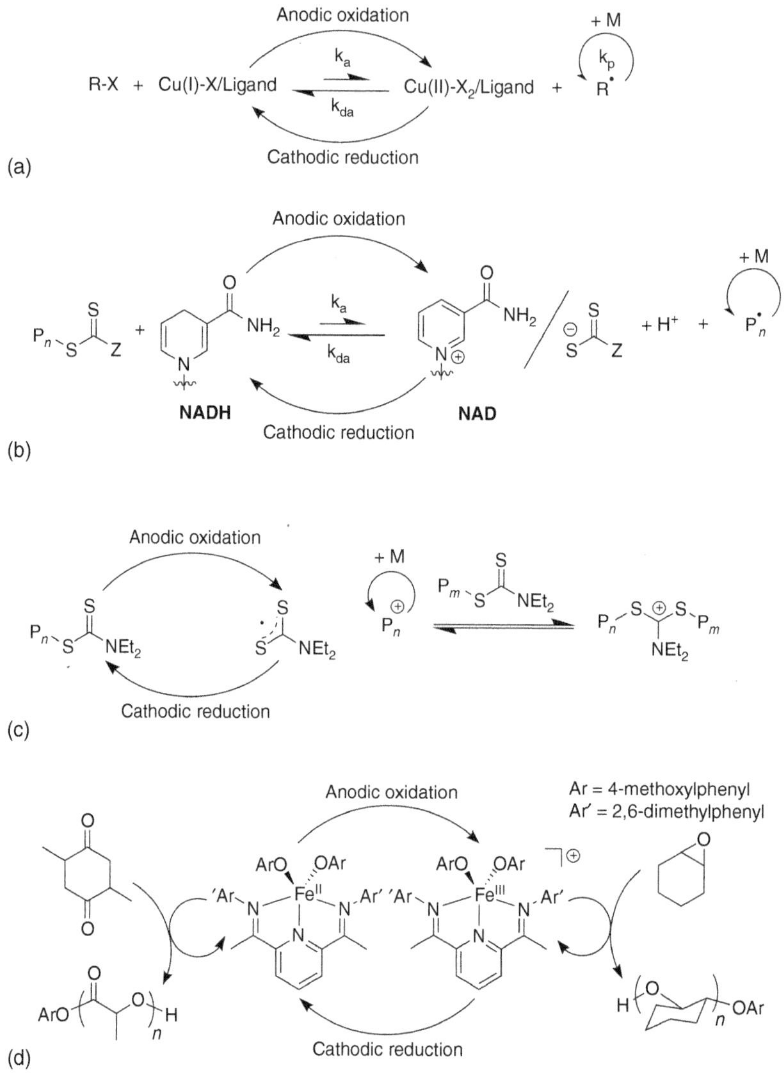

Figure 6.5 Recently developed polymerization methodologies to afford nonconjugated polymers under electrochemical control. (a) eATRP. Source: Modified from Magenau et al. [57], (b) eRAFT, (c) electrochemically controlled cationic polymerization. Source: Peterson et al. [58]. American Chemical Society, and (d) electrochemical switching of polymerization between lactide and epoxide. Source: Qi et al [59]. American Chemical Society.

(Figure 6.5a) [57]. They demonstrated that living polymerization proceeded by passing a cathodic current, whereas passing an anodic current completely stopped the polymerization.

After this report, other polymerization systems using electrochemical means were also reported. Matyjaszewski et al. [60] and Yan et al. [61] independently reported the electrochemical reversible addition-fragmentation chain transfer polymerization, i.e. electrochemical reversible addition-fragmentation chain transfer polymerization (*e*RAFT) (Figure 6.5b). Lin, Fors and coworkers reported the electrochemically controlled cationic polymerization of vinyl ether using dithiocarbamate as a chain transfer agent and (2,2,6,6-tetramethyylpiperidin-1-yl)oxyl as a mediator (Figure 6.5c) [58]. Byers and coworkers employed electrochemistry to switch the oxidation state of their iron alkoxide complex bearing bis(imino)pyridine ligands, which is active for the polymerization of lactide or epoxide, depending on the oxidation state of the iron center (Figure 6.5d) [59]. Based on this system, they successfully synthesized a poly(lactic acid-*b*-cyclohexene oxide) block copolymer and controlled the length of each segment electrochemically.

The idea of electrochemically controlled polymerization was also employed to achieve surface-initiated polymerization. Based on the *e*ATRP concept, Zhou and coworkers prepared an initiator-modified glass substrate and sandwiched this with a cathode. In this system, electrogenerated Cu(I) diffuses to the surface of the initiator-modified substrate to catalyze the polymerization. In this way, surface-grafted polymers, i.e. polymer brushes, were obtained in a facile manner [62]. Inagi and coworkers reported the fabrication of a gradient polymer brush by *e*ATRP using a bipolar electrode [63].

6.4.3 Electrochemically Induced Film Formation via Crosslinking

Electrogenerated catalysts have also been used for the crosslinking reaction of polymers. Boulmedais and coworkers reported the film deposition process using electrochemical click reaction [64]. They prepared two different polymers with azide and alkyne moieties at their respective side chains. The electrolyte solution contained these two polymers, a solvent and Cu(II). Electrochemical reduction of Cu(II) to Cu(I) catalyzed the azide-alkyne cycloaddition reaction (click reaction) to form the crosslinked polymer film on the electrode surface. The polymer film continued growing during the electrolysis, which was monitored using an electrochemical quartz crystal microbalance. Based on this technology, they also prepared a supermolecular polyrotaxane film [65]. Electrogenerated acid was used to induce hydrolysis of an anionic polymer to give the cationic poly(allylamine), which underwent complexation with anionic poly(styrenesulfonate) to give the polymer film on the electrode surface [66]. The same group also proposed a mussel-inspired film formation strategy based on the electrogeneration of *o*-benzoquinone derivative from the corresponding catechol, followed by reaction with poly(allylamine) to give an imine-linked polymer film [67].

6.5 Conclusion

In this chapter, we have summarized the general concept and various recent developments in electrochemical redox-mediated polymer synthesis. This field is relatively new and still not well explored compared to the electrosynthesis of small molecules. However, there is a significant trend to use electricity for the preparation of various polymeric materials, and new ideas and strategies are actively discussed.

We consider that these efforts will ultimately assist in the elimination of hazardous chemicals, harsh conditions, and harmful treatments from the industrial processes used for the synthesis of polymeric materials.

References

1 Heinze, J., Frontana-Uribe, B.A., and Ludwigs, S. (2010). *Chem. Rev.* 110: 4724–4771.
2 Beaujuge, P.M. and Reynolds, J.R. (2010). *Chem. Rev.* 110: 268–320.
3 Smela, E. (2003). *Adv. Mater.* 15: 481–494.
4 Ibanez, J.G., Rincón, M.E., Gutierrez-Granados, S. et al. (2018). *Chem. Rev.* 118: 4731–4816.
5 Sekiguchi, K., Atobe, M., and Fuchigami, T. (2002). *Electrochem. Commun.* 4: 881–885.
6 Chen, W. and Xue, G. (2005). *Prog. Polym. Sci.* 30: 783–811.
7 Guo, X., Baumgarten, M., and Müllen, K. (2013). *Prog. Polym. Sci.* 38: 1832–1908.
8 Huang, H. and Pickup, P.G. (1998). *Chem. Mater.* 10: 2212–2216.
9 Smith, E.L., Glidle, A., Mortimer, R.J., and Ryder, K.S. (2007). *Phys. Chem. Chem. Phys.* 9: 6098–6105.
10 Latonen, R., Kvarnström, C., and Ivaska, A. (1999). *Electrochim. Acta* 44: 1933–1943.
11 Roncali, J., Blanchard, P., and Frère, P. (2005). *J. Mater. Chem.* 15: 1589–1610.
12 Gohier, F., Frère, P., and Roncali, J. (2013). *J. Organomet. Chem.* 78: 1497–1503.
13 Kingsborough, R.P. and Swager, T.M. (1998). *Adv. Mater.* 10: 1100–1104.
14 Ruhlmann, L., Schulz, A., Giraudeau, A. et al. (1999). *J. Am. Chem. Soc.* 121: 6664–6667.
15 Saito, N., Kanbara, T., Nakamura, Y. et al. (1994). *Macromolecules* 27: 756–761.
16 Utley, J.H.P. and Gruber, J. (2002). *J. Mater. Chem.* 12: 1613–1624.
17 Miller, R.D. and Michl, J. (1989). *Chem. Rev.* 89: 1359–1410.
18 Shono, T., Kashimura, S., Ishifune, M., and Nishida, R. (1990). *J. Chem. Soc., Chem. Commun.*: 1160–1161.
19 Kashimura, S., Ishifune, M., Yamashita, N. et al. (1999). *J. Organomet. Chem.* 64: 6615–6621.
20 Atobe, M., Ohsuka, H., and Fuchigami, T. (2004). *Chem. Lett.* 33: 618–619.
21 Mizuno, M., Tateno, H., Matsumura, Y., and Atobe, M. (2017). *React. Chem. Eng.* 2: 642–645.
22 Gao, L., Mao, X., Zhu, H. et al. (2014). *Electrochim. Acta* 136: 97–104.
23 Matsushita, S., Yan, B., Yamamoto, S. et al. (2014). *Angew. Chem. Int. Ed.* 53: 1659–1663.
24 Cho, M.S., Yun, Y.Y., Nam, J.D. et al. (2008). *Synth. Met.* 158: 1043–1046.
25 Koizumi, Y., Shida, N., Ohira, M. et al. (2016). *Nat. Commun.* 7: 1–6.
26 Watanabe, T., Ohira, M., Koizumi, Y. et al. (2018). *ACS Macro Lett.* 7: 551–555.
27 Blasco, E., Sims, M.B., Goldmann, A.S. et al. (2017). *Macromolecules* 50: 5215–5252.

28 Inagi, S. and Fuchigami, T. (2014). *Macromol. Rapid Commun.* 35: 854–867.
29 Qi, Z.G. and Pickup, P.G. (1992). *J. Chem. Soc., Chem. Commun.*: 1675–1676.
30 Qi, Z. and Pickup, P.G. (1993). *Anal. Chem.* 65: 696–703.
31 Qi, Z., Rees, N.G., and Pickup, P.G. (1996). *Chem. Mater.* 8: 701–707.
32 Soudan, P., Lucas, P., Breau, L., and Bélanger, D. (2000). *Langmuir* 16: 4362–4366.
33 Li, Y., Kamata, K., Asaoka, S. et al. (2003). *Org. Biomol. Chem.* 1: 1779–1784.
34 Fabre, B. and Simonet, J. (1996). *J. Electroanal. Chem.* 416: 187–189.
35 Fabre, B., Kanoufi, F., and Simonet, J. (1997). *J. Electroanal. Chem.* 434: 225–234.
36 Inagi, S., Hayashi, S., Hosaka, K., and Fuchigami, T. (2009). *Macromolecules* 42: 3881–3883.
37 Fuchigami, T. and Inagi, S. (2011). *Chem. Commun.* 47: 10211–10223.
38 Inagi, S., Hayashi, S., and Fuchigami, T. (2009). *Chem. Commun.*: 1718–1720.
39 Hayashi, S., Inagi, S., and Fuchigami, T. (2009). *Macromolecules* 42: 3755–3760.
40 Shida, N., Okazaki, D., Kurioka, T. et al. (2017). *ChemElectroChem* 4: 1824–1827.
41 Yoshida, J., Shimizu, A., and Hayashi, R. (2018). *Chem. Rev.* 118: 4702–4730.
42 Atobe, M., Tateno, H., and Matsumura, Y. (2018). *Chem. Rev.* 118: 4541–4572.
43 Francke, R. and Little, R.D. (2014). *Chem. Soc. Rev.* 43: 2492–2521.
44 Kurioka, T., Nishiyama, H., Tomita, I., and Inagi, S. (2018). *ChemElectroChem* 5: 753–755.
45 Inagi, S., Koseki, K., Hayashi, S., and Fuchigami, T. (2010). *Langmuir* 26: 18631–18633.
46 Inagi, S., Nagai, H., Tomita, I., and Fuchigami, T. (2013). *Angew. Chem. Int. Ed.* 52: 6616–6619.
47 Han, C.C. and Jeng, R.C. (1997). *Chem. Commun.*: 553–554.
48 Han, C.C., Hong, S.P., Yang, K.F. et al. (2001). *Macromolecules* 34: 587–591.
49 Han, C.C. and Chen, H.Y. (2007). *Macromolecules* 40: 8969–8973.
50 Song, C. and Swager, T.M. (2009). *Macromolecules* 42: 1472–1475.
51 Hansen, T.S., Daugaard, A.E., Hvilsted, S., and Larsen, N.B. (2009). *Adv. Mater.* 21: 4483–4486.
52 Hansen, T.S., Lind, J.U., Daugaard, A.E. et al. (2010). *Langmuir* 26: 16171–16177.
53 Shida, N., Ishiguro, Y., Atobe, M. et al. (2012). *ACS Macro Lett.* 1: 656–659.
54 Abruña, H.D., Denisevich, P., Umana, M. et al. (1981). *J. Am. Chem. Soc.* 103: 1–5.
55 Zhong, Y.W., Yao, C.J., and Nie, H.J. (2013). *Coord. Chem. Rev.* 257: 1357–1372.
56 Matyjaszewski, K. and Tsarevsky, N.V. (2014). *J. Am. Chem. Soc.* 136: 6513–6533.
57 Magenau, A.J.D., Strandwitz, N.C., Gennaro, A., and Matyjaszewski, K. (2011). *Science* 332: 81–84.
58 Peterson, B.M., Lin, S., and Fors, B.P. (2018). *J. Am. Chem. Soc.* 140: 2076–2079.
59 Qi, M., Dong, Q., Wang, D., and Byers, J.A. (2018). *J. Am. Chem. Soc.* 140: 5686–5690.
60 Wang, Y., Fantin, M., Park, S. et al. (2017). *Macromolecules* 50: 7872–7879.
61 Sang, W., Xu, M., and Yan, Q. (2017). *ACS Macro Lett.* 6: 1337–1341.
62 Li, B., Yu, B., Huck, W.T.S. et al. (2013). *J. Am. Chem. Soc.* 135: 1708–1710.

63 Shida, N., Koizumi, Y., Nishiyama, H. et al. (2015). *Angew. Chem. Int. Ed.* 54: 3922–3926.
64 Rydzek, G., Jierry, L., Parat, A. et al. (2011). *Angew. Chem. Int. Ed.* 50: 4374–4377.
65 Rydzek, G., Garnier, T., Schaaf, P. et al. (2013). *Langmuir* 29: 10776–10784.
66 Dochter, A., Garnier, T., Pardieu, E. et al. (2015). *Langmuir* 31: 10208–10214.
67 Maerten, C., Garnier, T., Lupattelli, P. et al. (2015). *Langmuir* 31: 13385–13393.

7

Chemical Paired Transformations

Eiji Shirakawa

Kwansei Gakuin University, School of Science and Technology, Department of Applied Chemistry for Environment, 2-1 Gakuen, Sanda, 669-1337, Hyogo, Japan

7.1 Introduction

In the substitution reaction between an aromatic electrophile (Ar–X: X^- = (pseudo) halide) and a carbon nucleophile (m–R, H–R) to give Ar–R, neither oxidation nor reduction is usually included as the whole process. However, such a substitution reaction often includes both oxidation and reduction to complete the process. One of the most important substitution reactions that include both oxidation and reduction steps is the transition metal-catalyzed cross-coupling reaction of aryl halides, which have to be activated by reduction to be ready for the substitution reaction because they do not undergo S_N2 or S_N1 reaction [1]. In particular, the cross-coupling reaction of Ar–X with arylmetals (m–Ar′) is important as an indispensable method to prepare biaryl frameworks (Ar–Ar′), which are an important class of substructures appearing in diverse pharmaceuticals and materials (Scheme 7.1). Low-valent transition metals such as palladium and nickel are often used as catalysts for the cross-coupling reaction because they perform well in both the reductive activation of Ar–X and the oxidation in a later step to make ends meet. An Ar–X is an aryl cation equivalent and receives two electrons through oxidative addition to a low-valent transition metal, exemplified by Pd^0 thus far, being converted into an aryl anion equivalent as a form of Ar–Pd^{II}–X, which goes through transmetalation with m–Ar′. The resulting complex (Ar–Pd^{II}–Ar′), undergoes reductive elimination to give the coupling product (Ar–Ar′), and regenerates Pd^0, where Ar^- is regarded to be oxidized back to Ar^+, the oxidation state of the aryl halide. Transition metal catalysts had been considered to be indispensable to carry out the process until the electron-catalyzed cross-coupling reaction appeared (cf. 7.3). The utility of the transition metal-catalyzed cross-coupling reaction is authorized by 2010 Nobel Prize for chemistry, given for the palladium-catalyzed reaction of aryl halides with organoboron/zinc reagents (Suzuki–Miyaura/Negishi coupling) or alkenes (Mizoroki–Heck reaction).

On the other hand, aryl halides are known to be activated also by single-electron reduction to be ready for the substitution reaction with *in situ*-generated anionic

Organic Redox Chemistry: Chemical, Photochemical and Electrochemical Syntheses, First Edition.
Edited by Jun-Ichi Yoshida and Frédéric W. Patureau.

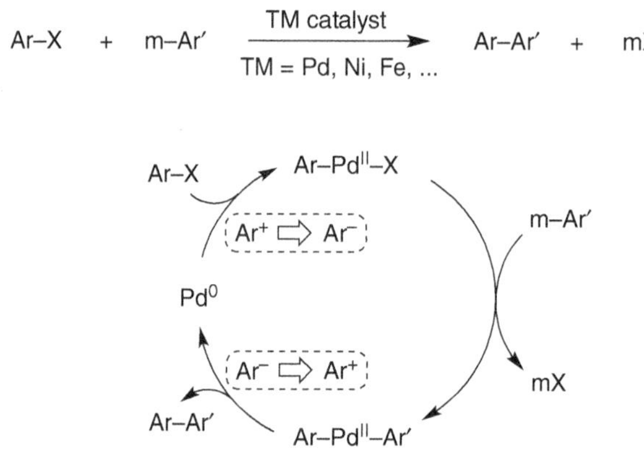

Scheme 7.1 Transition metal-catalyzed cross-coupling reaction of aryl halides with organometals.

nucleophiles such as enolates of ketones and thiolates with the aid of an electron donor often accompanied by photoirradiation (Scheme 7.2). The reaction of this type is called as $S_{RN}1$ (substitution radical nucleophilic unimolecular) reaction and has a history slightly longer than the transition metal-catalyzed cross-coupling reaction [2]. Single-electron transfer (SET) from an electron donor (D) to Ar–X gives an anion radical ($[Ar–X]\cdot^-$) to initiate a radical chain (step *a*). Elimination of the halide (X^-) from this anion radical gives the aryl radical (Ar·) (step *b*), which reacts with an anionic nucleophile (Nu^-), generated *in situ* by deprotonation (step *c*) from a pronucleophile (H–Nu), to give an anion radical ($[Ar–Nu]\cdot^-$) of the substitution product (step *d*). Finally, SET from $[Ar–Nu]\cdot^-$ to Ar–X gives the substitution product (Ar–Nu) and regenerates anion radical $[Ar–X]\cdot^-$ (step *e*), where the reduction to activate Ar–X and the oxidation to make ends meet simultaneously take place. It features intrinsic efficiency that such a tiny species as an electron, derived from an electron donor, is regarded to work as a catalyst, though the concept "electron catalysis" has just appeared quite recently. Transition metals are not required here, but the reaction had not been applicable to sp^2-carbon nucleophiles and consequently biaryls cannot be prepared by this method. Furthermore, the reaction procedure

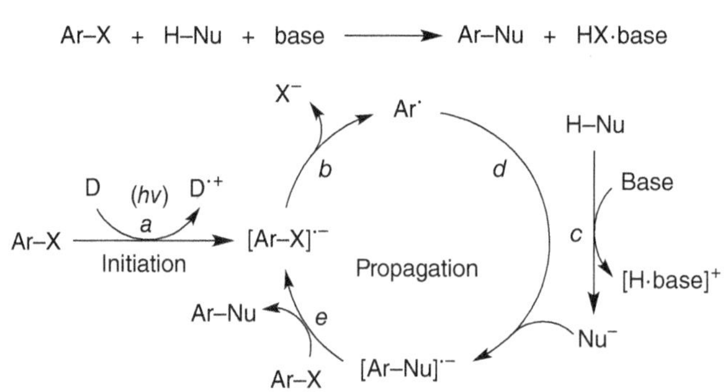

Scheme 7.2 $S_{RN}1$ reaction of aryl halides.

sometimes is bothersome, where Birch reduction conditions using an alkali metal in liquid ammonia are often required.

This chapter deals with the substitution reaction of aryl halides proceeding through a radical-chain mechanism containing a simultaneous single-electron reduction and oxidation step (such as step *e* in Scheme 7.2) in a transition metal-free manner: the $S_{RN}1$-type reaction is treated but only for so-called modified $S_{RN}1$ reactions developed after 2008 (Scheme 7.3) [3].

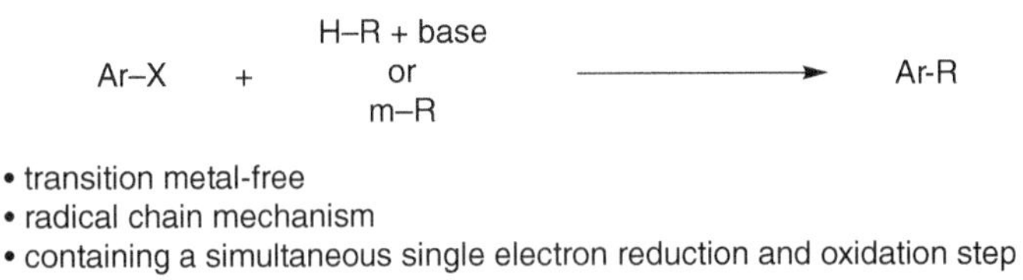

Scheme 7.3 Electron-catalyzed substitution reaction of aryl halides.

7.2 Direct Arylation of Arenes with Aryl Halides

Aryl radicals are unstable to react even with ordinary solvents having $C(sp^3)$–H bonds, and thus the scope of efficient reaction partners is extremely narrow. Arenes are one of the most suitable reaction partners, where the instability of aryl radicals allows themselves to add to benzene even though it breaks the aromaticity (Scheme 7.4). Abstraction of hydrogen ($H\cdot = H^+ + e^-$) from the resulting cyclohexadienyl radical gives the corresponding biaryl. In this process consisting of addition of a radical to an aromatic ring and elimination of another radical, called as homolytic aromatic substitution (HAS), the second step has a strong driving force to recover the lost aromaticity, being suitable for capturing high reactivities of aryl radicals [4]. However, when readily available aryl halides are used as aryl radical sources, compatibility between single-electron reduction and oxidation, required for aryl radical generation and hydrogen abstraction steps, respectively, has to be maintained [5]. As described above, $S_{RN}1$ reaction includes simultaneous single-electron reduction and oxidation in the final step (step *e* in Scheme 7.2), and this character has a potential to solve the problem of the conflicting demands mentioned above. And this holds true as follows.

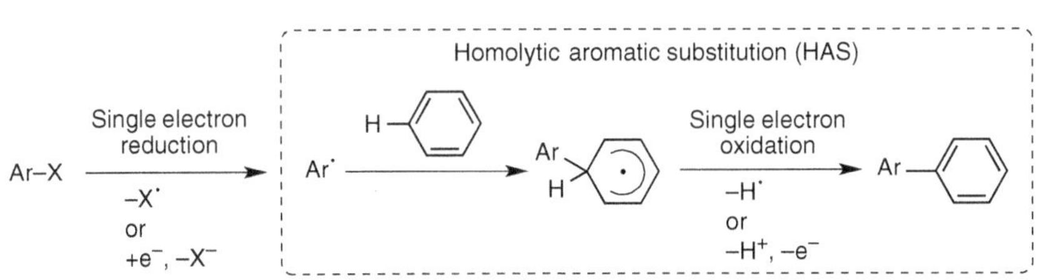

Scheme 7.4 Biaryl synthesis from aryl halides through homolytic aromatic substitution.

Until recently, transition metal catalysts were considered to be indispensable for biaryl synthesis from aryl halides and arenes, except for the cases that utilize rather bothersome methods such as ultraviolet light irradiation [6]. In 2008, Itami and coworkers reported that π-electron-deficient N-heteroarenes such as pyrazine and pyridine undergo the substitution reaction with aryl iodides to give the arylated heteroarenes with no aid of transition metal catalysis (Scheme 7.5) [7]. Use of a strong base such as potassium *tert*-butoxide is indispensable, where microwave irradiation accelerates the reaction. As radical scavengers inhibit the reaction, involvement of radical intermediates was mentioned, but the details were not clarified at that time.

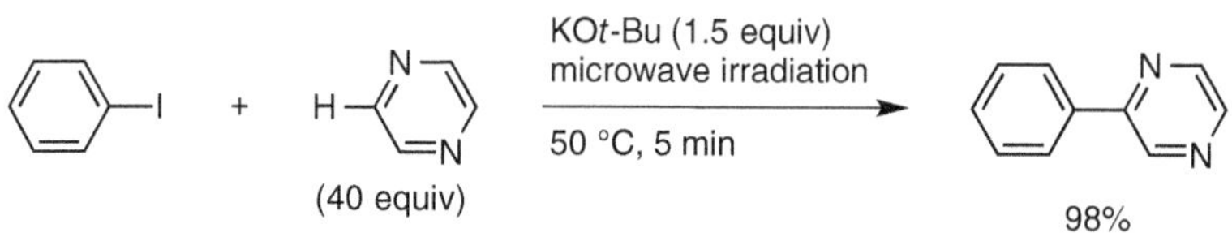

	98%
without microwave (120 °C, 13 h)	79%
NaO*t*-Bu instead of KO*t*-Bu	<1%
in the presence (1 equiv) of galvinoxyl	<1%
in the presence (1 equiv) of TEMPO	<1%
in the presence (1 equiv) of acrylonitrile	<1%
Ph–Br instead of Ph–I (80 °C, 0.5 h)	54%
Ph–Cl instead of Ph–I (80 °C, 0.5 h)	<1%

Scheme 7.5 Direct arylation of N-heteroarenes with aryl halides. Source: Modified from Yanagisawa et al. [7].

In 2010, three research groups reported, at almost the same time, the reaction of benzene with aryl halides using a stoichiometric amount of a *tert*-butoxide and a substoichiometric amount of a nitrogen bidentate ligand to give biaryls (Scheme 7.6) [8]. As expansion of the scope of arenes from π-deficient N-heteroarenes to benzene derivatives drew much attention as a transition metal-free coupling reaction, papers containing "transition metal free" in their titles keep increasing after 2010.

At the time of the early publications in 2010, the following clues were available on the reaction mechanism of the direct arylation of arenes with aryl halides. Shirakawa and Hayashi showed that SET from a mixture of sodium *tert*-butoxide and 4,7-diphenyl-1,10-phenanthroline (Ph-phen) to an aryl halide (Ar–X) gives anion radical $[Ar–X]^{\cdot -}$, which undergoes elimination of X^- to give the aryl radical intermediate ($Ar^{\cdot}$) [8a]. On the other hand, Kwong and Lei showed that the reaction proceeds through the anion radical ($[Ar–Ph]^{\cdot -}$ in the reaction with benzene) of the arylation product [8b]. From these observations in combination with the concept "base-promoted homolytic aromatic substitution (BHAS)" developed earlier by Russell [9], a modified $S_{RN}1$ mechanism, built-in with BHAS as shown in Scheme 7.7, was proposed by Studer and Curran [10]. A radical chain starts with SET from a single-electron donor (D) to an aryl halide (Ar–X), giving an anion radical ($[Ar–X]^{\cdot -}$) (step *a*). Elimination of X^- from $[Ar–X]^{\cdot -}$ gives the aryl radical ($Ar\cdot$) (step *b*), which adds to a benzene ring (step *c*). Deprotonation from

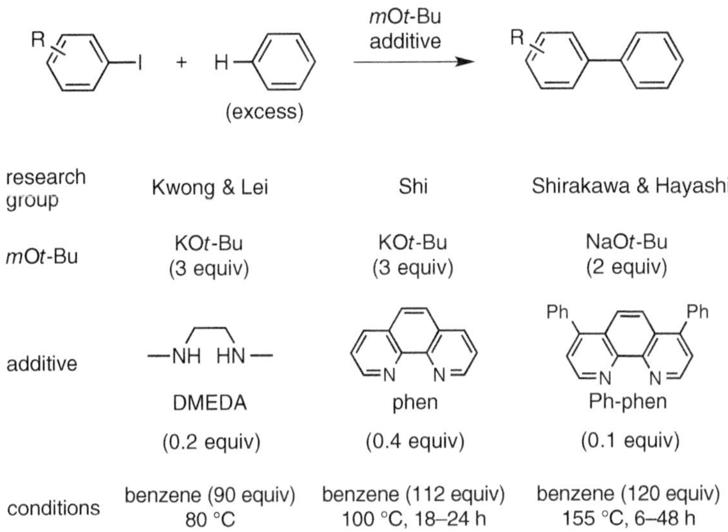

Scheme 7.6 Direct arylation of benzene derivatives with aryl halides. Source: Modified from Shirakawa et al. [8a], Liu et al. [8b] and Sun et al. [8c].

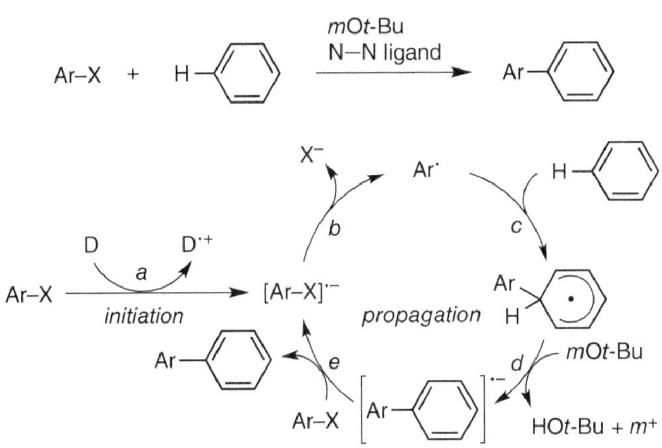

Scheme 7.7 Direct arylation through a base-promoted homolytic aromatic substitution.

the cyclohexadienyl radical recovers the lost aromaticity to give an anion radical ($[Ar–Ph]^{\cdot-}$) of the arylation product (step *d*). Finally, SET from $[Ar–Ph]^{\cdot-}$ to Ar–X gives the arylation product (Ar–Ph) and regenerates $[Ar–X]^{\cdot-}$ (step *e*). Here, the conflicting demands of single-electron reduction and oxidation are effectively solved in step *e*, and an electron derived from a *tert*-butoxide in combination with a nitrogen bidentate ligand acts as a catalyst in a manner similar to $S_{RN}1$ reaction. The major difference compared with $S_{RN}1$ reaction is the reversal of the sequence of steps *c* and *d*: deprotonation–addition in $S_{RN}1$ reaction *vs.* addition–deprotonation in the present reaction. Here a strong base promotes HAS consisting of steps *c*, *d*, and *e*, injecting electron-richness into benzene to make it work as a nucleophile toward aryl halides.

As described above, in the reaction using a substoichiometric amount of 4,7-diphenyl-1,10-phenanthroline (Ph-phen), the mixture with sodium *tert*-butoxide generates an electron catalyst. Murphy and coworkers proposed that additives work as single-electron donors after being converted into a so-called super electron donor upon reaction with a *tert*-butoxide [11], rather than additive–*tert*-butoxide complexes themselves work as single-electron donors [12]. Many reports are available for effective additives other than nitrogen bidentate ligands: butanol [13], phenylhydrazine [14], 2-pyridylmethanol [15], and indoline [16] are examples that have a relatively simple structure.

Instead of adding an additive that works as a single-electron donor in combination with a *tert*-butoxide, photoirradiation is used to accelerate SET in the initiation step. Rossi and coworkers reported that the arylation of benzene with aryl halides under photoirradiation proceeds at a room temperature, which is much lower than that in the reaction using an additive (Scheme 7.8, Eq. (i)) [17]. A method to get into the propagation cycle shown in Scheme 7.7 as an aryl radical also is available (Scheme 7.8, Eq. (ii)). The reaction of aryl iodides with benzene (a large excess as a solvent) in the presence of *t*-BuON=NO*t*-Bu (0.2 equiv), KO*t*-Bu (2 equiv) and dimethyl sulfoxide (DMSO) (10 equiv) proceeds at a moderate temperature (60 °C) through aryl radical intermediates generated from aryl iodides upon iodine atom abstraction by a methyl radical, which is produced, accompanied by acetone, from *t*-BuON=NO*t*-Bu through homolysis of the O–N bonds and the C–C bond cleavage of the resulting *t*-BuO· [18].

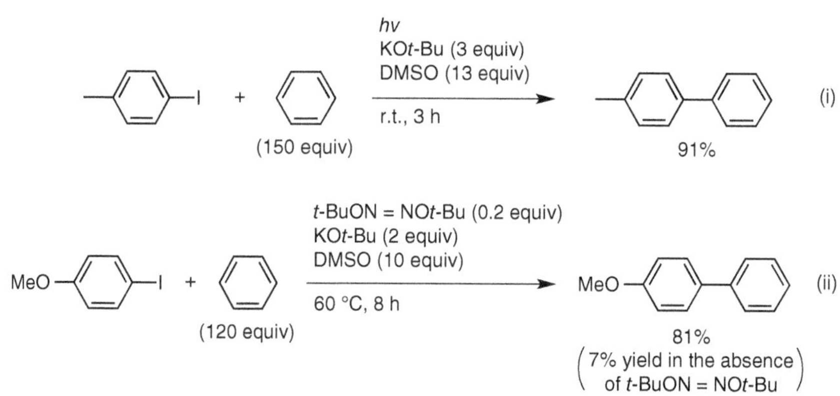

Scheme 7.8 Direct arylation initiated by photoirradiation or *tert*-butoxy radical. Source: Modified from Budén et al. [17a] and Kiriyama et al. [18].

So far, merely the examples using benzene as an arene have been treated, not mentioning the two major drawbacks of the electron-catalyzed arylation of arenes with aryl halides. The first problem is on the regioselectivity concerning arenes. In most cases, the arylation of substituted benzenes does not take place regioselectively in the same way as other reactions containing a HAS process. For example, the reaction of benzonitrile as an arene using NaO*t*-Bu and 4,7-diphenyl-1,10-phenanthroline (Ph-phen) results in production of a regioisomeric mixture ($o/m/p$ = 58/17/25)

(Scheme 7.9, Eq. (i)) [8a]. As intramolecular reactions sometimes do not suffer from the regioselectivity problem, this arylation can be suitable for preparation of aromatic compounds having more than three rings as shown in Eq. (ii) of Scheme 7.9 [19]. The second problem is that this arylation requires the use of a large excess amount of arenes to avoid the overarylation of the arylation products because radical addition to benzene rings is accelerated by any substituent on the benzene rings, and thus monoarylation products show much higher reactivities than arenes themselves toward aryl radicals. The use of unsubstituted benzene in a solvent amount is not problematic, and thus this arylation is especially beneficial for the preparation of biphenyls having substituents merely on the one of the two benzene rings derived from benzene and a substituted phenyl halide.

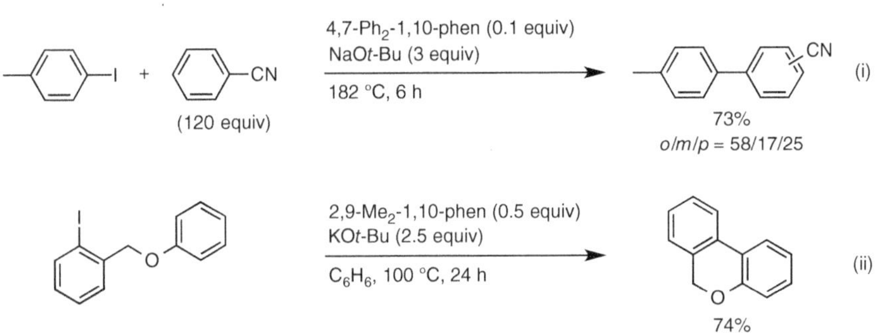

Scheme 7.9 Regiochemistry in the direct arylation. Source: Modified from Shirakawa et al. [8a] and Sun et al. [19].

The base-promoted arylation is applicable also to alkenes instead of arenes, though the scope of alkenes is limited to styrene derivatives, which are excellent radical acceptors to be converted into relatively stable benzylic radical intermediates (Scheme 7.10, Eq. (i)) [20]. The reaction of aryl iodides with carbon monoxide in the presence of KO*t*-Bu [21] or an amine [22] gives the corresponding benzoic

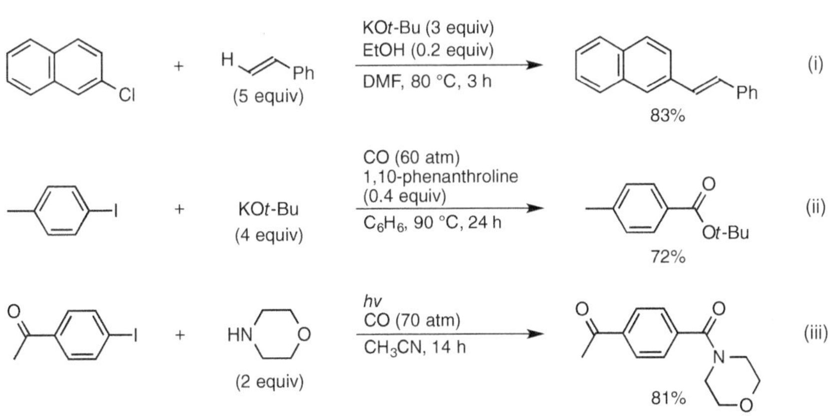

Scheme 7.10 Other base-promoted arylation reactions. Source: Modified from Shirakawa et al. [20], Zang et al. [21] and Kawamoto et al. [22].

acid derivatives, where carbon monoxide acts as an efficient aryl radical acceptor (Scheme 7.10, Eqs. (ii) and (iii)).

7.3 Electron-Catalyzed Cross-Coupling Reactions of Aryl Halides

As described in the Section 7.2, the use of arenes as nucleophiles toward aryl halides to construct biaryl frameworks is convenient because prior introduction of a functional group such as a metal moiety into arene nucleophiles is not required. However, the regioselective arylation of the arenes is not facile as the selection of an optional hydrogen on aromatic rings is often difficult. In 2012, the electron-catalyzed cross-coupling reaction of aryl halides was first reported to be applied to aryl Grignard reagents instead of arenes, where the regiochemistries concerning both aromatic rings are retained [23]. The reaction of aryl iodides with aryl Grignard reagents (1.5 equiv) in toluene in the presence of a small amount (6–12 equiv) of tetrahydrofuran (THF) at 110 °C for 24 hours gives the corresponding coupling products in high yields (Scheme 7.11). The presence of a small amount of THF is essential for the reaction to take place, but the use of THF as the sole solvent changes the reaction course to iodine–magnesium exchange.

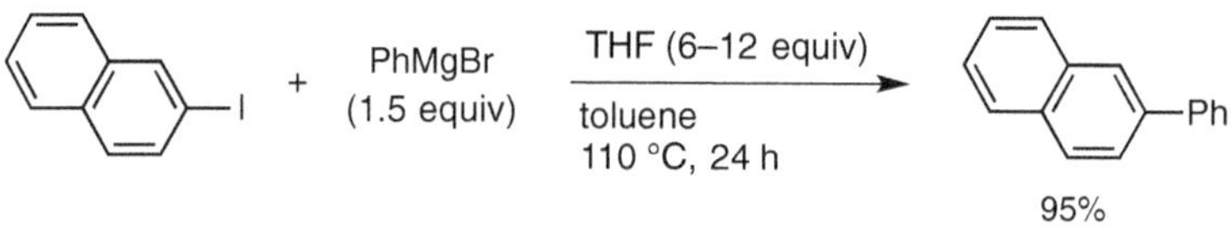

Scheme 7.11 Electron-catalyzed cross-coupling reaction of aryl halides with aryl Grignard reagents. Source: Adapted from Shirakawa et al. [23].

The Grignard cross-coupling reaction was disclosed to be catalyzed by an electron in a manner shown in Scheme 7.12 [23, 24]. The reaction is initiated by SET from an arylmagnesium bromide (exemplified by phenylmagnesium bromide: PhMgBr) to an aryl iodide (exemplified by 2-naphthyl iodide: Np–I), getting into the propagation steps in a form of anion radical $[Np–I]^{\cdot-}$. The anion radical is converted to the anion radical ($[Np–Ph]^{\cdot-}$) of the coupling product (Np–Ph) upon reaction with PhMgBr. Finally, SET from $[Np–Ph]^{\cdot-}$ to Np–I gives Np–Ph and regenerates $[Np–I]^{\cdot-}$, where the reduction to activate Np–I and the oxidation to give Np–Ph take place simultaneously as same as $S_{RN}1$ reaction. On the other hand, in the initiation steps, PhMgBr is converted into its cation radical ($[PhMgBr]^{\cdot+}$), which reacts with PhMgBr to give anion radical $[Ph–Ph]^{\cdot-}$. This anion radical also supplies an electron catalyst through SET toward Np–I, getting into the propagation steps again in the form of $[Np–I]^{\cdot-}$. Consequently, two electrons are provided from two molecules of PhMgBr to work as catalysts in the propagation steps. As shown in Scheme 7.13, the supply of an electron catalyst by adding an anion radical of a biaryl, lithium 4,4′-di-*tert*-butylbiphenylide (LDBB), which corresponds to

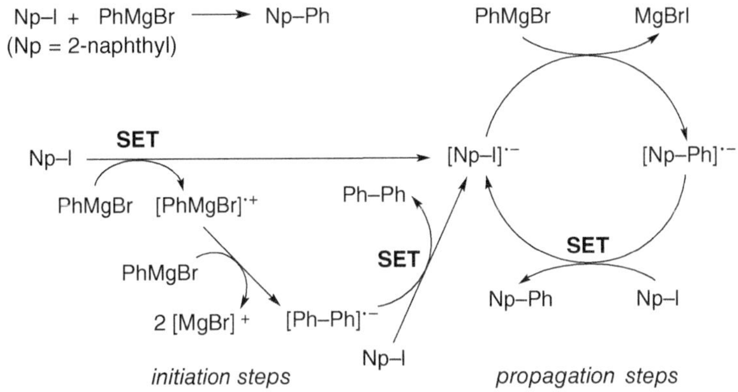

Scheme 7.12 Mechanism of the electron-catalyzed Grignard cross-coupling reaction. Source: Adapted from Uchiyama et al. [24a].

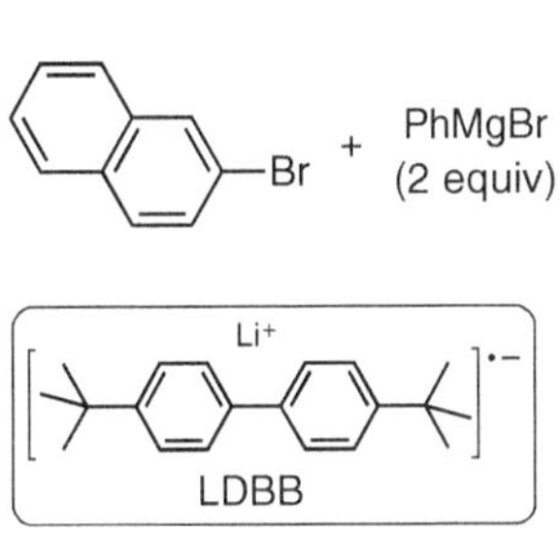

additive	conv	yield
none	2%	1%
LDBB (0.2 equiv)	98%	80%

Scheme 7.13 Acceleration of the Grignard cross-coupling reaction by addition of an electron equivalent. Source: Adapted from Shirakawa et al. [23].

[Np–Ph]$\cdot^-$ in the propagation steps in Scheme 7.12, drastically accelerated the coupling reaction of 2-bromonaphthalene (Np–Br) with PhMgBr at 80 °C, being consistent with operation of the electron catalysis and showing that high temperature is required for SET from PhMgBr to Np–X in the initiation steps. In contrast to the cases with $S_{RN}1$ mechanism and a modified one shown in Schemes 7.2 and 7.7, respectively, the propagation steps in Scheme 7.12 do not include any aryl radical intermediate. No involvement of an aryl radical intermediate derived from an aryl iodide is proved by a radical clock reaction using *o*-homoallylphenyl iodide, which reacts with PhMgBr to give the corresponding normal coupling product with no cyclization products (Scheme 7.14) [24a, 25]. The reactivity order of aryl halides is much different under the electron catalysis compared with that under a palladium catalysis (Scheme 7.15) [23]. A styryl-substituted bromobenzene showed a higher reactivity than a trifluoromethyl-substituted bromobenzene toward PhMgBr under electron catalysis, whereas the reactivity order of the aryl bromides is reversed under the palladium catalysis. It is rationally understood that, under electron catalysis, a lower lowest unoccupied molecular orbital (LUMO) level, brought about more efficiently by a highly conjugating styryl group, is important for the high reactivity. In contrast, a higher electrophilicity, brought about by a highly electron-withdrawing trifluoromethyl group, is critical under the palladium catalysis.

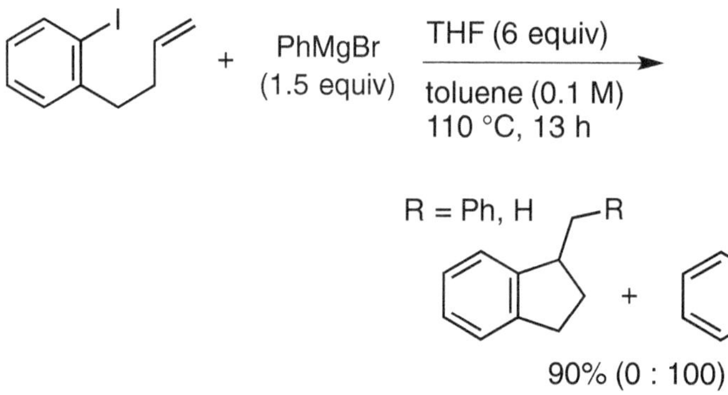

Scheme 7.14 Radical clock reaction in the Grignard cross-coupling reaction. Source: Adapted from Shirakawa et al. [23].

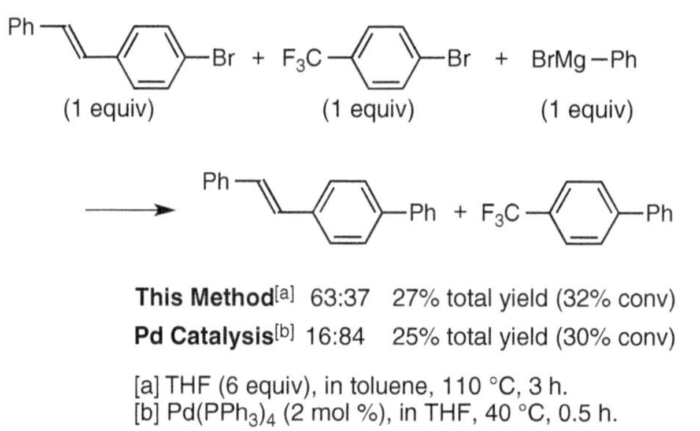

Scheme 7.15 Competition reaction between aryl halides in the Grignard cross-coupling reaction.

Aryl Grignard reagents react also with alkenyl iodides under the same conditions (Scheme 7.16, Eq. (i)) [26]. The stereochemistry of alkenyl iodides is retained in the coupling product, showing that alkenyl free radicals, which readily undergo *E*–*Z* isomerization, are not involved as intermediates. The scope of aryl/alkenyl halides is expanded from iodides to bromides by addition of lithium chloride (4 equiv) in combination with the use of a mixed solvent of toluene–THF in an increased THF ratio (Scheme 7.16, Eq. (ii)) [27]. The Grignard cross-coupling reaction is applied also to polymerization of *m*-iodophenylmagnesium chloride to give poly(*meta*-phenylene)s (Scheme 7.16, Eq. (iii)) [28].

Organozinc reagents, which show functional group tolerance higher than Grignard reagents, were found to undergo the cross-coupling reaction with aryl/alkenyl halides (Scheme 7.17). Diarylzinc reagents, prepared by transmetalation between an aryllithium and zinc chloride, react even with aryl iodides having a ketone moiety

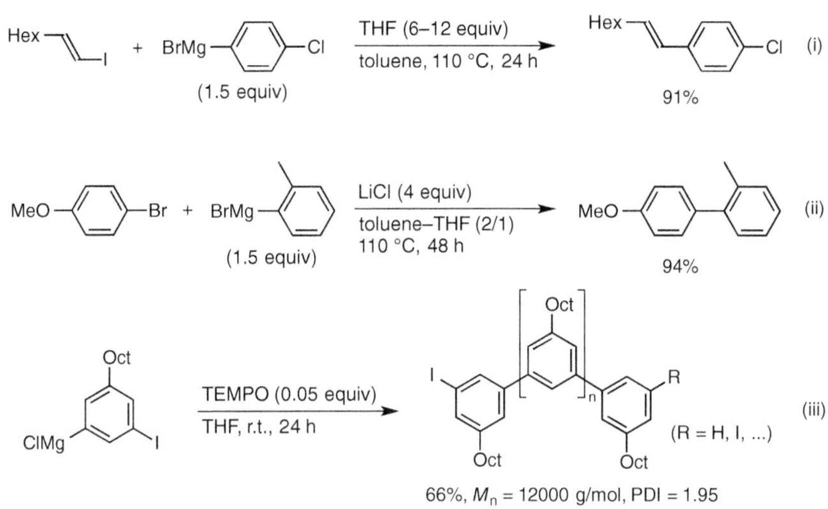

Scheme 7.16 Grignard cross-coupling reactions. Source: Modified from Shirakawa et al. [26], Shirakawa et al. [27], and Modified from Murarka and Studer [28].

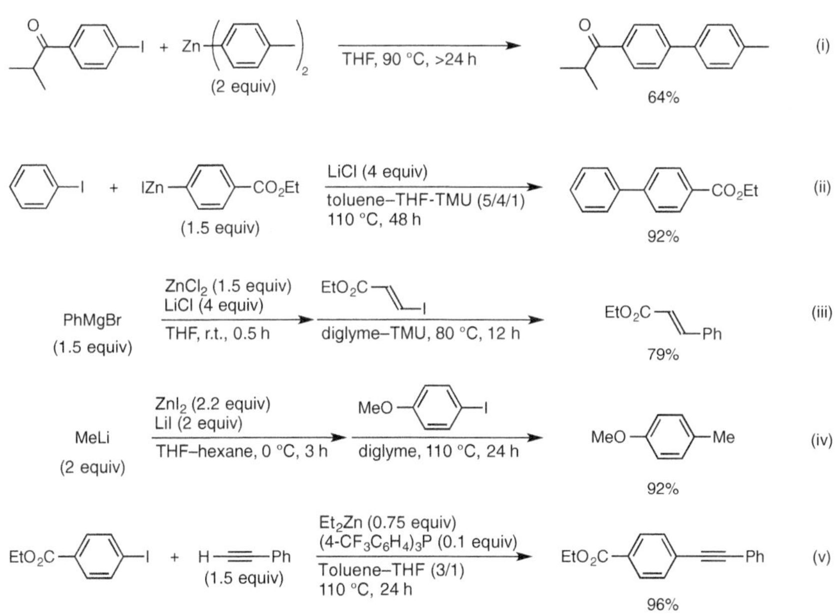

Scheme 7.17 Electron-catalyzed cross-coupling reaction of aryl and alkenyl halides with organozinc reagents. Source: Modified from Minami et al. [29], Shirakawa et al. [30], Okura and Shirakawa [31], and Okura et al. [32].

(Scheme 7.17, Eq. (i)) [29]. Use of arylzinc iodides, prepared through the redox reaction between an aryl iodide and zinc dust, in combination with lithium chloride is compatible with electrophilic substituents such as ester and nitrile on both aryl halides and arylzinc iodides (Scheme 7.17, Eq. (ii)) [30]. The cross-coupling reaction with aryl/alkenyl halides is applicable also to arylzinc reagents prepared by transmetalation of a readily available aryl Grignard reagent with zinc chloride in the presence of lithium chloride (Scheme 7.17, Eq. (iii)). Alkylzinc reagents, prepared through transmetalation between an alkyllithium and zinc iodide in the presence of lithium iodide, undergo coupling reaction with aryl iodides, where no reaction takes place by the use of zinc chloride and lithium chloride in contrast to the case with arylzinc reagents (Scheme 7.17, Eq. (iv)) [31]. As for alkynylzinc reagents, the use of a substoichiometric amount of a triarylphosphine is indispensable (Scheme 7.17, Eq. (v)) [32]. The reaction of a terminal alkyne (1.5 equiv) and diethylzinc (0.75 equiv), which are converted to the corresponding bisalkynylzinc *in situ*, with an aryl iodide in the presence of tris(4-trifluoromethylphenyl)phosphine (0.1 equiv) gives the corresponding alkynylarene.

In addition to organomagnesium/zinc reagents, the following nucleophiles undergo the electron-catalyzed cross-coupling reaction with aryl halides: arylboroxines with the aid of zinc chloride and potassium *tert*-butoxide [33], arylaluminum reagents [34], tetraarylstannanes [35], and magnesium diarylamides (Scheme 7.18) [36].

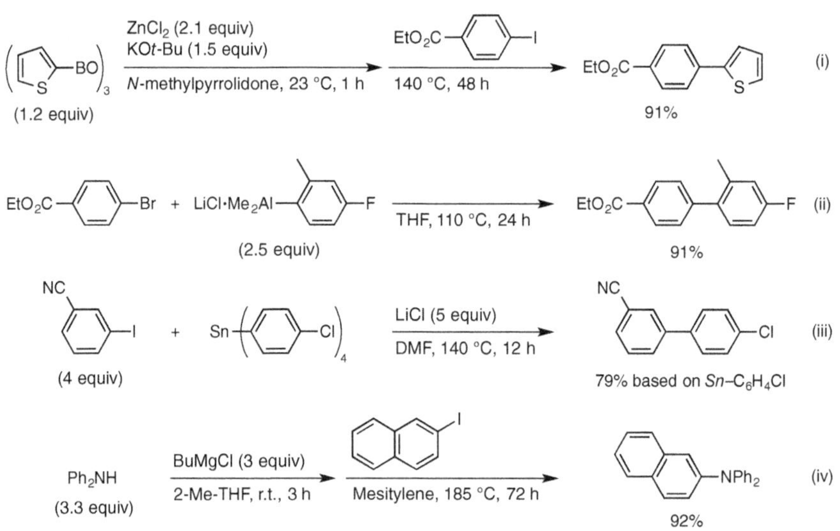

Scheme 7.18 Other electron-catalyzed cross-coupling reactions. Source: Modified from Okura et al. [33], Minami et al. [34], He et al. [35] and Kiriyama et al. [36].

7.4 Conclusions

The electron-catalyzed reactions of aryl halides with arenes–*tert*-butoxide or organometallic compounds were described in this chapter. These reactions employ an electron as a catalyst instead of a transition metal, which has been considered to be indispensable for this type of reactions. Here $S_{RN}1$-type mechanism is operative, sharing the efficiency that the reduction of aryl halides activating them to undergo the substitution reaction and the oxidation giving the substitution product to even redox accounts take place simultaneously. Expansion of the scope of $S_{RN}1$ reaction to arenes is brought about by *tert*-butoxides, which induce the BHAS, where the electron-richness of *tert*-butoxides is regarded to be transferred to arenes to be able to act as nucleophiles toward aryl halides, utilizing the high reactivity of aryl radicals produced through elimination of X^- from the anion radicals ($[Ar–X]^{\cdot -}$) of aryl halides (Ar–X). On the other hand, in the electron-catalyzed cross-coupling reactions with organometallic compounds, the anion radicals, $[Ar–X]^{\cdot -}$, directly react with organometallic compounds. No involvement of aryl radical intermediates, the high reactivity of which often induces diverse type of side reactions, contributes to high yields of the electron-catalyzed cross-coupling reactions. The new finding that electron-rich anion radicals $[Ar–X]^{\cdot -}$ act as electrophiles toward organometallic compounds is expected to be a clue for the development of new substitution reactions of aryl halides.

References

1 For reviews: (a) Stanforth, S.P. (1998). *Tetrahedron* 54: 263. (b) Miyaura, N. (ed.) (2002). *Cross-Coupling Reactions: A Practical Guide, Topics in Current Chemistry*, vol. 219. Berlin: Springer. (c) Hassan, J., Sévignon, M., Gozzi, C. et al. (2002). *Chem. Rev.* 102: 1359. (d) de Meijere, A. and Diederich, F. (eds.) (2004). *Metal-Catalyzed Cross-Coupling Reactions*, 2e, vol. 1–2. Weinheim: Wiley-VCH. (e) Corbet, J.-P. and Mignani, G. (2006). *Chem. Rev.* 106: 2651.

2 For an early example: (a) Kim, J.K. and Bunnett, J.F. (1970). *J. Am. Chem. Soc.* 92: 7463. For an account: (b) Bunnett, J.F. (1978). *Acc. Chem. Res.* 11: 413. For reviews: (c) Rossi, R.A., Pierini, A.B., and Peñéñory, A.B. (2003). *Chem. Rev.* 103: 71. (d) Banrdagí, J.I., Vaillard, V.A., and Rossi, R.A. (2012). *Encyclopedia of Radicals in Chemistry, Biology and Materials*, vol. 1 (eds. C. Chatgilialoglu and A. Studer), 333–364. Chichester: John Wiley & Sons, Ltd.

3 For $S_{RN}1$ reaction before 2008, see refs 2b–d.

4 Bolton, R. and Williams, G.H. (1986). *Chem. Soc. Rev.* 15: 261.

5 For an example that effectively solves the conflicting demand, see: Curran, D.P. and Keller, A.I. (2006). *J. Am. Chem. Soc.* 128: 13706.

6 Wolf, W. and Kharasch, N. (1965). *J. Org. Chem.* 30: 2493.

7 Yanagisawa, S., Ueda, K., Taniguchi, T., and Itami, K. (2008). *Org. Lett.* 10: 4673.

8 (a) Shirakawa, E., Itoh, K., Higashino, T., and Hayashi, T. (2010). *J. Am. Chem. Soc.* 132: 15537. (b) Liu, W., Cao, H., Zhang, H. et al. (2010). *J. Am. Chem. Soc.* 132: 16737. (c) Sun, C.-L., Li, H., Yu, D.-G. et al. (2010). *Nat. Chem.* 2: 1044.

9 (a) Russell, G.A., Chen, P., Kim, B.H., and Rajaratnam, R. (1997). *J. Am. Chem. Soc.* 119: 8795. (b) Wang, C., Russell, G.A., and Trahanovsky, W.S. (1998). *J. Org. Chem.* 63: 9956.

10 (a) Studer, A. and Curran, D.P. (2011). *Angew. Chem. Int. Ed.* 50: 5018. see also: (b) Shirakawa, E. and Hayashi, T. (2012). *Chem. Lett.* 41: 130.

11 (a) Zhou, S., Anderson, G.M., Mondal, B. et al. (2014). *Chem. Sci.* 5: 476. (b) Zhou, S., Doni, E., Anderson, G.M. et al. (2014). *J. Am. Chem. Soc.* 136: 17818. See also: (c) Cuthbertson, J., Gray, V.J., and Wilden, J.D. (2014). *Chem. Commun.* 50: 2575. (d) Patil, M. (2016). *J. Org. Chem.* 81: 632.

12 Yi, H., Jutand, A., and Lei, A. (2015). *Chem. Commun.* 51: 545.

13 Liu, W., Tian, F., Wang, X. et al. (2013). *Chem. Commun.* 49: 2983.

14 Dewanji, A., Murarka, S., Curran, D.P., and Studer, A. (2013). *Org. Lett.* 15: 6102.

15 Wu, Y., Choy, P.Y., and Kwong, F.Y. (2014). *Org. Biomol. Chem.* 12: 6820.

16 Yang, H., Chu, D.-Z., and Jiao, L. (2018). *Chem. Sci.* 9: 1534.

17 (a) Budén, M.E., Guastavino, J.F., and Rossi, R.A. (2013). *Org. Lett.* 15: 1174. For an example of the use of a photoredox catalysis for this purpose, see: (b) Cheng, Y., Gu, X., and Li, P. (2013). *Org. Lett.* 15: 2664.

18 Kiriyama, K. and Shirakawa, E. (2017). *Chem. Lett.* 46: 1757.

19 Sun, C.-L., Gu, Y.-F., Huang, W.-P., and Shi, Z.-J. (2011). *Chem. Commun.* 47: 9813.

20 Shirakawa, E., Zhang, X., and Hayashi, T. (2011). *Angew. Chem. Int. Ed.* 50: 4671.

21 Zhang, H., Shi, R., Ding, A. et al. (2012). *Angew. Chem. Int. Ed.* 51: 12542.

22 Kawamoto, T., Sato, A., and Ryu, I. (2015). *Chem. Eur. J.* 21: 14764.

23 Shirakawa, E., Hayashi, Y., Itoh, K. et al. (2012). *Angew. Chem. Int. Ed.* 51: 218.

24 (a) Uchiyama, N., Shirakawa, E., and Hayashi, T. (2013). *Chem. Commun.* 49: 364. *o*-Homoallylphenyl radical is known to readily cyclize ($k_c = 5 \times 10^8\ s^{-1}$ at 50 °C): (b) Abeywickrema, A.N. and Beckwith, A.L.J. (1986). *J. Chem. Soc. Chem. Commun.*: 464.

25 A report proposing, based on DFT calculations, that the Grignard cross-coupling reaction proceeds through an aryl radical intermediate, which exists under special conditions and thus does not show free radical character, is available. Haines, B.E. and Wiest, O. (2014). *J. Org. Chem.* 79: 2771.

26 Shirakawa, E., Watabe, R., Murakami, T., and Hayashi, T. (2013). *Chem. Commun.* 49: 5219.

27 Shirakawa, E., Okura, K., Uchiyama, N. et al. (2014). *Chem. Lett.* 43: 922.

28 Murarka, S. and Studer, A. (2012). *Angew. Chem. Int. Ed.* 51: 12362.

29 Minami, H., Wang, X., Wang, C., and Uchiyama, M. (2013). *Eur. J. Org. Chem.*: 7891.

30 Shirakawa, E., Tamakuni, F., Kusano, E. et al. (2014). *Angew. Chem. Int. Ed.* 53: 521.

31 Okura, K. and Shirakawa, E. (2016). *Eur. J. Org. Chem.*: 3043.

32 Okura, K., Kawashima, H., Tamakuni, F. et al. (2016). *Chem. Commun.* 52: 14019.

33 Okura, K., Teranishi, T., Yoshida, Y., and Shirakawa, E. (2018). *Angew. Chem. Int. Ed.* 57: 7186.

34 Minami, H., Saito, T., Wang, C., and Uchiyama, M. (2015). *Angew. Chem. Int. Ed.* 54: 4665.

35 He, Q., Wang, L., Liang, Y. et al. (2016). *J. Org. Chem.* 81: 9422.

36 Kiriyama, K., Okura, K., Tamakuni, F., and Shirakawa, E. (2018). *Chem. Eur. J.*: 4519.

8

Photochemical Paired Transformations

Takashi Koike and Munetaka Akita

Institute of Innovative Research, Tokyo Institute of Technology, Laboratory for Chemistry and Life Science, R1-27, 4259 Nagatsuta-cho, Midori-ku, Yokohama 226-8503, Japan

8.1 Introduction

Light is a unique stimulus to drive reactions. Photochemical processes start with absorption of light. Thus, even though plural components are placed in one vessel, irradiation of light with an appropriate wavelength, which is absorbed by one of them, selectively causes excitation of the particular component, leading to photochemical transformations (Grotthus–Draper law). The chemistry involved in the excited state is capable of inducing transformations, which cannot be achieved by thermal processes. For example, photosynthesis done by green plants is widely known as a representative photochemical transformation of CO_2 and H_2O, which are inorganic molecules. Artificial photosynthesis has been extensively studied as a key technology to solve the energy issue [1, 2]. On the other hand, photochemical strategies for synthesis of useful organic molecules such as pharmaceuticals, agrochemicals, and functional materials have been rapidly advancing for the past decade [3–21]. In particular, the emergence of well-designed photocatalytic reactions through hydrogen-atom abstraction from unactivated $C(sp^3)$—H bonds made a breakthrough in the research field.

While functionalization of inert $C(sp^3)$—H bonds under mild reaction conditions is desirable and still considered as "Holy Grail" in the field of synthetic organic chemistry, free-radical processes such as the Barton and Hofmann–Löffler–Freytag (HLF) reactions have been regarded as useful aliphatic C—H bond functionalization [22–24]. Hydrogen atom transfer (HAT) process, induced by an active radical species, is involved as a key process. But further extension to asymmetric transformations is one of the challenges. In general, asymmetric bond formation requires elaborate control over conformation of the involved intermediates near the chiral environment. But when cleavage of an unactivated bond is caused by thermal stimulus, the following bond formation is not always allowed to acquire high enantioselectivity because all plural components in one vessel are thermally activated. In contrast, photochemical strategies for HAT are promising from the viewpoint of access to the reactive radical species without thermal activation. Recently, photochemical generation

Organic Redox Chemistry: Chemical, Photochemical and Electrochemical Syntheses, First Edition.
Edited by Jun-Ichi Yoshida and Frédéric W. Patureau.

of the radical species has been renewed and, especially, photocatalytic protocols have been widely propagated. In this chapter, "photochemical paired transformations" are defined as asymmetric functionalization of inert $C(sp^3)$—H bonds triggered by photocatalytic HAT processes. Description of the two basic concepts for the photochemical HAT processes will be followed by selected examples and asymmetric transformations associated with each concept.

8.2 Basic Concepts for Photochemical Hydrogen Atom Transfer (HAT) Process

HAT process, given in Eq. (8.1), is regarded as a sort of proton-coupled electron transfer (PCET) process. HAT is now regarded as a useful strategy for functionalization of unactivated alkanes because it allows homolytic activation of aliphatic R—H bonds, often in a highly selective manner.

$$\mathrm{R-H} + \underset{\text{hydrogen abstractor}}{{}^{\cdot}\mathrm{A}} \rightarrow {}^{\cdot}\mathrm{R} + \mathrm{H-A} \tag{8.1}$$

Heteroatom-centered radicals such as O-, N-, S-, and halogen-based radicals can serve as the hydrogen abstractor ($^{\cdot}A$). Photochemical methods offer two basic concepts for HAT: (i) direct HAT by the excited photocatalyst ($^{\cdot}A = PC^*$, concept 1) and (ii) indirect HAT triggered by photocatalysis ($^{\cdot}A \neq PC^*$, concept 2).

8.2.1 Concept 1: Direct HAT by the Excited Photocatalyst

Carbonyl compounds, especially aromatic ketones, e.g. 4-benzoylpyridine (4-BzPy), 2-chloroanthraquinone (2-ClAQ), and anthraquinone-2-sodium sulfonate (2-SO_3NaAQ), undergo n–π^* transition upon light irradiation with appropriate wavelength to form the triplet excited species, which exhibits reactivity in a manner similar to alkoxy radicals ($^{\cdot}OR$) [25, 26]. In addition, some metal oxides (O=M species) such as the decatungstate anion, $[W_{10}O_{32}]^{4-}$ [27–30], and the uranyl cation, $[UO_2]^{2+}$ [31–33], also show highly electrophilic O-centered radical character under photoirradiation. Thus, the excited photocatalyst (PC*) can serve as a hydrogen abstractor ($^{\cdot}A$ shown in Eq. (8.1)) and is amenable to homolytic cleavage of a $C(sp^3)$—H bond (R—H) (Scheme 8.1). The PC is regenerated by a back HAT step of the reduced H–PC species to one of the intermediates (Y) formed in the reaction.

8.2.2 Concept 2: Indirect HAT Triggered by Photocatalysis

Recently, photoredox catalysis with transition-metal complexes such as Ir-pyridylphenyl derivatives and organic dyes such as 2,4,6-triphenylpyrylium tetrafluoroborate salt (TPT) has become a powerful tool for synthetic radical chemistry because they can promote single-electron-transfer (SET) processes under visible-light irradiation [3–21]. The excited photocatalyst (PC*) undergoes SET to/from electron-accepting/-donating precursors (A^+/A^-) to give the hydrogen abstractor $^{\cdot}A$.

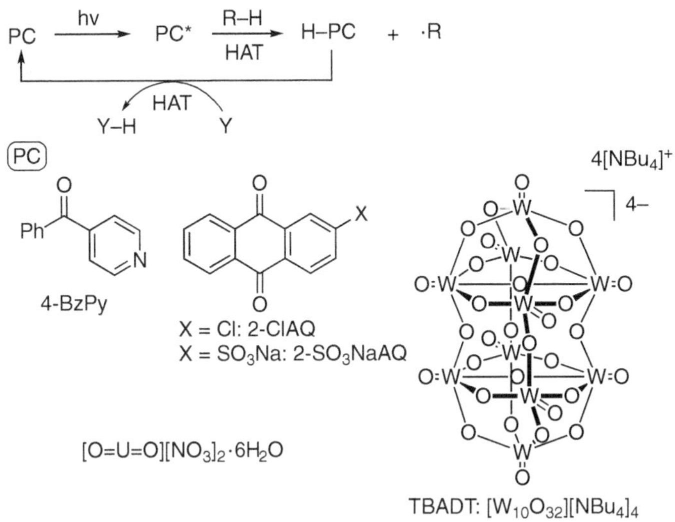

Scheme 8.1 Concept 1 where PC* = ·A highlighted in the present chapter.

Under appropriate circumstances, the generated radical ·A abstracts a hydrogen atom from an aliphatic C(sp^3)—H bond (R—H). In particular, alkanes bearing A^+ or A^- units in the molecules undergo highly regioselective radical translocation onto remote δ-positions (1,5-HAT). Highly oxidized/reduced photocatalysts (PC^+/PC^-) undergo successive SET from/to one of the intermediates in the reaction or sacrificial redox agents (Y), resulting in regeneration of PC (Scheme 8.2) [34–36]. In this chapter, photocatalytic 1,5-HAT reactions involving heteroatom-centered radicals will be discussed.

Both of the systems can produce alkyl radical (·R) from substrate (R–H). Thus, those systems associated with appropriate chiral environment can induce asymmetric functionalization of C(sp^3)—H bonds.

8.3 Asymmetric Radical Functionalization Associated with Direct HAT by Photocatalysts

In this section, recent examples of direct HAT by photoactivated catalysts are shown first. Herein transformations of inert C(sp^3)—H bonds to C—C bonds are dealt with. Then, applications to asymmetric functionalization are discussed.

8.3.1 Photocatalytic Functionalization of C(sp^3)−H Bonds Based on Concept 1

Chemistry induced by the excited state of the carbonyl group (^{3}(nπ)*) has been well studied. In particular, the Norrish type II reaction is a well-known photochemical process [37]. In the intermolecular HAT reaction, benzophenone and its derivatives

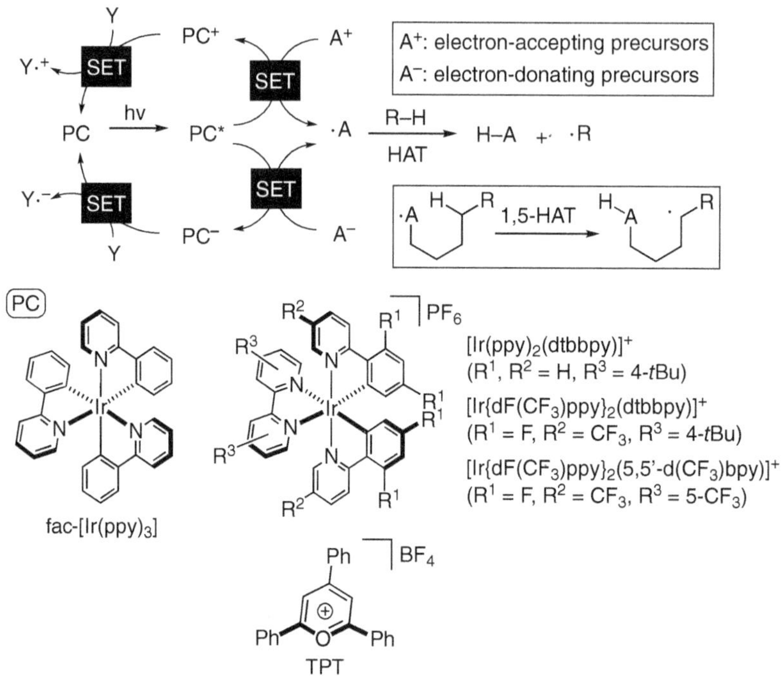

Scheme 8.2 Concept 2 where PC* ≠ ·A highlighted in the present chapter. Source: Adapted from Čeković [34], Curran et al. [35] and Curran and Shen [36].

are generally used as the carbonyl photocatalyst [25, 26]. In 2016, the group of Kamijo reported on introduction of the aldoxime functionality to C(sp^3)–H moieties, including cycloalkane, ether, azacycle, and cyclic sulfide (**1**) by employing the photoexcited 4-BzPy as a hydrogen abstractor under UV irradiation (Scheme 8.3) [38]. While a stoichiometric amount of 4-BzPy was used in the reaction, potentials for the photocatalytic reaction were also remarked in their work. The proposed mechanism is depicted in Scheme 8.3. The photoexcited 4-BzPy induces HAT to give the carbon radical intermediate **5** and α-oxy radical **4**. Addition of **5** to the sulfonyl oxime **2** provides N-centered radical intermediate **6**, elimination of the sulfonyl radical from which affords the product **3**. The released sulfonyl radical accepts the hydrogen atom from **4** to regenerate 4-BzPy.

Furthermore, they developed Giese-type reaction of cycloalkane, allyl and benzyl compounds, and heterocycle (**7**) by employing 2-ClAQ as a catalytic hydrogen abstractor under UV irradiation [39]. As shown in Scheme 8.4, the photoexcited 2-ClAQ abstracts a hydrogen atom from **7** to give the organyl radical **11**. The generated radical reacts with 1,1-di(phenylsulfonyl)ethylene **8**, an electron-deficient olefin, to provide the radical intermediate **12**, which abstracts the hydrogen atom in **10** generated by the initial HAT process to regenerate 2-ClAQ together with the product **9**.

Before the 2000s, the chemistry of the photoexcited decatungstate $[W_{10}O_{32}]^{4-}$ was well studied in terms of the mechanistic aspects of its photochemical processes [40],

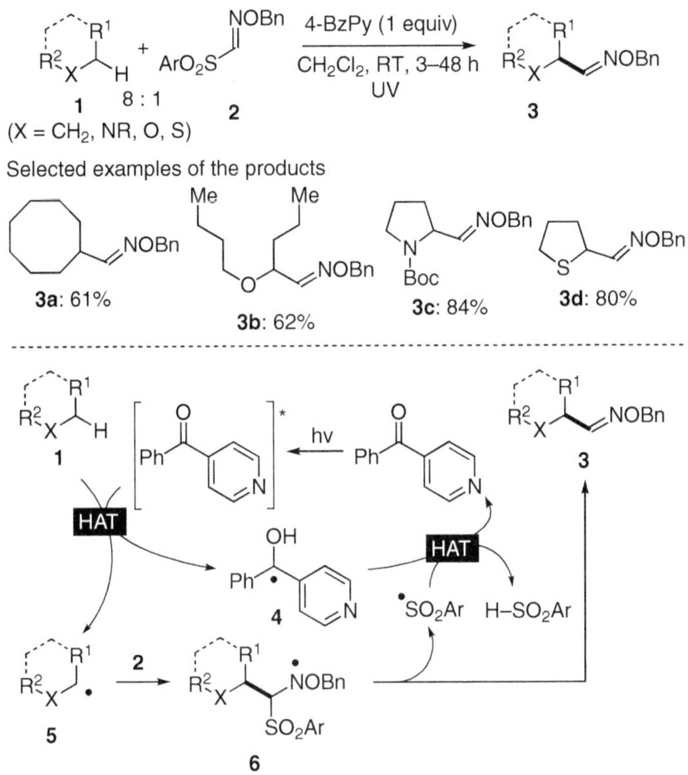

Scheme 8.3 Introduction of aldoxime functionality onto $C(sp^3)$–H bonds. Source: Modified from Kamijo et al. [38].

applications to photomicrolithography [41], and decontamination of toxic organic compounds [42]. Among them, the group of Hill reported on fundamental $C(sp^3)$–H activation of alkanes by the decatungstate [43–45]. Recently, the chemistry has been directed to organic synthesis because it contributes to a mild activation strategy for generation of alkyl radicals from simple alkanes. In 2004, the group of Fagnoni and Albini reported on the reaction of cycloalkane **13** with electron-deficient olefin **14** by employing tetrabutylammonium decatungstate (TBADT: $[W_{10}O_{32}][NBu_4]_4$) as a photocatalyst (Scheme 8.5) [46].

UV irradiation causes excitation of decatungstate, $[W_{10}O_{32}]^{4-}$, through ligand-to-metal charge transfer (LMCT). The excited decatungstate, $[W_{10}O_{32}]^{4-*}$, serves as a hydrogen abstractor from cycloalkane **13**, giving the alkyl radical **16** and $[W_{10}O_{32}]^{5-}H^+$. The generated radical **16** reacts with alkene **14** to provide the radical intermediate **17**, which is reduced by $[W_{10}O_{32}]^{5-}H^+$ to anionic intermediate **18**. Final protonation affords the product **15**. Later on, the group expanded the reaction system to those with aldehydes, ethers, and amides [47].

Furthermore, they and the group of Ryu cooperatively developed carbonylation and amidation with CO and azodicarboxylate **23** and β-alkylation of cyclopentanone **25** (Scheme 8.6) [48–50].

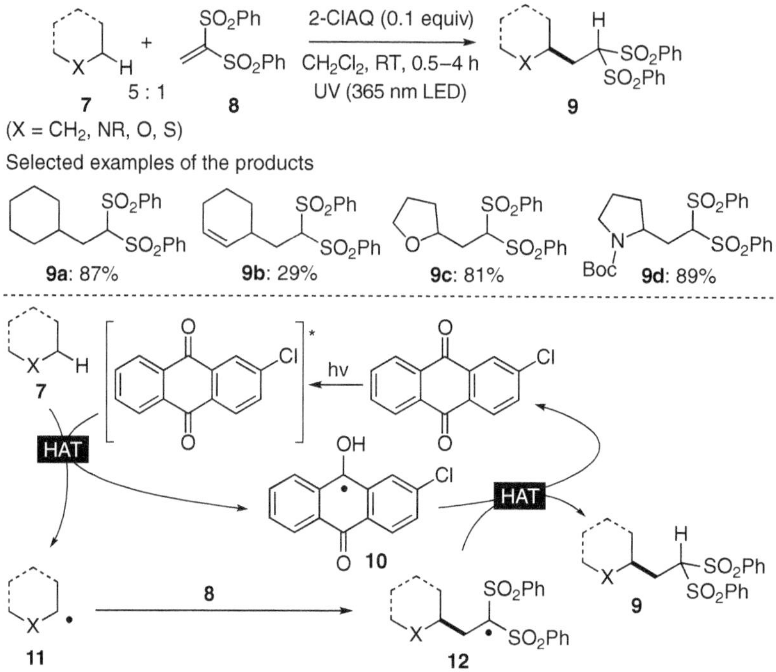

Scheme 8.4 Giese-type reaction of C(sp^3)–H bonds.

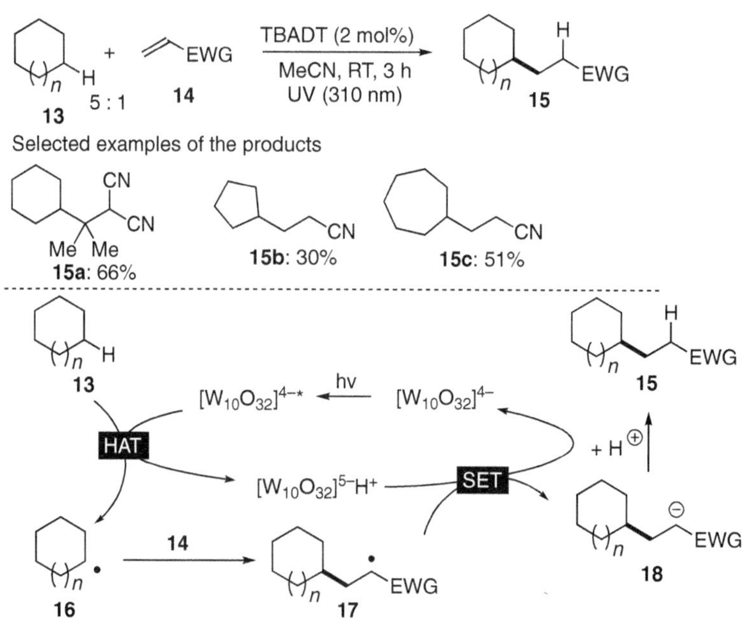

EWG: electron-withdrawing group

Scheme 8.5 Giese-type reaction of C(sp^3)–H bonds by decatungstate. Source: Modified from Dondi et al. [46].

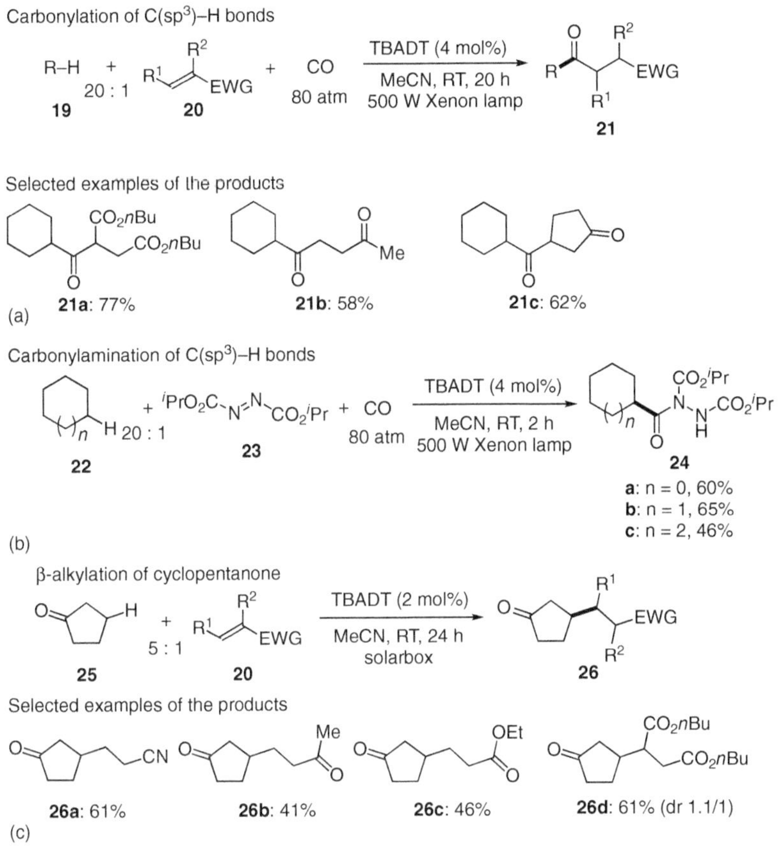

Scheme 8.6 Decatungstate photocatalysis. Source: Adapted from Ryu et al. [48] and Ryu et al. [49].

More recently, the group of MacMillan reported that a merger of TBADT photocatalysis and the Ni catalysis realized arylation of aliphatic C(sp^3)—H bonds (Scheme 8.7) [51]. Hydrogen abstraction from a C(sp^3)—H bond of alkane **27** by the excited TBADT, $[W_{10}O_{32}]^{4-*}$, provides the alkyl radical and $[W_{10}O_{32}]^{5-}$ H$^+$, a singly reduced species, which undergoes disproportionation to generate $[W_{10}O_{32}]^{4-}$ and a doubly reduced decatungstate, $[W_{10}O_{32}]^{6-}2H^+$. The $[W_{10}O_{32}]^{6-}$ species reduces the Ni precatalyst, NiBr$_2$(dtbbpy) (dtbbpy: 4,4′-di-*tert*-butyl-2,2′-bipyridine), to generate the catalytically active Ni^0 species. The Ni^0 species captures the alkyl radical mentioned above to give a Ni^I–R species. Subsequent oxidative addition of aryl halide (Ar–Br **28**) to the resultant Ni^I–R species affords the Ni^{III}(R)(Ar) species, and reductive elimination produces the cross-coupled product **29** and the Ni^I–Br species, which is reduced by the doubly reduced decatungstate, $[W_{10}O_{32}]^{6-}2H^+$, to regenerate the Ni^0 species. An alternative mechanism involving oxidative addition to the Ni^0 species to **28** could also be operative.

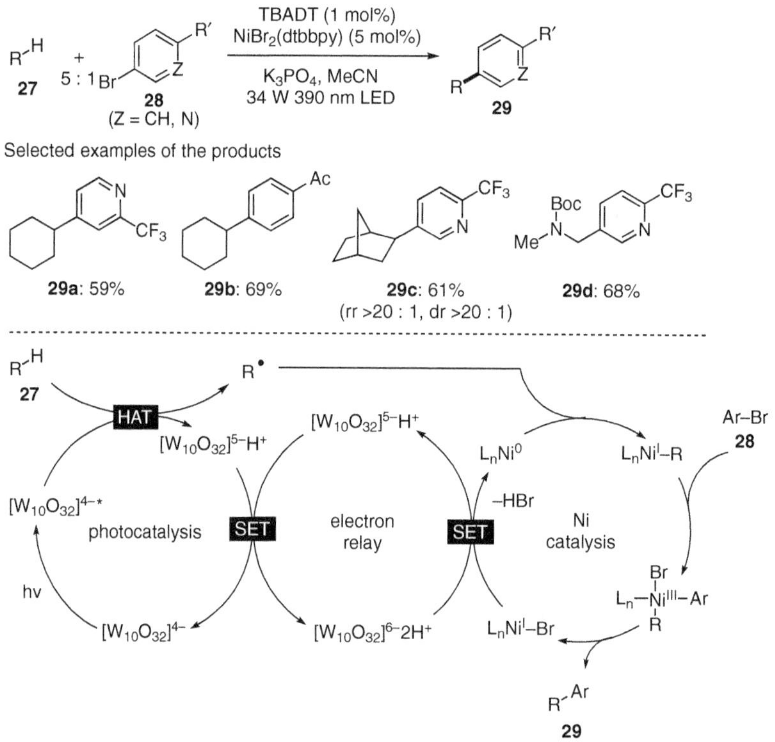

Scheme 8.7 Arylation of aliphatic C(sp^3)—H bonds. Source: Modified from Perry et al. [51].

8.3.2 Asymmetric Transformations Based on Concept 1

While photocatalytic transformations of C(sp^3)—H bonds to C—C bonds were described in the Section 8.3.1, pioneering works on asymmetric functionalization associated with photocatalytic HAT were reported recently. In 2015, the group of Itoh developed asymmetric aldol reaction preceded by HAT from a benzyl C—H bond in benzyl alcohol **30** [52]. Anthraquinone-2-sodium sulfonate (2-SO_3NaAQ) serves as a hydrogen abstractor under visible-light irradiation. The photocatalytic processes produce the corresponding aldehyde **33** through aerobic oxidation of the radical intermediate. Use of *trans*-4-*tert*-butyldiphenylsiloxy-L-proline as a chiral organocatalyst promotes asymmetric aldol condensation of **33** with ketone **31**, leading to stereoselective synthesis of β-hydroxy ketone **32** (Scheme 8.8). The reaction contains a HAT process, but the generated alkyl radical is not involved in asymmetric bond formation directly.

In 2016, the group of Melchiorre revealed that radicals were intrinsically suitable for connecting structurally congested carbon atoms. They developed an enantioselective method for construction of a quaternary stereogenic center, i.e. enantioselective radical conjugate addition to β-substituted cyclic enone **35**. Their electron-relay strategy, which is composed of TBADT photocatalysis and chiral cyclohexylamine catalysis bearing a redox-active carbazole moiety, is shown in Scheme 8.9 [53].

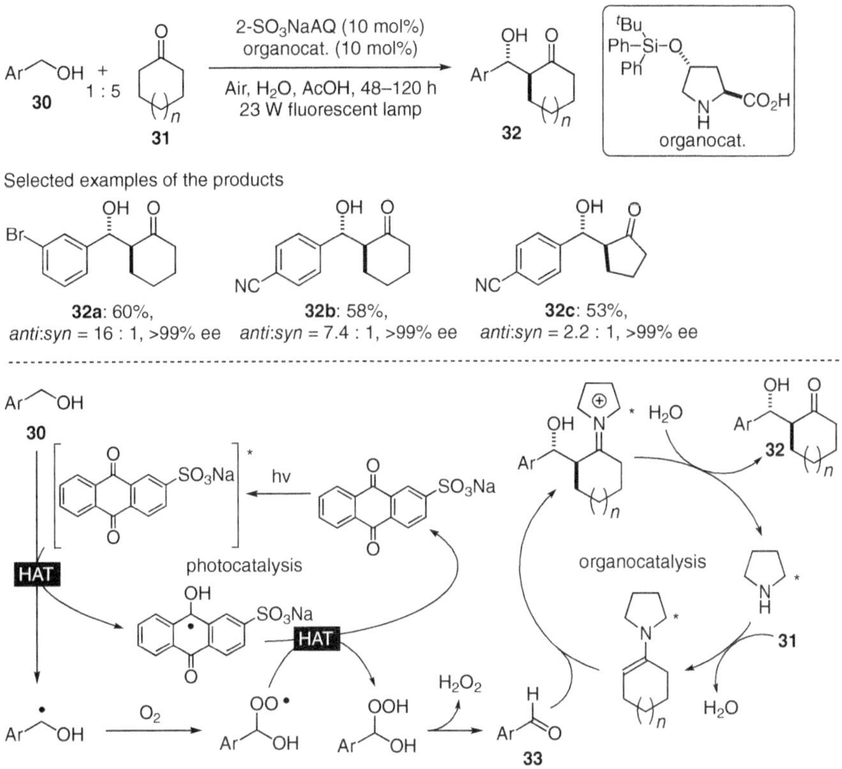

Scheme 8.8 Asymmetric aldol reaction of aldehydes generated through photocatalytic HAT processes.

The TBADT catalyst generates the carbon-centered radical **37** from benzodioxole **34** through a photocatalytic HAT process. On the other hand, a chiral amine catalyst reacts with **35** to form the iminium ion **38**. The bulky carbazole unit effectively shields the diastereotopic *Si* face of **38**, implying that the carbazole unit is located close to the olefinic moiety of **38**. Subsequent capture of **37** by **38** forms α-iminyl radical cation **39**, which is usually susceptible to β-scission. But the electron-rich carbazole unit transfers an electron to provide the enamine intermediate **40**, which is isomerized to the imine intermediate **41**. Final reduction of **41** by $[W_{10}O_{32}]^{5-}H^+$ followed by hydrolysis releases the product **36** together with the amine catalyst.

8.4 Asymmetric Radical Functionalization Associated with Indirect HAT Triggered by Photocatalysis

It is known that heteroatom-centered radicals serve as hydrogen abstractors. For the past decade, photocatalytic radical generation by photoredox SET processes has been extensively studied [3–21]. As a matter of course, generation and reaction

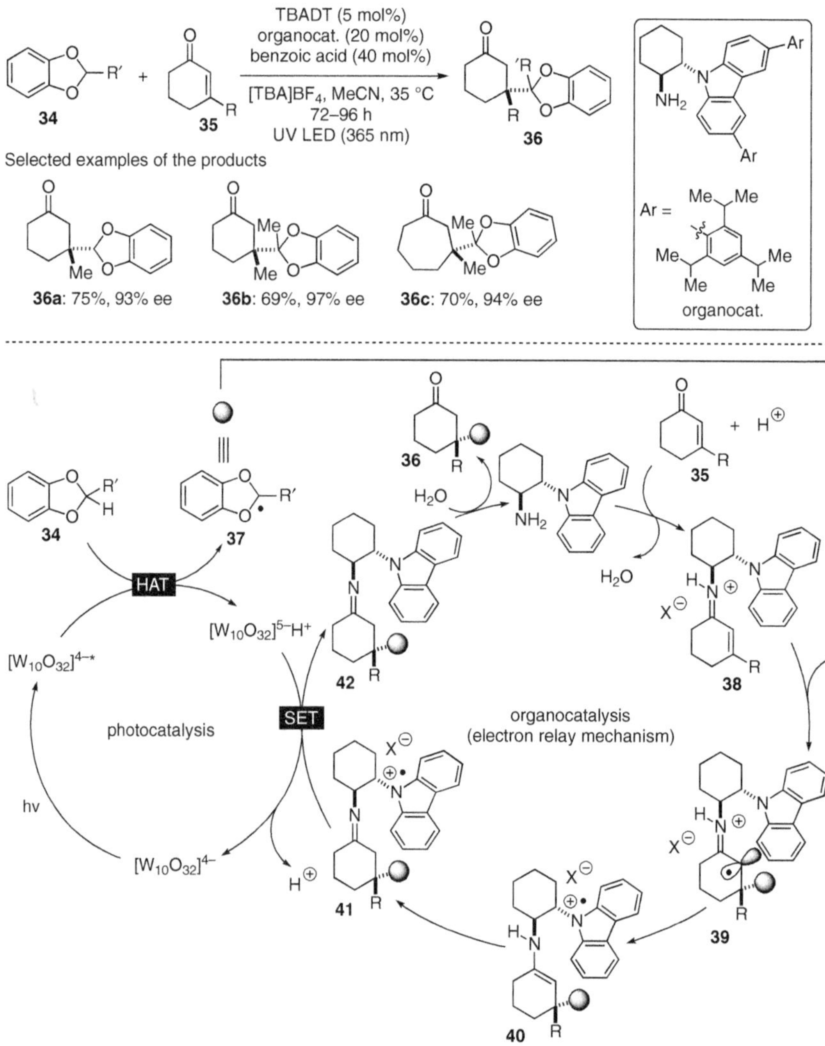

Scheme 8.9 Enantioselective construction of quaternary carbon center through photocatalytic HAT processes. Source: Modified from Murphy et al. [53].

of heteroatom-centered radicals have become research targets [54–57]. During the past few years, several seminal works on mild C(sp^3)–H functionalization promoted by catalytic amounts of hydrogen abstractors such as S-(thiols) [58, 59], *N*-(quinuclidines, sulfonamide) [60–63], *O*-(benzoate, *N*-hydroxylimide, phosphate) [64–66], and Br- or Cl-centered radicals [67–69] with cooperative action of photoredox catalysis were reported. On the other hand, intramolecular 1,5-HAT of well-designed substrates possessing heteroatom-centered radical precursors can become a powerful tool for highly regioselective radical functionalization of remote inert C(sp^3)—H bonds. Recently, asymmetric functionalization following 1,5-HAT processes triggered by photocatalysts was reported by the group of Meggers. In this section, pioneering works on photocatalytic functionalization of C(sp^3)—H bonds through 1,5-HAT processes will be introduced prior to asymmetric version.

8.4.1 Photocatalytic Functionalization of C(sp^3)—H Bonds Through 1,5-Hydrogen Atom Transfer Processes

The HLF reaction is a monumental reaction of N-centered radicals, but the reaction generally requires UV irradiation and strongly acidic conditions [24]. In 2015, the group of Qin and Yu reported on the HLF-type reaction under milder conditions (Scheme 8.10). They developed the photocatalytic method for generation of N-centered radicals from *N*-chlorosulfonamide **43** under visible-light irradiation (white LEDs) [70]. *N*-Chlorosulfonamide **43** undergoes 1e-reduction by the excited photocatalyst Ir^{III*} to give the N-centered radical **45**, which abstracts a hydrogen atom from a remote C—H bond. The generated C-radical **46** is transformed into the chlorinated intermediate **47** through either (i) ionic pathway involving the carbocationic intermediate **48** or (ii) radical chain pathway, resulting in production of cyclic amide **44** after treatment of **47** with NaOH.

Recently, the group of Reiher and Muñiz developed dual iodine/photoredox catalysis for the HLF-type amination of C(sp^3)—H bonds (Scheme 8.11) [71]. Their

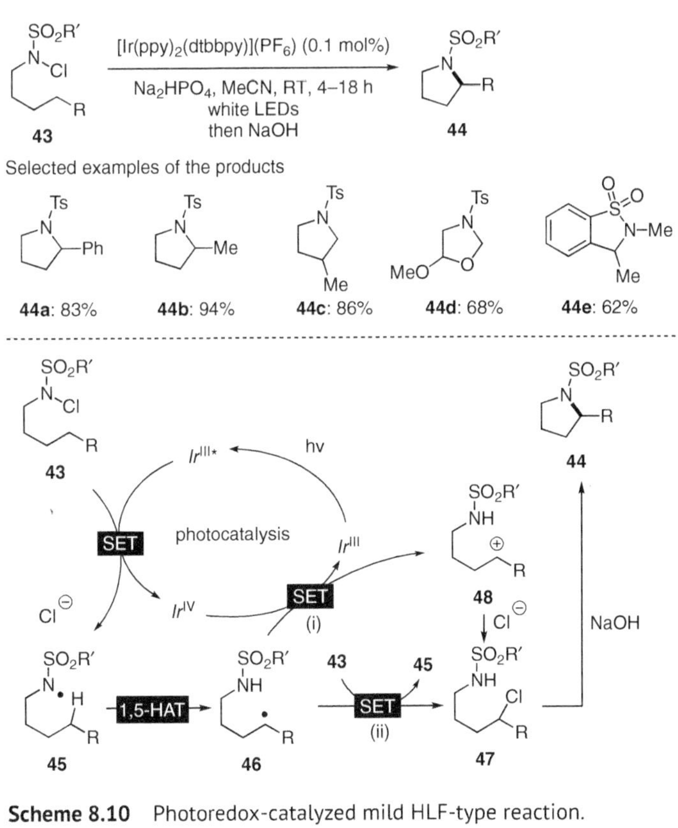

Scheme 8.10 Photoredox-catalyzed mild HLF-type reaction.

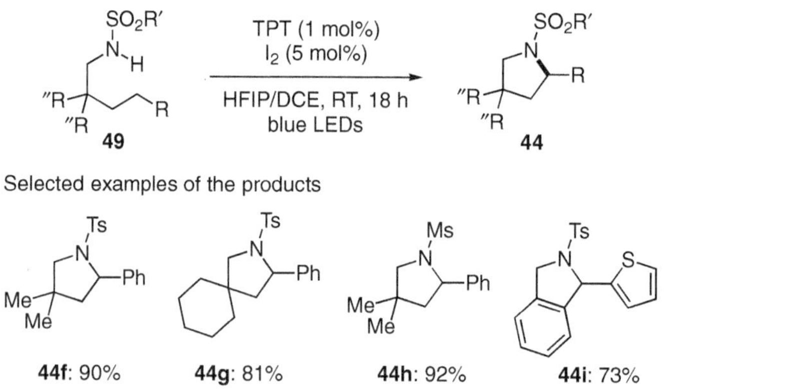

Scheme 8.11 Mild HLF-type reaction of sulfonamides. Source: Modified from Becker et al. [71].

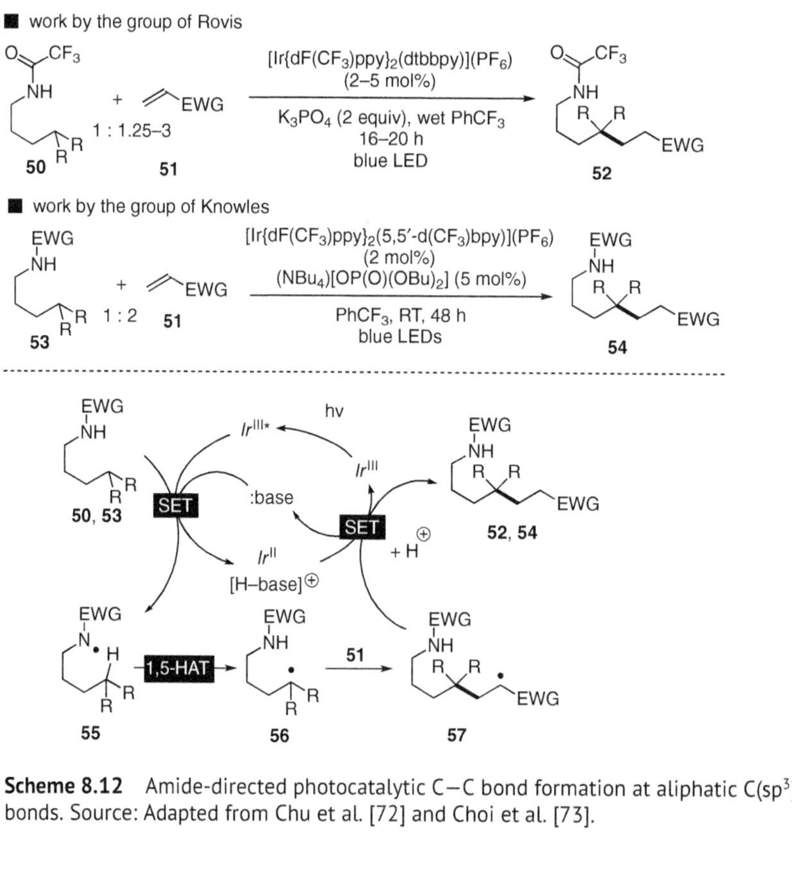

Scheme 8.12 Amide-directed photocatalytic C–C bond formation at aliphatic C(sp^3)–H bonds. Source: Adapted from Chu et al. [72] and Choi et al. [73].

method does not require the treatment of alkyl iodide corresponding to **47** with base. The *N*-iodosulfonamide is formed *in situ* from the reaction of sulfonamide **49** with hypoiodite, which is generated from the reaction of I_2 with H_2O. Upon irradiation with visible light, the *N*-iodosulfonamide collapses to generate the amidyl radical species corresponding to **45**. Subsequent radical-chain processes involving 1,5-HAT provide alkyl iodide, which undergoes spontaneous intramolecular cyclization to give the product **44** and HI. The highly oxidizing TPT photocatalyst converts HI to regenerate I_2.

In 2016, the reaction was expanded from the intramolecular C—N bond formation to intermolecular C—C bond formation. The groups of Rovis and Knowles independently reported on the photocatalytic alkylation of inert aliphatic C—H bonds (Scheme 8.12) [72, 73]. Their photocatalytic systems are based on the same concept. Generation of amidyl radical **55** from 1e-oxidation of amide substrate (**50** or **53**) by the excited Ir photocatalyst, Ir^{III}*, in the presence of a base is a key

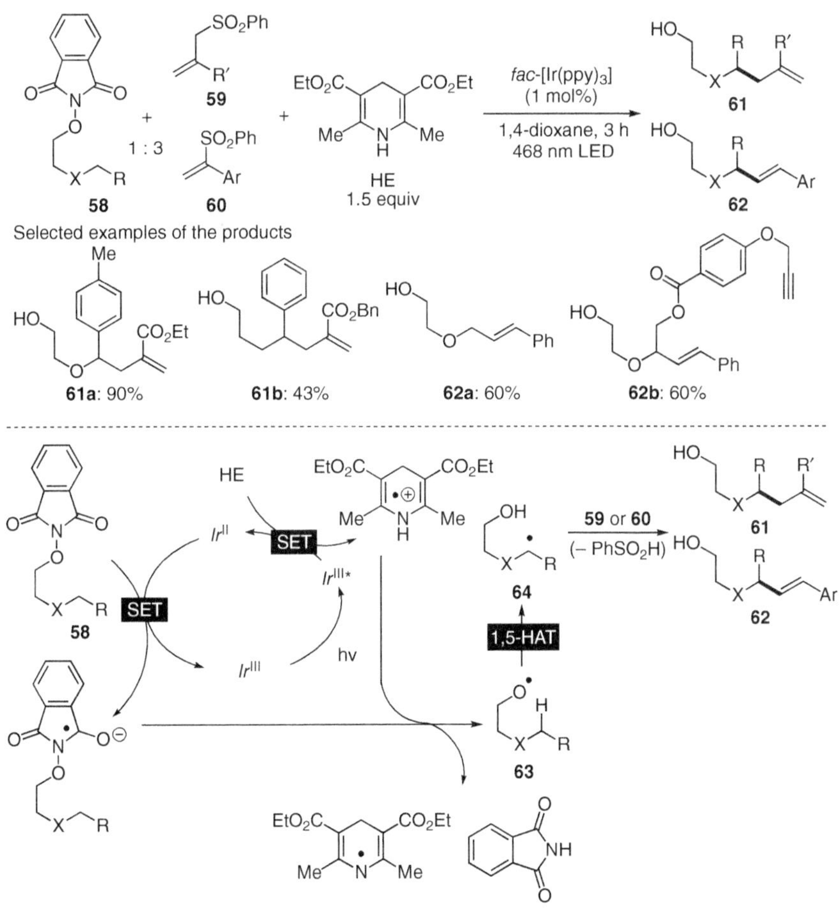

Scheme 8.13 Alkoxy radical-mediated photocatalytic allylation of alkenylation of aliphatic C(sp^3)—H bonds.

process. Subsequent 1,5-HAT forms a carbon radical **56**, which is captured by an electron-deficient alkene **51**. Finally, a radical intermediate **57** undergoes SET and protonation to give the product **52** or **54**.

In 2016, the Chen group also developed a selective allylation and alkenylation of inert C—H bonds by employing 1,5-HAT of O-centered radicals under mild photocatalytic conditions (Scheme 8.13) [74]. The alkoxy radical **63** generated from *N*-alkoxyphthalimide **58** by photoredox catalysis easily activates aliphatic C—H bonds by 1,5-HAT to give a C-radical **64**. Hantzsch ester (HE) serves as a sacrificial electron donor to yield the strongly reducing Ir active species (Ir^{II}) from the excited Ir species (Ir^{III*}). The radical intermediate **64** reacts with allyl or vinyl sulfone (**59** or **60**) to afford the olefinic product (**61** or **62**). More recently, the group of Zhu developed photocatalytic heteroarylation of aliphatic C(sp^3)—H bonds via direct generation of alkoxy radicals from alcohols (Scheme 8.14). While the mechanism is still unclear, a combination of an Ir photocatalyst and $K_2S_2O_8$ is a key to yield the alkoxy radical **67** from alcohol **65**. The C-radical **68** generated from remote HAT triggers the distal migration of heteroaryl group, providing the ketone product **66** heteroarylated at an aliphatic C—H bond [75].

8.4.2 Asymmetric Transformations Based on Concept 2

The combination of the above-mentioned concept 2-type remote activation of aliphatic C—H bonds with asymmetric catalysis has become one of the promising strategies for asymmetric functionalization of unactivated C(sp^3)—H bonds. In 2016, the group of Meggers reported on asymmetric C(sp^3)–H functionalization by dual photoredox/chiral Lewis acid catalysis (Scheme 8.15) [76].

They designed 1,5-HAT of *N*-alkoxyphthalimide **69** followed by stereocontrolled addition to alkene. The Rh catalyst (Δ-RhS), which was originally developed by themselves [77, 78], is used as a chiral Lewis acid and is coordinated by *N*-acylpyrazole substrate **70** through an N,O-chelating manner (**Rh–I**). The

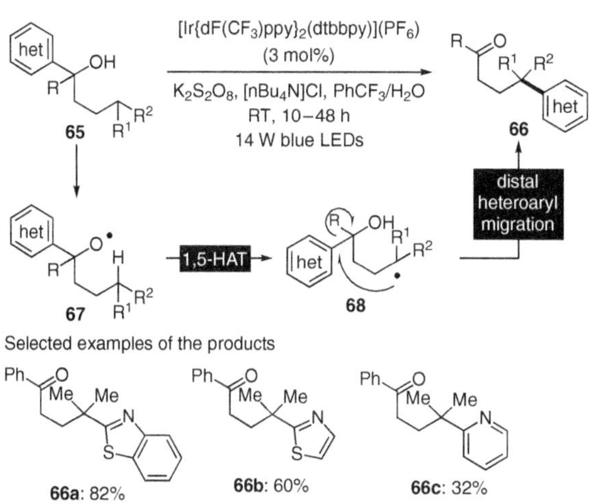

Scheme 8.14 Alcohol-directed photocatalytic heteroarylation of aliphatic C(sp^3)—H bonds.

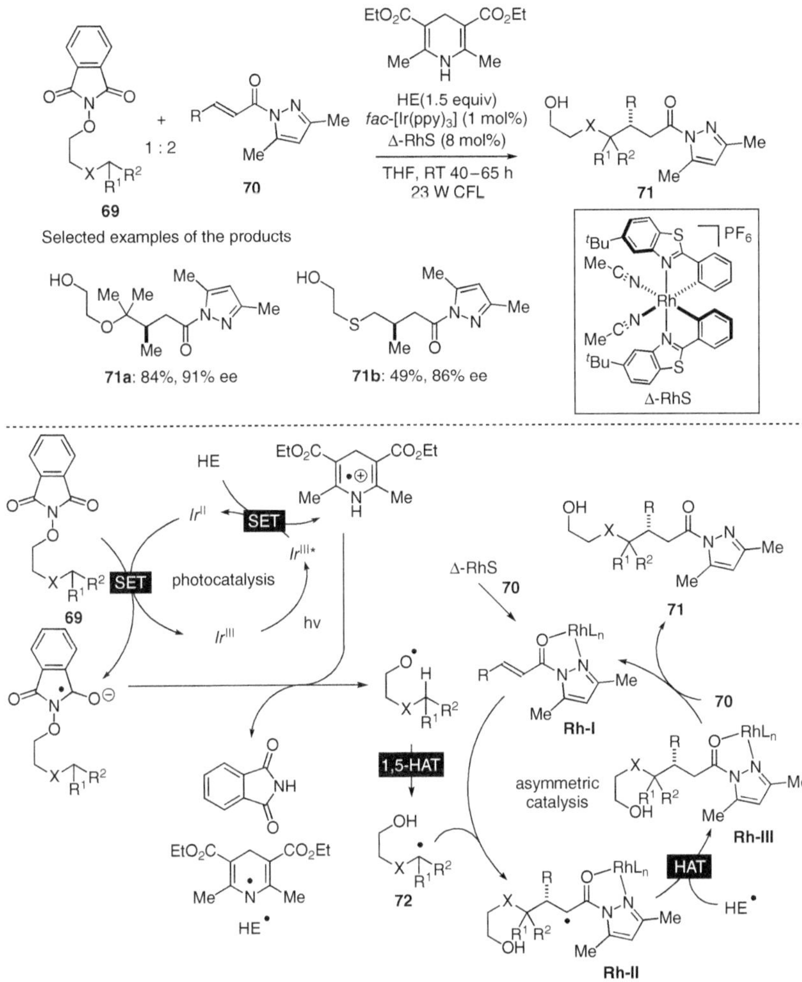

Scheme 8.15 Alkoxy radical-mediated asymmetric alkylation of aliphatic C(sp^3)—H bonds.

generated C-radical **72** via a 1,5-HAT process adds to the **Rh–I** intermediate to give the radical intermediate **Rh–II**, which undergoes HAT from the oxidized Hantzsch ester (HE$^•$). The Lewis acid significantly accelerates the radical addition so that a racemic background reaction can be negligible. Finally, ligand exchange of **Rh–III** produces the product **71**. More recently, they showed that the similar dual photo and Rh catalysis can be applied to amide-directed asymmetric alkylation of aliphatic C(sp^3)—H bonds (Scheme 8.16) [79].

8.5 Summary and Outlook

Photochemical HAT reaction is regarded as a classical chemistry. However, a merger of photocatalysis and another catalysis makes HAT processes useful modern

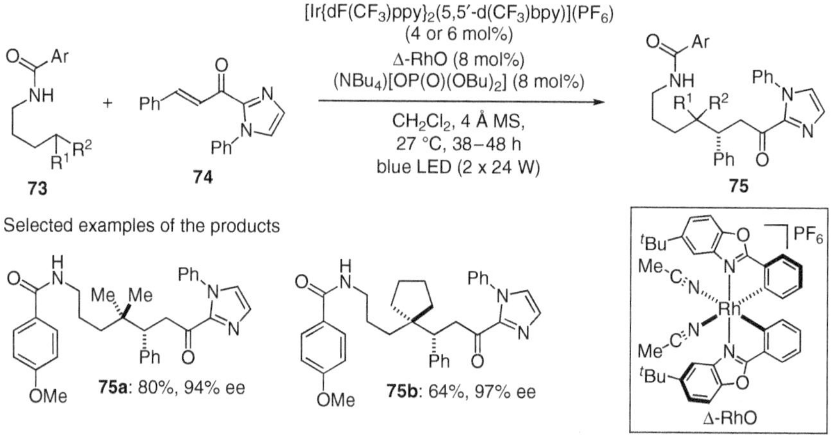

Scheme 8.16 Amide-directed asymmetric alkylation of aliphatic C(sp^3)–H bonds.

organic transformation. A development of elusive asymmetric functionalization of unactivated aliphatic C(sp^3)—H bonds is expected to be expanded. In this chapter, two strategies are highlighted: (i) direct HAT by the excited photocatalyst and (ii) HAT following photocatalytic generation of heteroatom radicals. Both strategies still have challenges with respect to the use of visible light as a light source in (i) and selective intermolecular reaction in (ii). Further development of photocatalytic systems merged with other chiral catalysts can overcome the problems and lead to novel photochemical-paired transformation technology.

References

1 Kudo, A. and Miseki, Y. (2009). Heterogeneous photocatalyst materials for water splitting. *Chem. Soc. Rev.* 38: 253–278.

2 Wang, Y., Suzuki, H., Xie, J. et al. (2018). Mimicking natural photosynthesis: solar to renewable H_2 fuel synthesis by Z-scheme water splitting systems. *Chem. Rev.* 118: 5201–5241.

3 Narayanam, J.M.R. and Stephenson, C.R.J. (2011). Visible light photoredox catalysis: applications in organic synthesis. *Chem. Soc. Rev.* 40: 102–113.

4 Xuan, J. and Xiao, W.-J. (2012). Visible-light photoredox catalysis. *Angew. Chem. Int. Ed.* 51: 6828–6838.

5 Reckenthäler, M. and Griesbeck, A.G. (2013). Photoredox catalysis for organic syntheses. *Adv. Synth. Catal.* 355: 2727–2744.

6 Fukuzumi, S. and Ohkubo, K. (2013). Selective photocatalytic reactions with organic photocatalysts. *Chem. Sci.* 4: 561–574.

7 Xi, Y., Yi, H., and Lei, A. (2013). Synthetic applications of photoredox catalysis with visible light. *Org. Biomol. Chem.* 11: 2387–2403.

8 Prier, C.K., Rankic, D.A., and MacMillan, D.W.C. (2013). Visible light photoredox catalysis with transition metal complexes: applications in organic synthesis. *Chem. Rev.* 113: 5322–5363.

9 Hopkinson, M.N., Sahoo, B., Li, J.-L., and Glorius, F. (2014). Dual catalysis sees the light: combining photoredox with organo-, acid, and transition-metal catalysis. *Chem. Eur. J.* 20: 3874–3886.

10 Koike, T. and Akita, M. (2014). Visible-light radical reaction designed by Ru- and Ir-based photoredox catalysis. *Inorg. Chem. Front.* 1: 562–576.

11 Hari, D.P. and König, B. (2014). Synthetic applications of eosin Y in photoredox catalysis. *Chem. Commun.* 50: 6688–6699.

12 Gong, L., Chen, L.-A., and Meggers, E. (2014). Asymmetric catalysis mediated by the ligand sphere of octahedral chiral-at-metal complexes. *Angew. Chem. Int. Ed.* 53: 10868–10874.

13 Protti, S., Fagnoni, M., and Ravelli, D. (2015). Photocatalytic C–H activation by hydrogen-atom transfer in synthesis. *ChemCatChem* 7: 1516–1523.

14 Ravelli, D., Protti, S., and Fagnoni, M. (2016). Carbon–carbon bond forming reactions via photogenerated intermediates. *Chem. Rev.* 116: 9850–9913.

15 Skubi, K.L., Blum, T.R., and Yoon, T.P. (2016). Dual catalysis strategies in photochemical synthesis. *Chem. Rev.* 116: 10035–10074.

16 Romero, N.A. and Nicewicz, D.A. (2016). Organic photoredox catalysis. *Chem. Rev.* 116: 10075–10166.

17 Shaw, M.H., Twilton, J., and MacMillan, D.W.C. (2016). Photoredox catalysis in organic chemistry. *J. Org. Chem.* 81: 6898–6926.

18 Matsui, J.K., Lang, S.B., Heitz, D.R., and Molander, G.A. (2017). Photoredox-mediated routes to radicals: the value of catalytic radical generation in synthetic methods development. *ACS Catal.* 7: 2563–2575.

19 Capaldo, L. and Ravelli, D. (2017). Hydrogen atom transfer (HAT): a versatile strategy for substrate activation in photocatalyzed organic synthesis. *Eur. J. Org. Chem.*: 2056–2071.

20 Marzo, L., Pagire, S.K., Reiser, O., and König, B. (2018). Visible-light photocatalysis: does it make a difference in organic synthesis? *Angew. Chem. Int. Ed.* 57: 10034–10072.

21 Silvi, M. and Melchiorre, P. (2018). Enhancing the potential of enantioselective organocatalysis with light. *Nature* 554: 41–49.

22 Barton, D.H.R., Beaton, J.M., Geller, L.E., and Pechet, M.M. (1960). A new photochemical reaction. *J. Am. Chem. Soc.* 82: 2640–2641.

23 Barton, D.H.R., Beaton, J.M., Geller, L.E., and Pechet, M.M. (1961). A new photochemical reaction. *J. Am. Chem. Soc.* 83: 4076–4083.

24 Wolf, M.E. (1963). Cyclization of N-halogenated amines (The Hofmann–Löffler reaction). *Chem. Rev.* 63: 55–64.

25 Ravelli, D., Fagnoni, M., and Albini, A. (2013). Photoorganocatalysis. What for? *Chem. Soc. Rev.* 42: 97–113.

26 Hoffman, N. (2016). Photochemical electron and hydrogen transfer in organic synthesis: the control of selectivity. *Synthesis* 48: 1782–1802.

27 Tzirakis, M.D., Lykakis, I.N., and Orfanopoulos, M. (2009). Decatungstate as an efficient photocatalyst in organic chemistry. *Chem. Soc. Rev.* 38: 2609–2621.

28 Ravelli, D., Protti, S., and Fagnoni, M. (2016). Decatungstate anion for photocatalyzed "window ledge" reactions. *Acc. Chem. Res.* 49: 2232–2242.

29 De Waele, V., Poizat, O., Fagnoni, M. et al. (2016). Unraveling the key features of the reactive state of decatungstate anion in hydrogen atom transfer (HAT) photocatalysis. *ACS Catal.* 6: 7174–7182.

30 Suzuki, K., Mizuno, N., and Yamaguchi, K. (2018). Polyoxometalate photocatalysis for liquid-phase selective organic functional group transformations. *ACS Catal.* 8: 10809–10825.

31 Burrows, H.D. and Kemp, T.J. (1974). The photochemistry of the uranyl ion. *Chem. Soc. Rev.* 3: 139–165.

32 Wang, W.-D., Bakac, A., and Espenson, J.H. (1995). Uranium(VI)-catalyzed photooxidation of hydrocarbons with molecular oxygen. *Inorg. Chem.* 34: 6034–6039.

33 West, J.G., Bedell, T.A., and Sorensen, E.J. (2016). The uranyl cation as a visible-light photocatalyst for C(sp^3)–H fluorination. *Angew. Chem. Int. Ed.* 55: 8923–8927.

34 Čeković, Z. (2003). Reactions of δ-carbon radicals generated by 1,5-hydrogen transfer to alkoxy radicals. *Tetrahedron* 59: 8073–8090.

35 Curran, D.P., Kim, D., Liu, H.T., and Shen, W. (1988). Translocation of radical sites by intramolecular 1,5-hydrogen atom transfer. *J. Am. Chem. Soc.* 110: 5900–5902.

36 Curran, D.P. and Shen, W. (1993). Radical translocation reactions of vinyl radicals: substituent effects on 1,5-hydrogen-transfer reactions. *J. Am. Chem. Soc.* 115: 6051–6059.

37 Yang, N.C. and Yang, D.-D.H. (1958). Photochemical reactions of ketones in solution. *J. Am. Chem. Soc.* 80: 2913–2914.

38 Kamijo, S., Takao, G., Kamijo, K. et al. (2016). Photo-induced substitutive introduction of the aldoxime functional group to carbon chains: a formal formylation of non-acidic C(sp^3)—H bonds. *Angew. Chem. Int. Ed.* 55: 9695–9699.

39 Kamijo, S., Takao, G., Kamijo, K. et al. (2016). Alkylation of nonacidic C(sp^3)—H bonds by photoinduced catalytic Michael-type radical addition. *Org. Lett.* 18: 4912–4915.

40 Tanielian, C. (1998). Decatungstate photocatalysis. *Coord. Chem. Rev.* 178–180: 1165–1181.

41 Okamoto, H., Ishikawa, A., and Kudo, T. (1989). Photoreaction mechanism for amorphous peroxopolytungstic acid as an inorganic photoresist material. *J. Photochem. Photobiol. A* 49: 377–385.

42 Mylonas, A., Hiskia, A., and Papaconstantinou, E. (1996). Contribution to water purification using polyoxometalates. Aromatic derivatives, chloroacetic acids. *J. Mol. Catal. A Chem.* 114: 191–200.

43 Renneke, R.F. and Hill, C.L. (1986). Homogeneous catalytic photochemical functionalization of alkanes by polyoxometalates. *J. Am. Chem. Soc.* 108: 3528–3529.

44 Renneke, R.F. and Hill, C.L. (1988). Homogeneous catalytic photochemical functionalization of alkanes by α-dodecatungstophosphate. Rate behavior, energetics, and general characteristics of the processes. *J. Am. Chem. Soc.* 110: 5461–5470.

45 Renneke, R.F. and Hill, C.L. (1988). Selective photochemical dehydrogenation of saturated hydrocarbons with quantum yields approaching unity. *Angew. Chem. Int. Ed. Engl.* 27: 1526–1527.

46 Dondi, D., Fagnoni, M., Molinari, A. et al. (2004). Polyoxotungstate photoinduced alkylation of electrophilic alkenes by cycloalkanes. *Chem. Eur. J.* 10: 142–148.

47 Protti, S., Ravelli, D., Fagnoni, M., and Albini, A. (2009). Solar light-driven photocatalyzed alkylations. Chemistry on the window ledge. *Chem. Commun.*: 7351–7353.

48 Ryu, I., Tani, A., Fukuyama, T. et al. (2011). Atom-economical synthesis of unsymmetrical ketones through photocatalyzed C–H activation of alkanes and coupling with CO and electrophilic alkenes. *Angew. Chem. Int. Ed.* 50: 1869–1872.

49 Ryu, I., Tani, A., Fukuyama, T. et al. (2013). Efficient C–H/C–N and C–H/C–CO–N conversion via decatungstate-photoinduced alkylation of diisopropyl azodicarboxylate. *Org. Lett.* 15: 2554–2557.

50 Okada, M., Fukuyama, T., Yamada, K. et al. (2014). Sunlight photocatalyzed regioselective β-alkylation and acylation of cyclopentanones. *Chem. Sci.* 5: 2893–2898.

51 Perry, I.B., Brewer, T.F., Sarver, P.J. et al. (2018). Direct arylation of strong aliphatic C—H bonds. *Nature* 560: 70–75.

52 Fujiya, A., Nobuta, T., Yamaguchi, E. et al. (2015). Aerobic photooxidative direct asymmetric aldol reactions of benzyl alcohols using water as the solvent. *RSC Adv.* 5: 39539–39543.

53 Murphy, J.J., Bastida, D., Paria, S. et al. (2016). Asymmetric catalytic formation of quaternary carbons by iminium ion trapping of radicals. *Nature* 532: 218–222.

54 Chen, J.-R., Hu, X.-Q., Lu, L.-Q., and Xiao, W.-J. (2016). Visible light photoredox-controlled reactions of N-radicals and radical ions. *Chem. Soc. Rev.* 45: 2044–2056.

55 Kärkäs, M.D. (2017). Photochemical generation of nitrogen-centered amidyl, hydrazonyl, and imidyl radicals: methodology developments and catalytic applications. *ACS Catal.* 7: 4999–5022.

56 Menigaux, D., Belmont, P., and Brachet, E. (2017). Light on unsaturated hydrocarbons-"Gotta heterofunctionalize them all". *Eur. J. Org. Chem.*: 2008–2055.

57 Jia, K. and Chen, Y. (2018). Visible-light-induced alkoxy radical generation for inert chemical bond cleavage/functionalization. *Chem. Commun.* 54: 6105–6112.

58 Qvortrup, K., Rankic, D.A., and MacMillan, D.W.C. (2014). A general strategy for organocatalytic activation of C—H bonds *via* photoredox catalysis: direct arylation of benzylic ethers. *J. Am. Chem. Soc.* 136: 626–629.

59 Cuthbertson, J.D. and MacMillan, D.W.C. (2015). The direct arylation of allylic sp^3 C—H bonds *via* organic and photoredox catalysis. *Nature* 519: 74–77.

60 Jeffrey, J.L., Terrett, J.A., and MacMillan, D.W.C. (2015). O–H hydrogen bonding promotes H-atom transfer from α C—H bonds for C-alkylation of alcohols. *Science* 349: 1532–1536.

61 Shaw, M.H., Shurtleff, V.W., Terrett, J.A. et al. (2016). Native functionality in triple catalytic cross-coupling: sp^3 C—H bonds as latent nucleophiles. *Science* 352: 1304–1308.

62 Le, C., Liang, Y., Evans, R.W. et al. (2017). Selective sp^3 C–H alkylation *via* polarity-match-based cross-coupling. *Nature* 547: 79–83.

63 Tanaka, H., Sakai, K., Kawamura, A. et al. (2018). Sulfonamides as new hydrogen atom transfer (HAT) catalysts for photoredox allylic and benzylic C–H arylations. *Chem. Commun.* 54: 3215–3218.

64 Mukherjee, S., Maji, B., Tlahuext-Aca, A., and Glorius, F. (2016). Visible-light-promoted activation of unactivvated $C(sp^3)$—H bonds and their selective trifluoromethylthiolation. *J. Am. Chem. Soc.* 138: 16200–16203.

65 Liu, X., Lin, L., Ye, X. et al. (2017). Aerobic oxidation of benzylic sp^3 C—H bonds through cooperative visible-light photoredox catalysis of *N*-hydroxyimide and dicyanopyrazine. *Asian J. Org. Chem.* 6: 422–425.

66 Margrey, K.A., Czaplyski, W.L., Nicewicz, D.A., and Alexanian, E.J. (2018). A general strategy for aliphatic C–H functionalization enabled by organic photoredox catalysis. *J. Am. Chem. Soc.* 140: 4213–4217.

67 Heitz, D.R., Tellis, J.C., and Molander, G.A. (2016). Photochemical nickel-catalyzed C–H arylation: synthetic scope and mechanistic investigations. *J. Am. Chem. Soc.* 138: 12715–12718.

68 Shields, B.J. and Doyle, A.G. (2016). Direct $C(sp^3)$–H cross coupling enabled by catalytic generation of chlorine radicals. *J. Am. Chem. Soc.* 138: 12719–12722.

69 Deng, H.-P., Fan, X.-Z., Chen, Z.-H. et al. (2017). Photoinduced nickel-catalyzed chemo- and regioselective hydroalkylation of internal alkynes with ether and amide α-hetero $C(sp^3)$—H bonds. *J. Am. Chem. Soc.* 139: 13579–13584.

70 Qin, Q. and Yu, S. (2015). Visible-light-promoted remote $C(sp^3)$–H amidation and chlorination. *Org. Lett.* 17: 1894–1897.

71 Becker, P., Duhamel, T., Stein, C.J. et al. (2017). Cooperative light-activated iodine and photoredox catalysis for the amination of ${C_{sp}}^3$—H bonds. *Angew. Chem. Int. Ed.* 56: 8004–8008.

72 Chu, J.C.K. and Rovis, T. (2016). Amide-directed photoredox-catalyzed C—C bond formation at unactivated sp^3 C—H bonds. *Nature* 539: 272–275.

73 Choi, G.J., Zhu, Q., Miller, D.C. et al. (2016). Catalytic alkylation of remote C—H bonds enabled by proton-coupled electron transfer. *Nature* 539: 268–271.

74 Zhang, J., Li, Y., Zhang, F. et al. (2016). Generation of alkoxyl radicals by photoredox catalysis enables selective $C(sp^3)$–H functionalization under mild reaction conditions. *Angew. Chem. Int. Ed.* 55: 1872–1875.

75 Wu, X., Wang, M., Huan, L. et al. (2018). Tertiary-alcohol-directed functionalization of remote $C(sp^3)$—H bonds by sequential hydrogen atom and heteroaryl migrations. *Angew. Chem. Int. Ed.* 57: 1640–1644.

76 Wang, C., Harms, K., and Meggers, E. (2016). Catalytic asymmetric ${C_{sp}}^3$–H functionalization under photoredox conditions by radical translocation and stereocontrolled alkene addition. *Angew. Chem. Int. Ed.* 55: 13495–13498.

77 Ma, J., Shen, X., Harms, K., and Meggers, E. (2016). Expanding the family of bis-cyclometalated chiral-at-metal rhodium(III) catalysts with a benzothiazole derivative. *Dalton Trans.* 45: 8320–8323.

78 Wang, C., Chen, L.-A., Huo, H. et al. (2015). Asymmetric Lewis acid catalysis directed by octahedral rhodium centrochirality. *Chem. Sci.* 6: 1094–1100.

79 Yuan, W., Zhou, Z., Gong, L., and Meggers, E. (2017). Asymmetric alkylation of remote $C(sp^3)$—H bonds by combining proton-coupled electron transfer with chiral Lewis acid catalysis. *Chem. Commun.* 53: 8964–8967.

9

Paired Electrolysis

Kouichi Matsumoto[1] *and Toshiki Nokami*[2]

[1] *Kindai University, School of Science and Engineering, Department of Chemistry, 3-4-1 Kowakae, Higashi-Osaka, 577-8502, Osaka, Japan*
[2] *Tottori University, Department of Chemistry and Biotechnology, 4-101 Koyamacho-minami, Tottori city, 680-8552, Tottori, Japan*

9.1 Introduction

There are numbers of reviews and book chapters of organic electrochemistry featuring paired electrolysis [1–6]. The authors ask readers to follow some pioneering examples of paired electrolysis by reading those previous articles. Here, we introduce some recent and selected examples of paired electrolysis. Electrochemical reactions are always performed in a paired manner, because anodic oxidation and cathodic reduction occur at the same time. Therefore, in principle, all electrochemical reactions can be categorized as a paired electrolysis; however, we excluded examples of electrochemical reactions such as systems with the reaction on the counter electrode only neutralizing the reaction media.

We divide this chapter of paired electrolysis into three sections (Figure 9.1). In the first section, we focus on a paired electrolysis for sequential reactions at both electrodes (Figure 9.1a). For example, a substrate **A** is converted to an intermediate **B** at an anode (a cathode) and the intermediate **B** is sequentially converted to the product **C** at the counter electrode. These types of reactions are normally performed using undivided cell or flow cell without a separator, because a chemical species generated at the electrode should reach to the counter electrode for the subsequent reaction in the same pot or flow as fast as possible. In the second section, we focus on a paired electrolysis with two different reactions at each electrode (Figure 9.1b). For example, substrates **A** and **C** are converted to products **B** and **D** at each electrode, respectively. These types of reactions are normally performed using a divided cell with a separator, which prevents overreactions of products **C** and **D**. Although a separator has a negative influence such as high cell voltage, it isolates two different species in both chambers. In the third section, we focus on a paired electrolysis for generation of two intermediates to afford the final

Organic Redox Chemistry: Chemical, Photochemical and Electrochemical Syntheses, First Edition.
Edited by Jun-Ichi Yoshida and Frédéric W. Patureau.

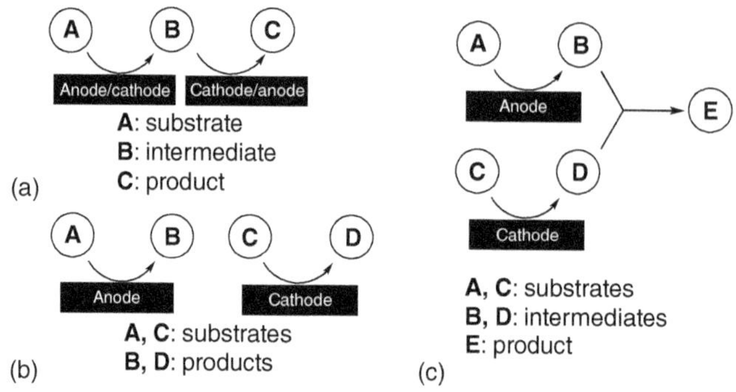

Figure 9.1 Three types of paired electrolysis.

product through the sequential reaction (Figure 9.1c). For example, substrates **A** and **C** are converted to intermediates **B** and **D** at each electrode respectively, and the subsequent reaction of the intermediates **B** and **D** affords the final product **E**. These types of reactions are performed in *in-cell* or *ex-cell* method by using undivided cell or divided cell, respectively. Flow cells are also ideal systems for these reactions.

9.2 Paired Electrolysis for Sequential Reactions at both Electrodes

9.2.1 Using an Undivided Cell

Paired electrolysis using an undivided cell is one of the most simple and practical procedures for the organic transformations. One of the representative example is the Monsanto process for the synthesis of adiponitrile from acrylonitrile by electrochemistry, which can be established in the industry [6, 7]. The cathodic reduction of acrylonitrile followed by the dimerization gives the dianionic species. Then, it can be allowed to react with $2H^+$, which is prepared by the anodic oxidation of H_2, to form the corresponding adiponitrile, effectively. This process can be regarded as the paired electrolysis, and one of the historical and successful reactions (Figure 9.2).

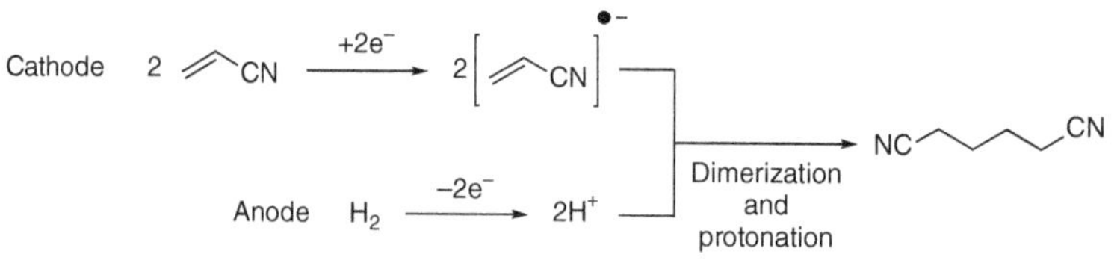

Figure 9.2 Monsanto process for the synthesis of adiponitrile from acrylonitrile in an undivided cell by using paired electrolysis.

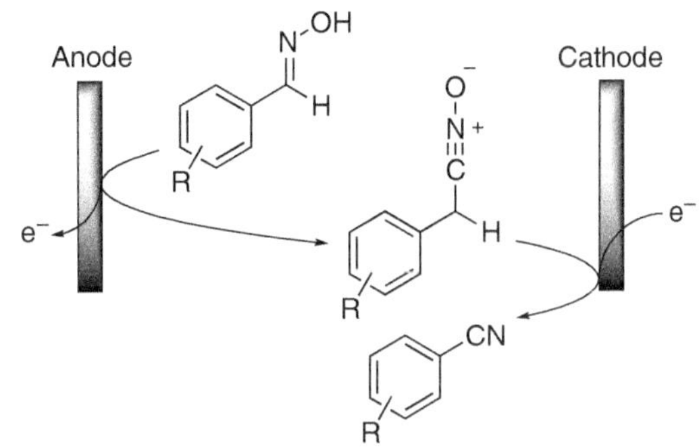

Figure 9.3 Sequential oxidation and reduction of oximes by electrochemical reaction using an undivided cell. Source: Modified from Hartmer and Waldvogel [8].

Among the recent examples of paired electrolysis using an undivided cell, Waldvogel and coworker demonstrated simple but important electrochemical transformations from oximes to nitrile by the sequential oxidation and reduction using electrochemistry (Figure 9.3) [8]. This procedure is valuable from a viewpoint of the metal-free and halogen-free reaction in addition to the wide range of substrate scope. In addition, benzonitrile and related compounds are the versatile building blocks in an industry process. Therefore, the current process might be attractive.

Both electrodes in an undivided cell are useful for the homo-coupling reactions. For example, imidazopyridine-based heterocycles undergo oxidation and homo-coupling reaction followed by the reduction to neutralize the intermediate (Figure 9.4) [9]. The plausible mechanism suggests the reaction of radical cation intermediate with starting material to give a homo-coupled intermediate, which can be neutralized by cathodic reduction. Another possibility of the homo-coupling of two radical cation intermediates followed by the reduction of the dicationic species cannot be ruled out.

9.2.2 Using a Flow Cell

The important example of paired electrolysis using the simple flow cell was reported by Karakus and Zuman (Figure 9.5) [10]. They have demonstrated the sequential reduction of nitrobenzene followed by oxidation of their intermediates to obtain nitrosobenzene. Since then, various types of electrochemical flow devices have been developed.

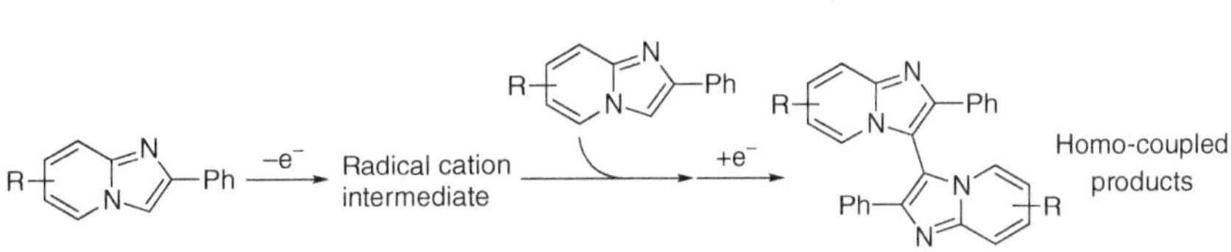

Figure 9.4 The homo coupling reaction of imidazopyridine-based heterocycles via the sequential oxidation and reduction using an undivided cell. Source: Modified from Gao et al. [9].

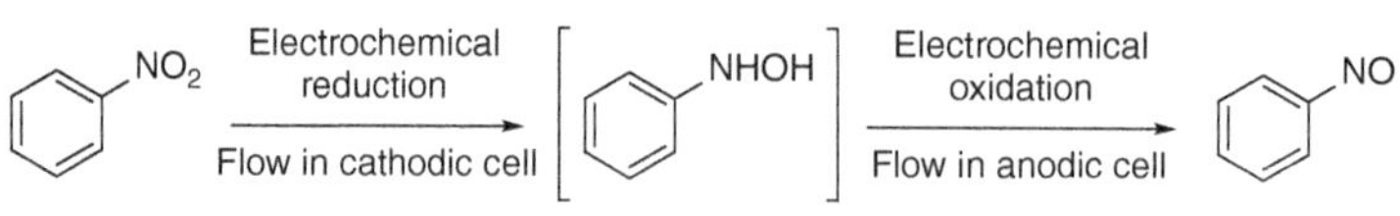

Figure 9.5 Electrosynthesis of nitrosobenzene from nitrobenzene by using the simple flow cell. Source: Modified from Karakus and Zuman [10].

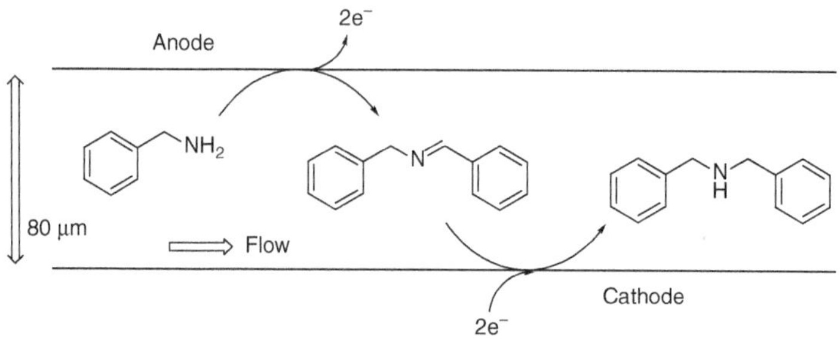

Figure 9.6 Anodic oxidation followed by cathodic reduction of benzylamine in an electrochemical micro reactor. Source: Modified from Amemiya et al. [11].

Among them, the use of the electrochemical flow-microreactor can realize the sequential oxidation and reduction for the completion of the desired reactions. For example, Atobe, Fuchigami, and coworkers have applied the effective paired electrolysis in an electrochemical microreactor, in which oxidation of benzylamine to *N*-benzylidenebenzylamine, followed by the reduction to afford dibenzylamine as a product (Figure 9.6) [11]. In the micro space, diffusion of the intermediate can realize the effective paired electrolysis. In fact, the selectivity of the desired product is increased by using an electrochemical microreactor, comparing with bach-type electrolysis. The same phenomenon is also found in the ether formation of methyl 2-(3,5-dichloro-4-hydroxyphenyl)acetate by electrochemistry as reported by Nishiyama, Atobe, and coworkers (Figure 9.7) [12]. If the electrochemical reaction of methyl 2-(3,5-dichloro-4-hydroxyphenyl)acetate is performed in a batch-type reactor, the desired dimer is produced in lower yields. However, the same reaction in an electrochemical microreactor enables the selective formation of the dimer in good yield. This result is indicating that the sequential

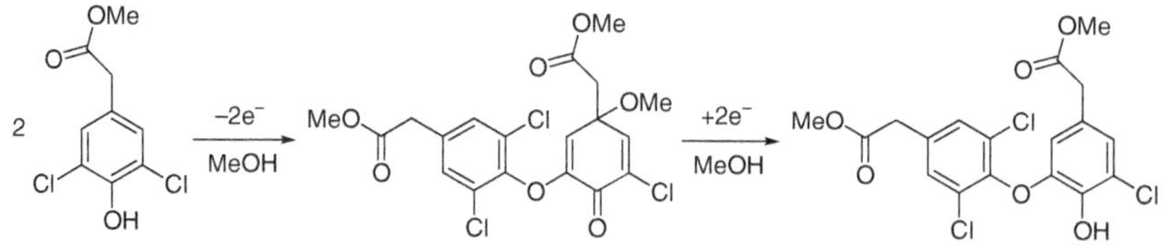

Figure 9.7 Anodic oxidation followed by the cathodic reduction of the phenol derivative in an electrochemical micro reactor. Source: Kashiwagi et al. [12].

oxidation followed by the reduction in micro spaces is essential for short living intermediates.

9.3 Paired Electrolysis with Two Different Reactions at both Electrodes

9.3.1 Using an Undivided Cell

If two products can be easily separated like gas and liquid, an undivided cell is useful enough to produce two different products in the same pot. For example, hydrogen (H_2) gas generated at the cathode can be transferred to a flask containing another substrate and Pd/C as a catalyst, which is used for hydrogenation (Figure 9.8) [13, 14]. In this case, oxidation of (3,4-dimethoxyphenyl)methanol to the corresponding aldehyde, reduction of methanol and hydrogenation of olefin can be carried out at the same time.

Another interesting reaction has also been established by Inagi et al. (Figure 9.9) [15]. They carried out the electrochemical polymer reactions at both electrodes, in which the polymer unit containing 9-fluorenol is cast on the anode and cathode. Under the electrolysis condition using an undivided cell, anodic oxidation of the 9-fluorenol unit in the polymer affords the corresponding 9-fluorenone unit. In contrast, the cathodic reduction of the 9-fluorenol unit in the polymer affords the corresponding fluorene unit. Because the thin layer polymer on both electrodes is not dissolved in $Et_4NOTs/^iPrOH$ during the electrolysis, the undesired electrochemical reactions on the counter electrode can be avoided. In addition, it is easy

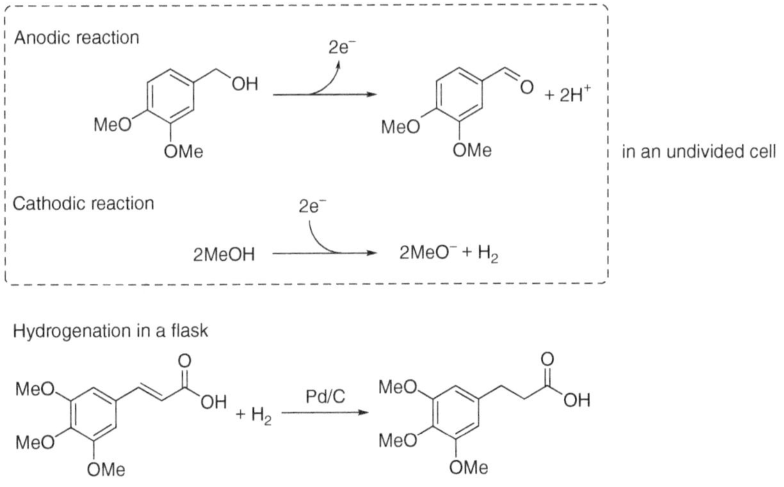

Figure 9.8 Formation of two different products at both electrodes using an undivided cell and utilization of thus-generated H_2 gas for hydrogenation in another flask. Source: Adapted from Nguyen et al. [13] and Wu et al. [14].

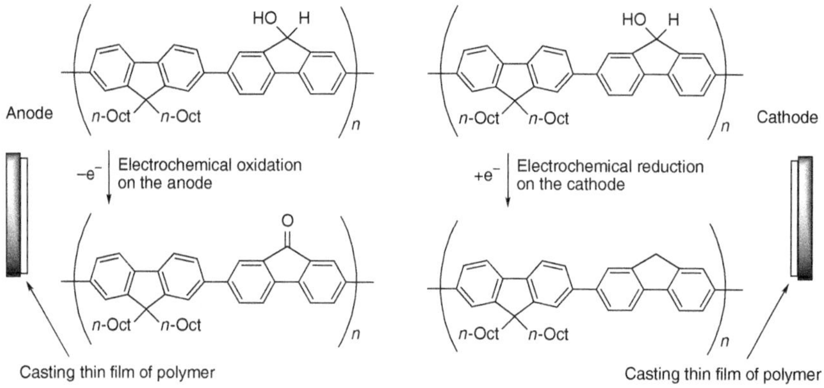

Figure 9.9 Paired polymer electrolysis on both electrodes in an undivided cell. Source: Modified from S. Inagi et al. [15].

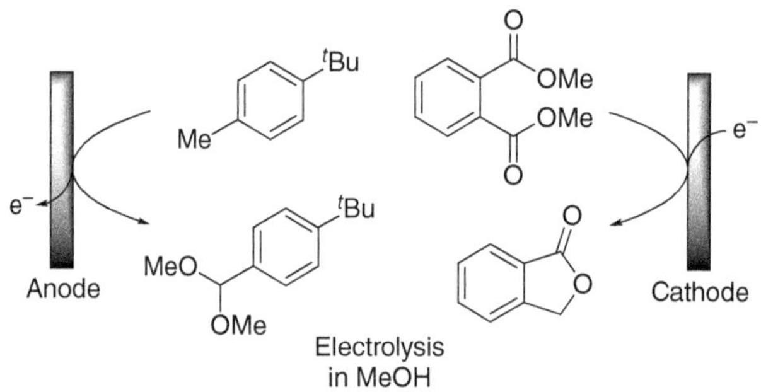

Figure 9.10 Paired electrolysis in an industrial scale. Source: Modified from Steckhan et al. [16].

to separate and obtain the final products, because the casting materials onto the electrodes is solid, and can be dissolved in $CHCl_3$.

In an industrial scale, one of the successful applications was achieved by BASF (Figure 9.10) [16]. The electrochemical oxidation of 1-(*tert*-butyl)-4-methylbenzene and the electrochemical reduction of dimethyl phthalate are combined in the same batch-type electrochemical reactor, in which the corresponding products such as 1-(*tert*-butyl)-4-(dimethoxymethyl)benzene and isobenzofuran-1(3*H*)-one (phthalide) can be obtained efficiently. These products are important building blocks in the chemical industry.

9.3.2 Using a Divided Cell

A divided cell enables two entirely different reactions at the same time with low energy. For example, Kubiak, Moeller, and coworkers nicely reported synthesis of benzimidazole and generation of carbon monoxide using two different mediator systems (Figure 9.11) [17].

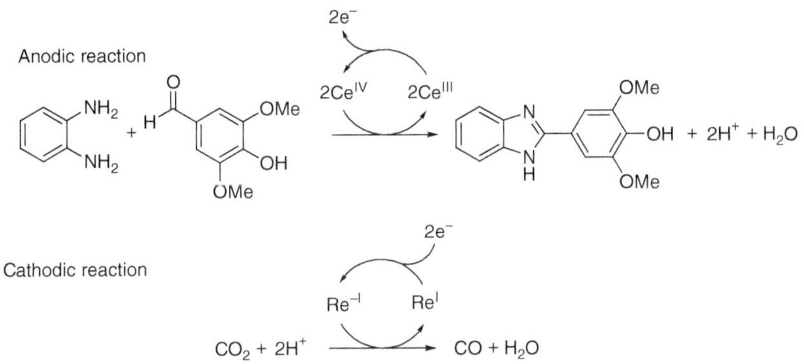

Figure 9.11 Synthesis of two different products in a divided cell. Source: Modified from Llorence et al. [17].

9.3.3 Using a Flow Cell

One of the features in a paired electrolysis using flow cells is no use of supporting electrolyte, in the case that an electrochemical flow cell bearing close distance between anode and cathode is used. For example, Atobe, Fuchigami, and coworkers have demonstrated that the anodic oxidation of 1-phenylethyl alcohol and the cathodic reduction of benzyl chloride in the electrochemical flow cell can be performed without a separator (Figure 9.12) [18]. In this electrochemical process, H^+ around the anode and Cl^- around the cathode are generated at the same time, and these ionic species can serve as the role of a supporting electrolyte. Thus, there was no need to add a supporting electrolyte in the solvent. They named this system as *self-supported paired electrolysis*. The methoxylation such as 2,5-dimethoxy-2,5-dihydrofuran from furan and methanol as well as the acetoxylation in the thin layer flow cell has also been demonstrated using the self-supported paired electrolysis by Atobe, Fuchigami, and Marken [19, 20]. In these cases, electro-generated reactive intermediates can be served as a supporting electrolyte.

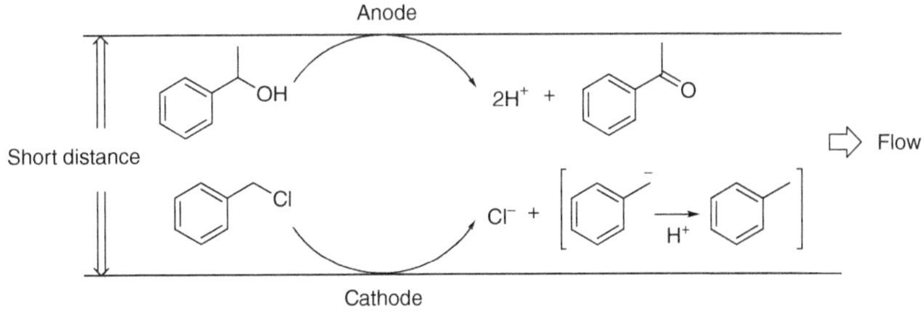

Figure 9.12 Self-supported paired electrolysis for anodic oxidation of 1-phenylethyl alcohol and cathodic reduction of benzyl chloride in the absence of added electrolyte. Source: Modified from Amemiya et al. [18].

9.4 Paired Electrolysis for Generation of Two Intermediates to Afford a Final Product by the Sequential Reaction

9.4.1 Using an Undivided Cell

There has been various types of reactions reported in this classification. As one of the valuable applications, the generation of alkoxycarbenium ion from tetrahydrofuran (THF) is developed (Figure 9.13). Kashimura, Ishifune, and coworkers found that the reductive generation of alkoxide ion from the corresponding esters can be coupled with the carbocation such as alkoxycarbenium ion from THF, to afford the desired acetal products in good yields [21–23]. The similar reaction has also been developed by Hilt, in which the anodically generated carbocation from THF can be reacted with alcoholate intermediate to give acetal product in good yield, shown in Figure 9.14 [24]. This result can be also recognized as the three components coupling reactions. In this process, the dissolved catalytic amount of In^{III} can be reduced to In^0, which is used for the next cycle with allyl bromide to form allylindium intermediate. It can be reacted with an aldehyde to give alcoholate intermediate together with In^{III}. Thus, catalytic cycles can be turned by the electrochemical reduction.

Another type of three components coupling reaction can be found in the field of electrochemical carboxylation, developed by Senboku et al. (Figure 9.15) [25]. The

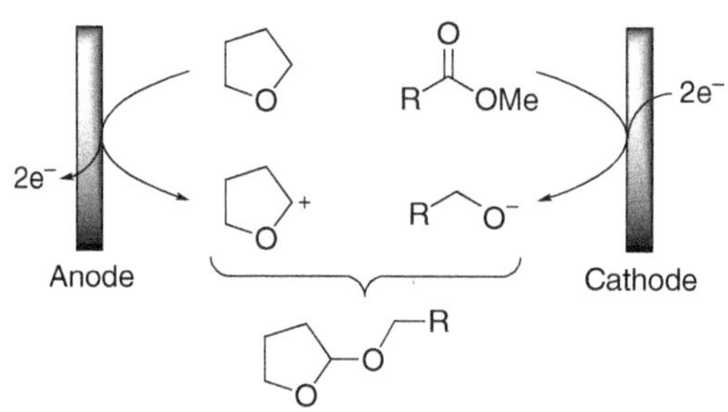

Figure 9.13 Generation of two intermediates such as alkoxycarbenium ion and alkoxide ion followed by the coupling reaction in an undivided cell.

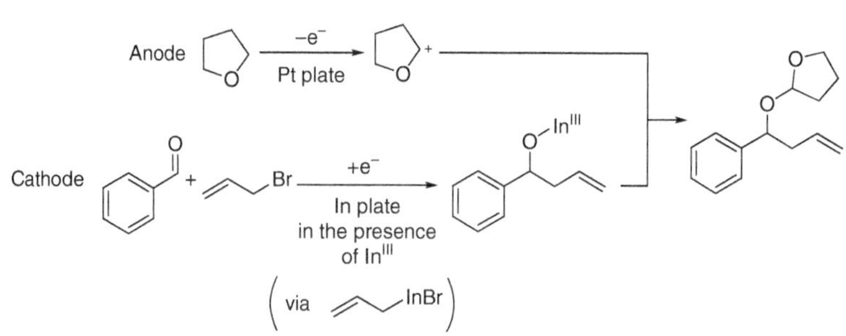

Figure 9.14 Generation of two intermediates such as alkoxycarbenium ion and indium alkoxide in an undivided cell. The formation of acetal via three components reaction. Source: Modified from Hilt [24].

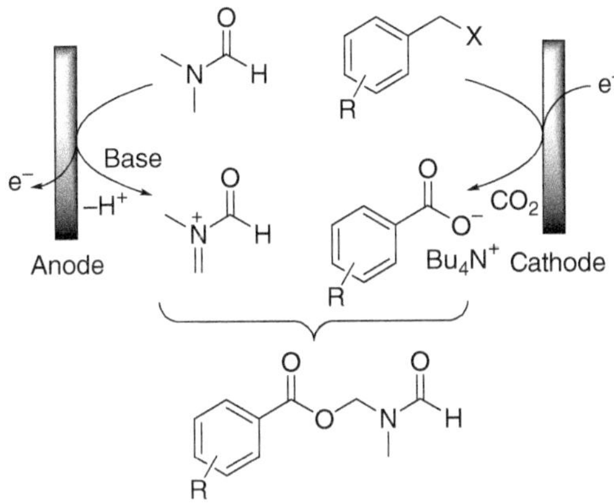

Figure 9.15 Generation of two intermediates in an undivided cell followed by the coupling reaction via carboxylation. Source: Modified from Senboku et al. [25].

cathodic reduction of aryl halide in the presence of CO_2 gives the corresponding aryl carboxylate ion intermediate. In contrast, the anodic oxidation of DMF (*N*,*N*-dimethylformamide), which has a role of solvent, generates the iminium ion intermediate. In the final stage, the aryl carboxylate ion intermediate can be allowed to react with the iminium ion intermediate to afford the three components coupling product in moderate to good yields. The success of the current reaction seems to be the use of Pt plates for both anode and cathode instead of the use of Mg electrodes. If the Mg electrode is used in the anode, magnesium ion (Mg^{2+}) is dissolved from Mg electrode. Because the counter cation of aryl carboxylate ion intermediate is Mg^{2+}, the reactivity of the anionic species might be decreased. However, in the current case the counter cation of aryl carboxylate ion intermediate is Bu_4N^+, which is derived from the supporting electrolyte, because a Pt plate is employed as an electrode. The reactivity of aryl carboxylate ion intermediate might be increased, due to the bulky cation such as Bu_4N^+ [26, 27].

In the above-described reactions, the reactive intermediates are derived from the oxidation of organic solvents such as THF and DMF. As another example for reductive generation of reactive intermediate from solvents, Yoshida et al. demonstrated the pioneering work shown in Figure 9.16 [28]. In general, Bu_4NBr is the first choice for the source of Br^-. However, the electrochemical condition can easily oxidize Br^- to form Br_2, preventing the oxidation of the substrate due to the order of the oxidation potentials. In order to overcome this challenge, Br^- is supplied by the reduction of CH_2Br_2, which is coupled with Prins cyclized carbocations derived from the oxidation of the corresponding materials, to form the brominated pyran rings.

In the viewpoint of carbon–carbon bond formation in a paired electrolysis for the generation of two intermediates, an important work was demonstrated by Chiba and coworkers (Figure 9.17) [29]. The benzylic sulfide is oxidized on the surface of an anode to give the benzylic cation intermediate. In this case, the phenylthio substituent (PhS) works as an electro-auxiliary [30]. In contrast, CH_3NO_2 is reduced to give $^-CH_2NO_2$. Overall, the reaction of two intermediates affords the corresponding coupling products via C—C bond formation. The key factor of this reaction is probably the difference of the oxidation potentials between the sulfide of the starting

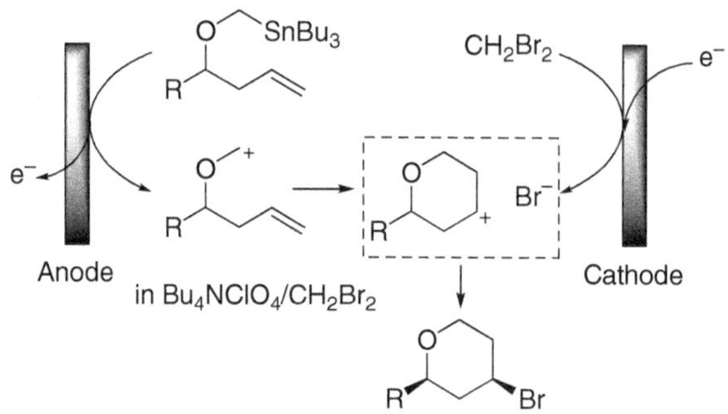

Figure 9.16 Paired electrolysis of bromination in Prins cyclization involving reductive generation of Br^- from CH_2Br_2 as a key step. Source: Modified from Yoshida et al. [28].

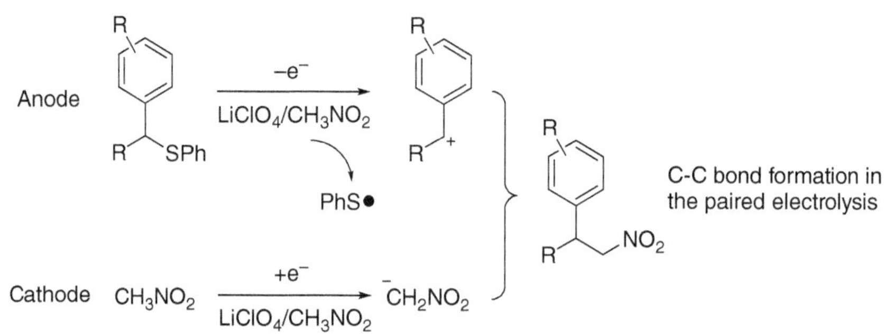

Figure 9.17 Paired electrolysis in an undivided cell via the carbon–carbon bond formation. Source: Modified from Kim et al. [29].

material and the final product. The sulfide compounds have generally lower oxidation potentials than the benzylic compounds.

As another excellent example of a carbon–carbon bond formation via the paired electrolysis, the reaction of carbon–carbon double bond with CF_3 radical has been developed to form oxytrifluoromethylated products by Kappe and Cantillo (Figure 9.18) [31]. The anodic oxidation of CF_3SO_2Na generates CF_3 radical, which can react with the styrene derivatives followed by sequential one-electron oxidation to give the benzylic cation intermediates. The hydroxide ion (OH^-) is generated by the cathodic reduction of H_2O. Thus, the carbocation can react with OH^- to give the corresponding products. This procedure is a useful method, because the source of CF_3 radical is cheap CF_3SO_2Na. In addition, the CF_3-containing compounds have attracted much attention in the field of biologically active compounds as well as drug materials.

Zeng and coworkers have also demonstrated the reaction of carbon–carbon double bonds with electro-generated CF_3 radical from CF_3SO_2Na, which undergoes an intramolecular cyclization [32]. The electrochemical reaction is carried out in an undivided cell by using Br^- as source of Br_2 and mediator. The reaction system is the paired electrolysis. In the proposed mechanism, the CF_3 radical is generated by the reduction of CF_3SO_2Br. This reaction is similar to the classification of the Section 2.1, but rigidly somewhat different.

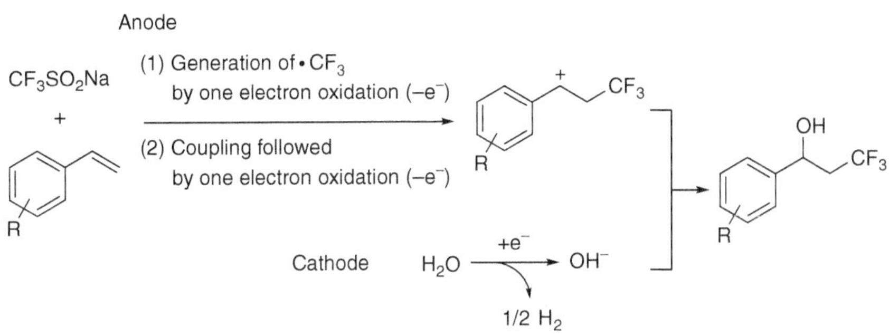

Figure 9.18 Oxytrifluoromethylation by using a paired electrolysis. Source: Modified from Jud et al. [31].

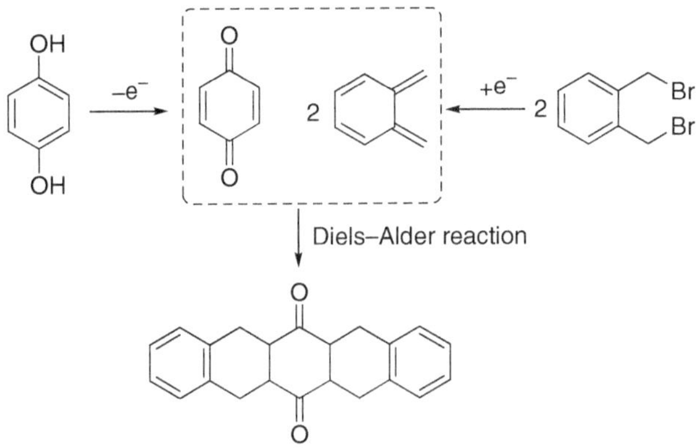

Figure 9.19 Sequential Diels–Alder reaction promoted by paired electrolysis in an undivided cell. Source: Modified from Habibi et al. [33].

Sequential Diels–Alder reaction has been demonstrated by Habibi et al. (Figure 9.19) [33]. They have focused on the combination of the anodic oxidation of 1,4-hydroquinone and the cathodic reduction of 1,2-bis(bromomethyl)benzene in an undivided cell. In-situ generated *p*-benzoquinone from 1,4-hydroquinone can be allowed to react with *in-situ* generated *o*-quinodimethane from 1,2-bis(bromomethyl)benzene to lead to the formation of the adduct via the sequential Diels–Alder reaction, which gives an interesting coupling product.

9.4.2 Using a Divided Cell

Generation of two intermediates by using a divided cell can achieve the allylation to aldehydes with a catalytic amount of metal catalyst. For example, Wang, Zha, and coworker demonstrated the anodic oxidation of benzyl alcohol to form benzaldehyde [34]. At the same time, allylic bromide and a catalytic amount of Sn^0 was dissolved in the cathodic chamber. After the electrolysis, the anodic solution was transferred into the cathodic chamber, and then the continuous electrochemical reduction was carried out. Thus, tin mediated electrochemical allylation of aldehydes has been

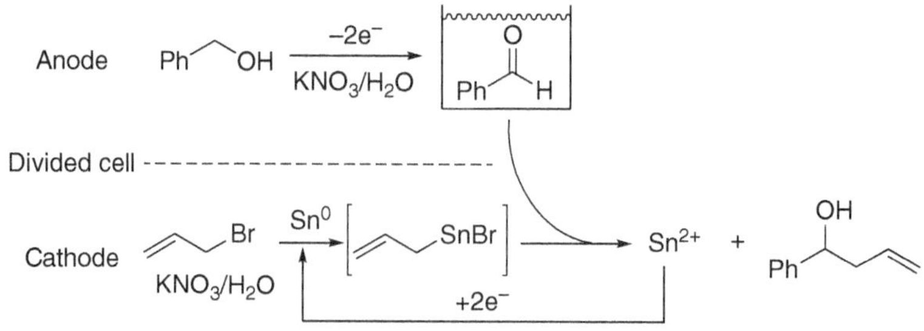

Figure 9.20 Synthesis of homoallylic alcohol by using a divided cell. Thus-accumulated anodic solution of an aldehyde was transferred to the cathodic chamber.

developed as shown in Figure 9.20. They also reported a similar type of electrochemical reactions [34–36].

9.4.3 Using a Flow Cell

The electrochemical flow cell is an effective and powerful tool for the preparation of metal catalyst. Thus, Kapur, Nguyen, Willans, and coworkers demonstrated the efficient synthesis of a copper-*N*-heterocyclic carbene complex using the electrochemical flow cell (Figure 9.21) [37]. Namely, the anodic conversion of Cu metal to Cu^{2+}, and the cathodic reduction of imidazolium ion to the corresponding carbene can be combined in the flow cell. The feature of this reaction is effective and selective formation of metal-catalyst. The application of thus synthesized catalyst to the hydrosilylation was also performed.

Another aspect of electrochemical reaction using flow cells is rapid generation of intermediates in both electrodes, followed by the coupling reaction in the outside of electrolysis cell. Therefore, the overreaction can be avoided. Yoshida, Suga, and coworkers reported the paired electrochemical microflow system [38]. The anodic oxidation of silyl-substituted pyrrolidinecarboxylate generates the iminium ion intermediate (Figure 9.22). The trimethylsilyl (Me_3Si) group can serve as an electro-auxiliary. The cathodic reduction of cinnamyl chloride in the presence of

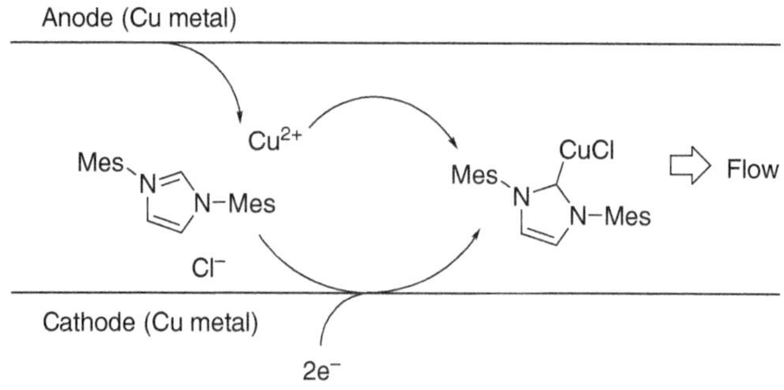

Figure 9.21 Paired electrochemical synthesis of the metal catalyst in flow. Source: Modified from Chapman et al. [37].

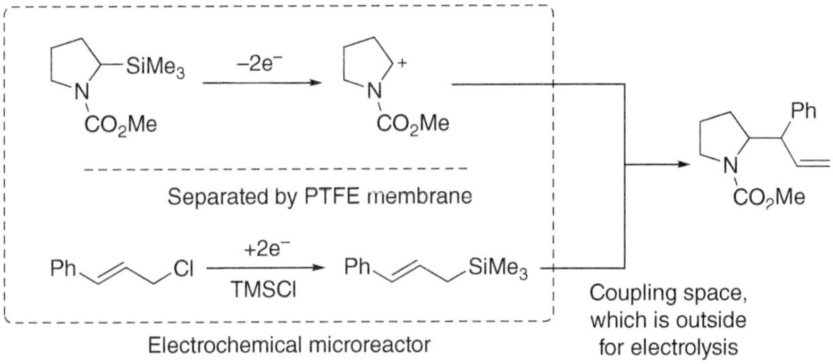

Figure 9.22 Electrochemical generation of two intermediates in a microreactor, followed by the coupling reaction. PFTE = polytetrafluoroethylene.

TMSCl (trimethylsilyl chloride) gives the allylic silanes as the nucleophiles. They can be reacted in the subsequent space of the flow system. Thus, the overoxidation of the desired product can be avoided. As another consideration for the current reaction, the same electrons are used for both electrodes for oxidation and reduction. This fact means the effective electrolysis for both intermediates and generation of the same amount of intermediates.

9.5 Conclusion

In this chapter, we categorized paired electrolysis into three types of processes, and selected examples have been introduced according to these three types. When we think benefits of electrolysis compared with conventional chemical methods, electrolysis is the only method, which enables oxidation and reduction reactions at the same time. In most cases products of organic reactions are neutral. In other word, oxidation should be followed by reduction and vice versa. Therefore, electrolysis especially paired electrolysis is an ideal process to transform organic molecules effectively. The choice of reaction system is important to perform the desired process with high efficiency. Although the numbers of examples are still limited, recent progress of flow microreactors opens new era of electrolysis of organic molecules.

References

1 Yoshida, J., Kataoka, K., Horcajada, R., and Nagaki, A. (2008). *Chem. Rev.* 108: 2265–2299.
2 Frontana-Uribe, B.A., Little, R.D., Ibanez, J.G. et al. (2010). *Green Chem.* 12: 2099–2119.
3 Yan, M., Kawamata, Y., and Baran, P.S. (2017). *Chem. Rev.* 117: 13230–13319.
4 Moeller, K.D. (2018). *Chem. Rev.* 118: 4817–4833.
5 Cardoso, D.S.P., Šljukić, B., Santos, D.M.F., and Sequeira, C.A.C. (2017). *Org. Process Res. Dev.* 21: 1213–1226.

6 Paddon, C.A., Atobe, M., Fuchigami, T. et al. (2006). *J. Appl. Electrochem.* 36: 617–634.

7 Baizer, M.M. and Danly, D.E. (1980). *Chem. Tech.* 10: 161–164.

8 Hartmer, M.F. and Waldvogel, S.R. (2015). *Chem. Commun.* 51: 16346–16348.

9 Gao, Y., Wang, Y., Zhou, J. et al. (2018). *Green Chem.* 20: 583–587.

10 Karakus, C. and Zuman, P. (1995). *J. Electrochem. Soc.* 142: 4018–4025.

11 Amemiya, F., Kashiwagi, T., Fuchigami, T., and Atobe, M. (2011). *Chem. Lett.* 40: 606–608.

12 Kashiwagi, T., Fuchigami, T., Saito, T. et al. (2014). *Chem. Lett.* 43: 799–801.

13 Nguyen, B.H., Perkins, R.J., Smith, J.A., and Moeller, K.D. (2015). *J. Organomet. Chem.* 80: 11953–11962.

14 Wu, T., Nguyen, B.H., Daugherty, M.C., and Moeller, K.D. (2019). *Angew. Chem. Int. Ed.* 58: 3562–3565.

15 Inagi, S., Nagai, H., Tomita, I., and Fuchigami, T. (2013). *Angew. Chem. Int. Ed.* 52: 6616–6619.

16 For example, Steckhan, E., Arns, T., Heineman, W.R. et al. (2001). *Chem. Tech.* 43: 63–73.

17 Llorence, M.J., Nguyen, B.H., Kubiak, C.P., and Moeller, K.D. (2016). *J. Am. Chem. Soc.* 138: 15110–15113.

18 Amemiya, F., Horii, D., Fuchigami, T., and Atobe, M. (2008). *J. Electrochem. Soc.* 155: E162–E165.

19 Horii, D., Atobe, M., Fuchigami, T., and Marken, F. (2005). *Electrochem. Commun.* 7: 35–39.

20 Horii, D., Atobe, M., Fuchigami, T., and Marken, F. (2006). *J. Electrochem. Soc.* 153: D143–D147.

21 Ishifune, M., Yamashita, H., Matsuda, M. et al. (2001). *Electrochim. Acta* 46: 3259–3264.

22 Kashimura, S., Yamashita, H., Murai, Y. et al. (2002). *Electrochim. Acta* 48: 7–10.

23 Ishifune, M., Yamashita, H., Matsuda, M. et al. (2003). *Electrochim. Acta* 48: 1879–1884.

24 Hilt, G. (2003). *Angew. Chem. Int. Ed.* 42: 1720–1721.

25 Senboku, H., Nagakura, K., Fukuhara, T., and Hara, S. (2015). *Tetrahedron* 71: 3850–3856.

26 Miyamoto, Y., Yamada, Y., Shimazaki, H. et al. (2015). *Electrochemistry* 83: 161–164.

27 Fuchigami, T., Awata, T., Nonaka, T., and Baizer, M.M. (1986). *Bull. Chem. Soc. Jpn.* 59: 2873–2879.

28 Yoshida, J., Takada, K., Ishichi, Y., and Isoe, S. (1994). *J. Chem. Soc., Chem. Commun.*: 2361–2362.

29 Kim, S., Uchiyama, R., Kitano, Y. et al. (2001). *J. Electroanal. Chem.* 507: 152–156.

30 For example, Yoshida, J. and Nishiwaki, K. (1998). *J. Chem. Soc., Dalton Trans.*: 2589–2596.

31 Jud, W., Kappe, C.O., and Cantillo, D. (2018). *Chem. Eur. J.* 24: 17234–17238.

32 Jiang, Y., Dou, G., Xu, K., and Zeng, C. (2018). *Org. Chem. Front.* 5: 2573–2577.

33 Habibi, D., Pakravan, N., and Nematollahi, D. (2014). *Electrochem. Commun.* 49: 65–69.

34 Zhang, L., Zha, Z., and Wang, Z. (2010). *Synlett*: 1915–1918.

35 See also, Zhang, L., Zha, Z., Wang, Z., and Fu, S. (2010). *Tetrahedron Lett.* 51: 1426–1429.

36 The reaction using an undivided cell was also reported. Zhang, L., Zha, Z., Zhang, Z. et al. (2010). *Chem. Commun.* 46: 7196–7198.

37 Chapman, M.R., Shafi, Y.M., Kapur, N. et al. (2015). *Chem. Commun.* 51: 1282–1284.

38 Suga, S., Okajima, M., Fujiwara, K., and Yoshida, J. (2005). *QSAR Comb. Sci.* 24: 728–741.

Index

a

absorption 52, 53–57, 59, 61–66, 70, 71, 85, 88, 187
absorption spectroscopy 54, 61, 62, 65, 66
absorption spectrum 54, 62–64, 66, 85
aerobic C–H bond functionalization approach 122
aerobic linear allylic amination method 121
alkoxylation 105, 160
alkylzinc reagents 182
4-allyl-2-methoxyphenols 30
allylsilanes 32
amino alkanes 107, 108
8-aminoquinoline-derived benzamides 20
aminyl anions 143
anodic C–C homo-coupling reactions 29–30
anodic coupling reactions 31, 33
anodic oxidation 135, 156, 161–163, 209, 210, 212, 213, 215, 217–220
anthraquinone 136, 145, 188, 194
anthraquinone-2-sodium sulfonate ($2\text{-}SO_3NaAQ$) 188, 194
anti-Stokes shift 57
aryl C–N coupling reactions 81–82
aryl C–S couplings 81–82
aryl diazonium salts 46, 47
aryl Grignard reagents 178, 180, 182
aryl halides 171, 173
  direct arylation of arenes 173–178
  electron-catalyzed cross-coupling reactions of 178–182
aryl-alkynyl coupling 16, 17
arylacetylenes 15, 16, 18
arylsilanes 12
atom transfer radical polymerization (ATRP) 78–80, 165
1,3-azoles 10, 11, 15, 16

b

back electron transfer (BET) 60, 86, 87
Beer's Law 54, 62–64
benzamides 12, 15, 17, 18, 20, 105, 106, 115
benzene 5, 7, 10–14, 17, 18, 76, 77, 84, 173–177, 214, 219
benzonitrile 143, 176, 211
benzothiophenes 34, 35
4-benzoylpyridine (4-BzPy) 188, 190
benzylamine 212
1-benzyl-1,4-dihydronicotinamide 46
binaphthol 164
2,2′-biphenols 29, 33
bipyridine 72, 73, 156, 165, 193
4,4′-bipyridine 156, 193
Birch reduction conditions 173
1,2-bis(bromomethyl)benzene 219
2,5-bis(2-hydroxyphenyl)thiophenes 34

Organic Redox Chemistry: Chemical, Photochemical and Electrochemical Syntheses, First Edition.
Edited by Jun-Ichi Yoshida and Frédéric W. Patureau.

bithiophene 13, 164
boronic acids 2, 4, 10, 12, 20, 106, 111
bromination 160, 161, 218
tert-butoxide 174–176, 182, 183
1-(*tert*-butyl)-4-(dimethoxymethyl) benzene 214
1-(*tert*-butyl)-4-methylbenzene 214

c

carbon-sulfur bond formation 49
cathodic reduction
  activated olefin and conjugated olefin 142
  active hydrogen compound 143
  alcohol and carboxylic acid 143
  alkyl halides 142
  14-Family and 15-Family Element Species 144
  gem- and *vic*-dihalogeno compounds 143
  ketone and imine 142
cation-pool method 36
C3-fluorinated oxindoles 39
charge transfer (CT) complexes 50
o-chloranil 12
2-chloroanthraquinone (2-ClAQ) 188, 190
cobalt-catalyzed dehydrogenative cross-coupling 105
concurrent reduction and substitution (CRS) method 162
consecutive photoinduced electron transfer (ConPET) 82, 83
copolymer synthesis 155–157
copper salts 106, 108, 114, 120
copper-catalyzed decarboxylative oxidative coupling 20
cross-coupling reactions 9, 10, 18, 20, 29–37, 49, 171, 172, 178, 180–183
cross dehydrogenative amination method 109, 120
cross-dehydrogenative coupling (CDC) 18–20, 39, 103, 104, 108–110, 113, 119, 120, 124
cyclic voltammetry (CV) measurement 59, 67, 68, 155, 159, 160, 165
cycloadditions 32, 37, 46–48, 77
1,4-cyclohexadiene 90

d

dehydrogenative C–O bond formation 105–107
dehydrogenative couplings 18, 39, 103, 104, 108–111, 113, 119, 124–125
deprotonation 148, 154, 172, 174, 175
2,2′-diaminobiaryls 35
diarylzinc reagents 180
diastereoisomer α-diisoeugenol 30
diastereoselective [2+2] photocycloadditions of bis(enones) 47
dibenzosiloles 12
1,3-dicarbonyls 32
1,9-dicyanoanthracene (DCA) 83
Diels–Alder reaction 219
N,*N*-dihalo compounds 143
dihydrophenazines 79, 81
(3,4-dimethoxyphenyl)methanol 213
2,4-dimethylphenol 29
dinitrobenzenes 136
4,7-diphenyl-1,10-phenanthroline (Ph-phen) 174, 176
1,1-di(phenylsulfonyl)ethylene 190

e

early photoredox catalysis 46
El Sayed's rules 58
electroauxiliaries
  coordination effects 136
  electron-transferable functional groups 135
electrochemical micro reactor 212
electrochemical oxidative arylation

of alkynes 39
of olefins 36–39
electrochemical oxidative benzyl-aryl cross-coupling reaction 35–36
electrochemical oxidative cross-dehydrogenative coupling
of $C(sp^3)$—H and $C(sp^2)$—H bonds 39
electrochemical post-functionalization 153, 159
electrochemical reactions
characteristics of 129–130
organic electrochemical reactions 130–131
electrochemical redox
by anodic substitution 159–162
aromatic monomers
oxidative electropolymerization 154–155
reductive electropolymerization of 157
cathodic reduction and paired reactions 162
conducting polymers 154–158
copolymer synthesis 155–157
electroactive polymers 164–165
electrochemically induced film formation 167
electrogenerated transition-metal catalysts 164
electropolymerization processes 158
oxidation-induced intramolecular cyclization 163–164
polyaniline functionalization 162–163
polymerization 165–167
electrochemical reductive transformations
direct and indirect electrochemical reductions 144–145
electrogenerated bases 148–150
reductive mediators 145–147
electrolysis 31, 33, 133, 148, 155, 156, 158, 159, 161–163, 165, 167, 209–221
electromagnetic radiation 50
electromagnetic spectrum 50–51
electron catalysis 172, 179
electron donor-acceptor (EDA) complexes 49
catalytic EDA reactions 91
definition 85
electron transfer 86
enantioselective reactions of 91–92
environmental factors 86–87
light intraction 85–86
photoredox catalysis 87–88
sacrificial donors and acceptors 89–90
stoichiometric EDA reactions 88–89
electron transfers
Marcus theory 60–61
mechanisms of 59–60
photoinduced electron transfer 58–59
electrophiles 9, 22, 131, 143, 183
electropolymerization 153–155, 157, 158, 160, 163–165
electroreduction method 144
enolate-enolate oxidative cross-coupling reaction 20
excitation spectrum 64
excited state absorption (ESA) 66
π-extended poly aromatic hydrocarbons (PAHs) 12
extinction coefficient 54

f

flow cell 209–213, 215, 220–221
fluorene-containing conjugated polymer 160, 161, 163
9-fluorenol 162, 213
fluorescence 52, 56–58, 64, 92
fluorophore 64, 65

Fourier-transform infrared (FT-IR) spectroscopy 160
Frank–Condon principle 53, 57, 60
Fujiwara–Moritani reaction 2, 4–10, 104

g
gel permeation chromatography (GPC) 160
Grignard compounds 32
Grignard cross-coupling reaction 178–181
Grignard reaction 132
ground state bleach (GSB) 66

h
heteroaromatics 5, 17
heterobiaryls 33
1,1,1,3,3,3-hexafluoroisopropanol (HFIP) 30, 31, 33, 35, 36
3-hexylthiophene 158, 160
Hofmann–Löffler–Freytag (HLF) reaction 187, 197, 198
homo-coupling reactions 29, 30, 33, 211
homolytic aromatic substitution (HAS) 173–175
hydrogen atom transfer (HAT) process
  asymmetric transformations 194–195, 200–201
  direct HAT, of excited photocatalyst 188–189
  indirect HAT 188–195
  photocatalytic functionalization of $C(sp^3)$–H bonds 189–194
hydrogen bonding 31, 33, 136
hydrogenated graphene derivative 113
1,4-hydroquinone 219
β-hydroxy ketone 194
9-hydroxyphenanthrene 30

i
indole carboxylic acids 7
inner sphere electron transfer (ISET) 59, 86, 130
inner-filter effect 65
internal conversion 52, 56, 58
internal oxidants
  C–H bond functionalization methods 115–116
  postulated mechanism 117–118
intersystem crossing (ISC) 52, 57–58, 64, 80, 82
intramolecular Pschorr cyclization 47
iridium polypyridyl complexes 46
isobenzofuran-1(3H)-one (phthalide) 214

k
Kipping method 157
Klinger, Heinrich 45

l
Laporte selection rule 51
leucoemeraldine base (LB) 162
Lewis acid 136, 200, 201
low-valent transition metal 171

m
Marcus theory 60–61
metal-based oxidants
  copper salts 108–109
  silver salt oxidants 105–108
methyl 2-(3,5-dichloro-4-hydroxyphenyl) acetate 212
10-methyl-9,10-dihydroacridine 46
2-methoxynaphthalene 30
2-methylquinoline 46
microwave irradiation 174
Mizoroki–Heck reaction 1
  arylmetal reagents 2–4
molar absorptivity 54, 55, 62, 63, 71

n

2-naphthol 30
nitrenes 143
nitrobenzene 46, 145, 211, 212
nitrosobenzene 211, 212
non-conductive polymers 153, 154, 159
non-crossed dimerization 140
nonradiative processes 55, 58
non-selective oxidation processes 31

o

O_2, use of 119–124
octafluoronaphthalene 13
open shell photoredox catalysts 82–84
optical density (OD) 54, 62
organic electrochemical reductions
  addition type 138
  asymmetric synthesis type 141
  chemoselectivity 133
  cleavage type 141
  crossed dimerization 140
  cyclization type 140
  dimerization type 139–140
  elimination type 139
  insertion type 138
  metalation type 141
  polymerization type 141
  polymorphism formation type 140
  reaction pathway selectivity 133
  regioselectivity 133–134
  stereoselectivity 134–135
  substitution type 139
  substitutive exchange type 139
  transformation type of functional group 137–138
  umpolung 131–132
organic oxidants
  organic peroxides 110–112
  quinones 112–115
organic peroxides 110–112
organocatalysis 47
organozinc reagents 180, 181
outer sphere electron transfer (OSET) 59–60, 86, 130
oxidative aryl–aryl bond formation
  aniline derivatives 17
  oxidative C–C bond formation 18–22
  oxidative C–H/C–M biaryl cross-coupling 10–12
  phenyloxazine-containing benzamides 17
  2-(pyridine-2-yl)isopropylamine 18
  robust palladium-based catalyst system 17
  Sonogashira coupling 16
oxidative C–C bond formations 1–23, 29–39
oxidative C–H bond functionalization techniques 113
oxidative electropolymerization 153–155
oxidative strategy 1, 22

p

paired electrolysis
  divided cell 214–215, 219–220
  flow cell 211–213, 215, 220–221
  undivided cell 210–211, 213–214, 216–219
palladium catalysts 1–2, 31
palladium catalyzed C–H functionalizations 107
palladium-catalyzed reaction 171
persulfates 109
phenol-arene cross-coupling products 33
phenols 29, 31–34, 38, 89, 109, 112, 120
phenoxazine PC 79, 82
phenoxazines 80–82
phosphorescence 52, 58, 63–64
photocatalytic 1,5-HAT reactions 189

photochemical reactions
  choice of light source 69
  factors 68
  photoreactor design 68–69
photoinduced electron transfer (PET) 58–60, 82, 85, 91
photoirradiation 172, 176, 188
photophysical processes
  absorption 53–55
  fluorescence 56–57
  internal conversion 56
  intersystem crossing 57–58
  Jablonski diagrams 52–53
  phosphorescence 58
  vibrational relaxation 55–56
photoredox 47
  dehalogenation reactions 49
  [2+2] enone cycloadditions 48
photoredox catalysis 46
  effective absorption of light 70–71
  favorable thermodynamics 71–72
  long lived excited state 71
  mechanisms of 69–70
  quantum yield 71
photoredox processes
  cyclic voltammetry 67–68
  emission spectroscopy 63–65
  transient absorption spectroscopy 65–67
  UV-visible spectroscopy 61–63
pinacol-type couplings 46
classical tin-mediated dehalogenations 48
poly(3-(2-ethylhexyl)thiopehene) 161
poly(3-(4-fluorophenyl)thiophene) 160
poly(3-alkylselenophene) (P3AS) 160
poly(3-alkylthiophene) (P3AT) 160
poly(3-hexylselenophene) 161
poly(3-methylthiophene) (P3MT) 159–160
poly(9,9′-dioctylfluorene-*alt*-thiophene) 160, 162
poly(*p*-phenylenevinylene) (PPV) 157
polymer reaction 153, 160, 161–162, 213
polypyrrole 160
Priestley, Joseph 45
product selectivity 134
proton-coupled electron transfer (PCET) process 188
pump-probe spectroscopy 65
N-pyrimidylindoles 15

q
quantum yield 64, 71
quasi-reversible 67
quinone derivatives 110, 114, 167

r
radical-cation pool method 31–32
redox reversibility
  inorganic photocatalysts 72–75
  open shell photoredox catalysts 82–84
  organic excited state oxidants 75–78
  organic excited state reductants 78–82
reductive quenching 70, 72–73, 83
regioselectivity 3, 6, 10, 13, 133–134, 176–177
reorganization energy 60–61
reoxidation system 7
rod–rod block copolymer 161
ruthenium 6, 46, 165

s
self-supported paired electrolysis 215
sequential Diels–Alder reaction 219
silver salt oxidants 105–108
silver trifluoroacetate (AgTFA) 107
silylenol ethers 32
single electron transfer (SET) processes 46, 105, 110, 172, 174–176, 178–179, 180, 188, 189, 195, 200
single wavelength kinetics 66

solvatochromism 57
Sonogashira coupling 16
Stark–Einstein Law 54
Stern–Volmer quenching 64, 66
stimulated emission (SE) 66
styrenes 2, 5, 37, 74, 177, 218
sulfonium salts 46
symmetric OCO-pincer ligands 33

t
tetrabutylammonium decatungstate 191
tetrahydrofuran (THF) 58, 111, 178–182, 216–217
tetrahydroisoquinoline 18–19, 73
tetrahydrothiophen-3-one 113
transient absorption spectroscopy (TAS) 65–67
transition metal catalysts 16, 75, 108, 164, 171, 174
transition metal photocatalysts 46
transition metal-catalyzed cross-coupling reaction 160, 171–172
2,2,2-trifluoroacetic acid 35
trifluoromethylation 81, 87
2,4,6-triphenylpyrylium tetrafluoroborate salt (TPT) 188, 199
2,5,5-trisubstituted oxazoles 122
twisted intramolecular charge transfer (TICT) 58

u
α,β-unsaturated δ-lactams 4
α,β-unsaturated carbonyls 2
UV-visible spectroscopy 61–63

v
vibrational relaxation 52, 55–56, 58

x
X-ray photoelectron spectroscopy (XPS) analysis 51, 159–160

z
zinc-porphyrin 156–157